Fine Homebuilding
Construction Techniques 2

Fine Homebuilding
Construction Techniques 2

The Taunton Press

International Standard Book Number 0-918804-47-7
Library of Congress Catalog Card Number 84-50164
Printed in the United States of America

A FINE HOMEBUILDING Book

FINE HOMEBUILDING ® is a trademark of
The Taunton Press, Inc.,
registered in the U.S. Patent and Trademark Office.

The Taunton Press, Inc.
63 South Main Street
Box 355
Newtown, Connecticut 06470

Contents

Introduction

Fine Homebuilding Construction Techniques 2 includes the best articles on building methods from issues 16 through 25 of *Fine Homebuilding* magazine. Whether you're an architect, professional builder or serious beginner, if you want to learn about materials and construction techniques, you'll find the articles here helpful.

—The editors

General Construction

Making a Structural Model

Fine-tuning your design and working out construction details is a lot less expensive at ¹⁄₂₄ scale

by Mark Feirer

If you've ever passed a construction site and paused to watch a framing crew, you may have marveled at the precise choreography of their work. A glance at the blueprints, and these carpenters are back to fast, precise cutting and nailing as if genetically encoded with the right instructions. Not all of us have such a sense of how a building's many structural parts fit together, and this explains the value of building a model before you're facing full-scale construction.

The framer's source of information is a set of blueprints, but even the best-drawn blueprints can't match the comprehensive impression a scale model conveys. A novice builder can be intimidated by a relatively simple house in the same way that an experienced contractor can be befuddled by an unusual design. Both can benefit from building a scale model. Models don't eliminate the need for working drawings, but they do increase the builder's understanding of a structure. If you believe that practice makes perfect, consider this: How else can you build your house twice and benefit from your mistakes without having to live with them?

The design model—This is a sort of three-dimensional sketch. Its purpose is to let you briefly explore a design before you make final decisions that will be translated onto your working drawings. It often takes several models before you can make a decision, so these models are characteristically inexpensive and quick to construct. They're usually built from sheet materials like mat board or cardboard at a scale of ¼ in. = 1 ft. (1⁄48 size).

The design model replicates only the shell of the building and doesn't attempt to answer structural questions. Nevertheless, because it's a scale replica of the real thing, it lets you test for shading and sun penetration quite accurately. Solar designers can place a model in a device called a heliodon, which simulates the path the sun takes at different latitudes and different times of the year (see *FHB* #13, p. 18). A design model can also be used to test the livability of a floor plan by adding interior partitions to it. With the roof removed, you can "walk" through the house.

Structural models—While design models show the skin of a house, structural models show a building's skeleton, and are therefore more useful from a builder's point of view. Constructing this kind of model enables you to trouble-shoot the structural design of a house, eliminating or revising framing techniques that are difficult or wasteful. As any builder knows, just because you can draw it doesn't mean you should build it that way. Complex or experimental structures can be examined at the modelmaking stage for flaws that might otherwise make site work a nightmare, or drive the cost beyond reason.

Engineers can test a structural model by subjecting it to scaled-down stresses of gravity, wind shear and seismic activity. Models given this type of testing have to be specially built. However, if your craftsmanship is tolerable and your finished model collapses when you accidentally bump the table, you might want to re-examine your design.

Another benefit of structural modelmaking is that once the model is finished, it can help you make a materials estimate for the building. All you have to do is count the number of studs, joists, rafters and other structural members. You can also determine the most efficient layout for sheathing and subflooring.

Scale and materials—The level of detail that you can achieve in a model depends largely on its scale. The detail can increase as the scale gets larger, but at some point the model can get unwieldy and expensive without really offering more advantages than you'd have at a smaller scale. For structural models that replicate an entire house, a scale of ½ in. = 1 ft. (1⁄24 size) is a good compromise: it's large enough to show joist cross-bridging but small enough for tabletop work. At this scale, individual pieces of the model are usually large enough to handle without tweezers. If you want more detail, 1⁄12 scale (1 in. = 1 ft.) is the logical next step up. You can buy modelmaking materials that are predimensioned to both these scales.

Modelmaking materials and tools are available at most hobby shops (sometimes in the guise of model-railroad or model-aircraft supplies) and some stores that supply architects and architecture students. Materials may be purchased by mail order as well, from Northeastern Scale Models, Inc. (P.O. Box 425, Methuen, Mass. 01844) and Eugene Toy and Hobby (32 E. 11th Ave., Eugene, Ore. 97401).

Spruce, basswood and balsa are the most popular modelmaking woods, and you can also find predimensioned plastic extrusions. Balsa is the lightest, weakest and easiest to work, and it comes in the widest variety of forms: sheets, slabs, scaled-down 2x lumber and blocks. Unless the work calls for extra strength or stiffness, most of the structural members in our models are balsa.

Model lumber is scaled to the nominal dimension of the full-size item; a 2x4 will scale to 2 in. by 4 in., not 1½ in. by 3½ in. The difference is barely noticeable at 1⁄24 size. Model woods are usually found in lengths of 2 ft. or 3 ft. (48 ft. and 72 ft. in 1⁄24 scale) so a considerable amount of cutting to length is required before you can begin gluing.

You can, of course, dimension and prepare your own materials from scrap stock, but I wouldn't recommend doing so unless you're long on time and short on funds. However, if you need an off-sized piece and you can find balsa sheets in the thickness you want, it's easy to rip individual framing members to width with a razor knife and straightedge. You could also use redwood, cedar, yellow poplar or fir, as long as the wood is dry.

Adhesives take the place of mechanical fasteners in modelmaking. White or yellow wood glues make a strong bond and get tacky quickly. It's usually only 10 seconds before a piece dabbed with a pinprick of glue will stay in position without being held, and after 10 minutes the bond is fairly secure. Sometimes balsa requires two coats of glue because it's so porous, particularly in end-grain joints.

Testor Corp. (620 Buckbee St., Rockford, Ill. 61101) makes an extremely strong and quick-drying adhesive called Cement for Wood Models (modelers sometimes call it Green Label). Though its unpleasant odor and relatively high cost ($.59 for a ⅝-oz. tube) count against it, wood glued with Testor's develops a strong bond in just 15 to 20 seconds. For an even faster bond, I've had success using the cyanoacrylate "super glues" that sell for about $3 per ½-oz. tube. They're expensive, but they bond almost instantly.

Tools—The razor knife is the portable circular saw of model building—you can't do without it. I use two: a lightweight knife with a scored blade that can be snapped off when it gets dull, baring a fresh edge; and a standard mat knife, like the one Stanley makes, for heavier cutting. A straightedge is the companion to the razor knife. I use a metal one 18 in. long and 1½ in. wide. The third essential tool is an architect's scale, so that you can dimension structural members for cutting. The standard triangular-section scale works fine.

A miter box and a fine-tooth modeler's saw aren't essential, but they are very useful. The miter box we use is an aluminum, hobby-shop model with slots for both 45° and 90° cuts. A pair of dividers or a compass with two metal tips is great for laying out dimensions and marking wood to be cut. Even thin pencil lines are usually too wide to use in marking small pieces, so the pinprick left by dividers is useful. A large hat pin comes in handy for this same kind of scribing and for applying glue.

In addition to these basic tools, most modelmakers end up making or improvising a number of their own. When you're working at such small scale, it's amazing what household or workshop items get pressed into service: straight pins for temporarily fastening wood to wood, paper clips (which can be bent to form fine clamps) and long-nosed tweezers. The X-acto Co. (4535 Van Dam St., Long Island City, N. Y. 11101) makes a variety of useful blades to fit their basic razor knife. Dremel Co. (4915 21st St., Racine, Wis. 53406) makes such exotic modelmaking tools as miniature power saws and lathes.

Estimating materials—If you already have a materials estimate for the full-scale house, just use the lumber portion of it as your shop-

Building a structural model gives you a manageable replica to study before and during full-scale construction. Facing page: the author positions a roof truss on a model built to 1⁄24 scale (½-in. = 1 ft.). White or yellow wood glue, a razor knife or two, and a fine-toothed hobbyist's saw are some of the tools you'll need. Model construction follows the same sequence as actual construction, although it's possible to skip the foundation and start with floor framing, as was done here. Floor sections were only partially sheathed so that the floor-framing details would remain visible.

Using a cut-off jig like the one shown above enables you to mass-produce studs or joists of uniform length. The jig has a thin plywood bottom that can be wedged in different positions in the metal miter box, and a glued-down stop for the stock to butt against.

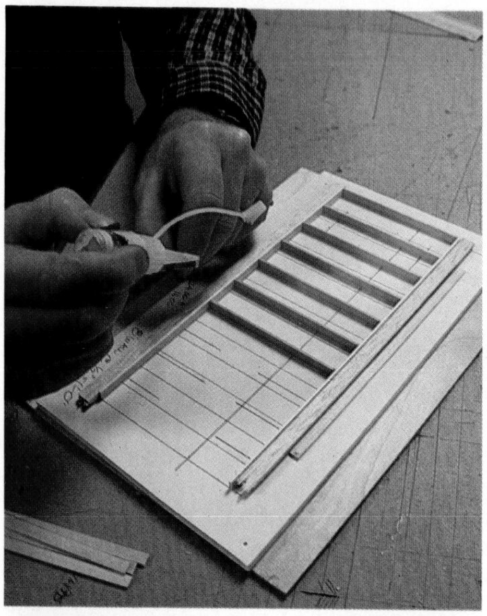

The wall-assembly jig, top, consists of two fixed, parallel guideboards that are spaced the same distance apart as the full plate-to-plate height of the wall. To use the jig, top and bottom plates are aligned against the guides and then studs are positioned over perpendicular layout lines and glued at top and bottom to both plates. Once the glue is dry, completed wall sections can be stood up and glued in place or glued to scaled-down sheathing or siding, as shown above. A razor knife is your ripsaw at 1⁄24 scale.

ping list for model materials—with one adjustment. Convert any listing of individual pieces into a total lineal-foot measurement for each dimension category. For example, a listing of 200 lineal feet of 2x4s is more helpful than a listing of 20 10-ft. 2x4s because model lumber usually isn't scaled to length. I add a waste factor of 5% to 8% to my materials list.

If you don't have a materials take-off for the plan yet, make a quick estimate for the model. You can figure liberally in this case. Carrying an inventory of 1⁄24-scale lumber over to your next model won't create a cash-flow problem. Once the model is complete, it will help you make an accurate list of framing and sheathing materials for the real house.

Before you buy materials, you have to decide how closely you intend to simulate the full-scale construction process. If your plans call for a plate on one wall composed of two 16-ft. 2x4s, will you use that combination on the model or will you use a continuous plate? Using the continuous plate will save time, but there's a lot to be said for building to the plans, especially if this is your first house. The more you can learn from your model, the better. As you work, imagine yourself doing the same thing at the site. It's easy to forget how heavy a 24-ft. 6x10 beam really is when you're lifting its counterpart with tweezers.

The model base and foundation—You need a flat, rigid working platform to build on, preferably one that's easy to mark up so you can outline your foundation. Particleboard and plywood (with one A face) both work well. Use ¾-in. material to get the rigidity you'll need to take your scale creation on site for reference. I make the base 2 in. to 3 in. larger than the overall size of the model, and sometimes allow a few extra inches on one side to serve as a cutting area for materials. Lumberyard remainder bins are a fairly reliable, low-cost source of base stock.

For the sake of authenticity, it's best to construct your model in the same sequence you'd follow in building a real house: foundation, floor system, first-floor walls, second floor and so on. Taking the measurements from the plans, draw an outline of the foundation on the base, representing the outside face of the foundation wall. Include the location of any interior support walls, piers or other structural members.

If I'm planning to build on a level site, I usually consider the surface of the model base to represent the top of the footing, and build foundation walls up from there. For hillside sites, it's possible to create a scaled-down slope by layering mat board (or something similar) on a flat base to correspond with actual site contours. If you don't foresee any problems with the foundation because it's straightforward, you can skip this step by using the top surface of the base as the top of the foundation. This way you'll begin by laying out the mudsill.

Double-check your foundation or perimeter-wall layout against your plans—square corners and accurate dimensions are just as important now as they are at full scale. If you elect to build a foundation for your model, particleboard or plywood are good materials to work with. At 1⁄24 scale, thicknesses of ¼ in., ⅜ in. and ½ in. correspond to full-size wall thicknesses of 6 in., 8 in. and 12 in., respectively. Cut the foundation walls to proper height and length with a saw and then glue them in place on your layout lines. Butt-joined corners are fine, but mitered corners give the model a cleaner look.

Floor systems—Whether your plans call for 2x6 T&G decking over girders or ¾-in. plywood over 2x joists, you should be able to duplicate this construction at 1⁄24 scale. To represent decking over girders, you can either lay down individual pieces or use a thin (1⁄16-in.) sheet of balsa or model-aircraft plywood, which is far stronger. Sometimes called wingskin, model-aircraft plywood is available in thicknesses ranging from 1⁄64 in. to 1⁄16 in. It cuts easily with several passes of a sharp razor knife, but to make short work of cutting individual sheets to consistent size I use a paper cutter.

Joisted floor systems take a bit more work. Once you've cut the joists to length, mark their locations on the plates or mudsills with dividers set to the center-to-center spacing of the joists. Prick a small hole with the point of the dividers at each joist location, and don't forget to indicate doubled joists, headed-off joists and other specials that are shown on your blueprints. Now dab the joist with a pinhead of glue and set it on its mark. Fight the temptation to use more glue than you need; a little goes a long way. Once you've got several joist runs glued down, eyeball them to make sure that they are running true.

To sheathe a floor or wall, you can use either sheet balsa or model-aircraft plywood. You might not want to sheathe your model completely at the floor, walls or roof, since this will cover many of the structural parts that you'll want to look at later.

Two jigs—To build our models, I use two jigs to speed construction and ensure accuracy: a cut-off jig for cutting studs to length, and a wall-assembly jig for putting all the parts together quickly and accurately.

The fastest way to cut modeling lumber to length is with a cut-off jig. The one I use is essentially a small, adjustable table that fits inside my model-maker's miter box, and has a stop block fastened permanently to one end (photo top left). The table is made from 1⁄16-in. aircraft plywood; it also protects the sawblade by preventing it from touching the metal base of the miter box.

To use the jig, mark a single stick of material to the length you want and position it with one end against the stop block and one edge against the side of the miter box. Slide the jig forward or back until the stick's cut-off mark lines up exactly with the blade-guide slot in the box. Then shim the plywood table firmly against the side of the miter box to hold its position—a business card or two should be

Rafters as well as roof trusses are modeled flat on a template or assembly jig and then glued in place atop the top plates, as shown at right. A time-saving alternative is to model the trusses from solid pieces of plywood or cardboard.

enough if the table has been sized correctly. Now cut your test piece and check its length. If it's right, you can load the box with five or six random lengths of material and gang-cut them to size. Just make sure that the stock is aligned straight and butted against the stop block as you cut. Scribing registration marks on both jig and miter box allows you to return the table to a given position without having to re-measure. Variations of this jig allow you to make repetitive cuts at different angles.

The wall-assembly jig is basically a platform about 8 in. long and an inch or so wider than the finished wall height. Two parallel wood guides are glued to the top and bottom of the platform; the distance between them should equal the wall height, from bottom plate to top plate. Perpendicular layout lines drawn on the jig between the guides correspond to the on-center spacing of the studs (photo facing page, center). Horizontal lines can be added to locate blocking or headers.

Using the jig is not much different from building a full-scale stud wall on a completed subfloor. Cut a bottom plate and a top plate to the length of your wall and align them on edge against the parallel guides. Now glue in your studs, dabbing a pinhead of glue on both ends and aligning each one on its proper layout line. The fit of the studs in the jig is important, so I cut them after it is built. I do this because it's easier to adjust the studs to the jig than vice versa, and getting the height of the wall correct is the most critical factor. The fit of the studs should be snug enough that they stay in place (before they're glued) when the jig is turned upside down, but if they bow (bowing is particularly easy with balsa), they are too tight. Keep a piece of 150-grit sandpaper nearby to make minor adjustments. Walls of most common lengths will fit within the 8-in. length of the jig. But you can also build longer walls simply by sliding completed portions through the jig.

Installing walls—A wall can usually be installed on the subfloor as soon as it leaves the assembly jig, though I like to let one wall assembly dry while I'm building the next one. When the second one is finished I install the first, and so on through all the first-floor walls and partitions.

The first step in installation is to mark the location of interior-partition walls because this is easiest to do before perimeter walls go up. Exterior walls can be installed using the outer edge of the subfloor as a guide; I usually begin with a pair of adjoining exterior walls.

To attach a wall to the subfloor, simply hold the wall in position after dabbing the sole plate (bottom plate) with glue. Walls are usually straight and square as they come from the assembly jig, but you can use a straightedge to take out the bow in a plate. Once two exterior

walls are up I install any partitions that attach to them. This construction sequence does two things: The partitions brace the exterior walls and help to hold them in place should I accidentally bump one. And I'm also spared the tricky maneuver of reaching between completed exterior walls to install interior partitions, which can demand the patience and finesse of a surgeon.

Other installations—The second and third floors are built in the same way, using whatever scaled-down components the plans call for. If walls of a different height are required, you have to modify your wall-assembly jig, or build a new one.

Builders often use floor trusses to frame the second story, because they allow long clear spans. You can model these in two ways: as a solid member or as a truss. The quickest and easiest method is to model the truss as a solid, using a ¹⁄₂₄ scale 2x member with the same depth as the truss. To model it as a truss with all its webbing, I use a modified wall jig.

You can prepare rafters with an angled cut-off jig similar to the straight cut-off jig mentioned earlier. A fairly accurate bird's mouth can be cut in each rafter with a razor knife,

though you can't lay it out with a framing square. The best approach is to make a template rafter by setting a single piece of rafter stock in position against the ridge and the top plate and marking it for plumb and seat cuts. Cut and adjust the fit with a razor knife until you get a close-fitting template, then use it to pattern the other rafters.

Roof trusses (photo above) can be fairly complex, and you might not want to spend the extra time modeling them exactly if you're planning to use prefab units. If this is the case, you can use one of four options: 1) Cut a solid piece of aircraft plywood or balsa to represent each truss; 2) build just the outer chords of each truss, leaving the webbing out; 3) ignore the roof framing structure altogether and build a tent-type roof out of mat board with cleats glued to the underside to anchor this sheathing to the top plate; 4) or build one or two trusses, webbing and all, as illustrations of how they would look, and build the rest of the roof using one of the other techniques. Any of these approaches will save time, particularly if at this point you're anxious to see the overall shape of the house. □

Mark Feirer lives in Eugene, Ore.

Floor Framing

With production techniques and the right materials, a solid, squeakless floor is a day's work

by Don Dunkley

Heading off the joists to make room for DWV lines, as shown here, is a lot easier if plumbing runs are considered before framing begins.

ing walls will have to be spread for waste lines that run parallel to the joists, or headed off with solid blocking to accommodate perpendicular plumbing runs (photo above). Doubling joists under a heavy cast-iron bathtub is also a good idea.

Next, lay out the joist spacing shown on your plans. This layout has to be adjusted so that the butting edges of the decking will fall in the center of a joist, allowing two sheets to join over a single joist. Stretch your tape from

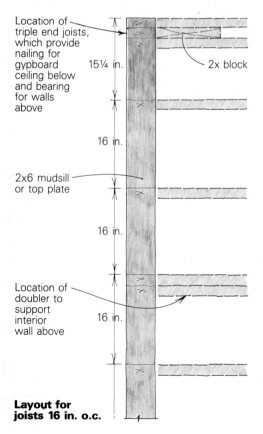

Layout for joists 16 in. o.c.

- Location of triple end joists, which provide nailing for gypboard ceiling below and bearing for walls above
- 15¼ in.
- 2x block
- 16 in.
- 2x6 mudsill or top plate
- 16 in.
- Location of doubler to support interior wall above
- 16 in.

the outside edge of the sill and make a mark ¾ in. (or half the thickness of a 2x joist) shy of the joist-spacing dimension—15¼ for 16 in. o. c. Put your X on the leading side (far side) of this mark for the first joist. The rest of the layout can be taken from this first joist at the required centers without any further adjustment. If your floor has joists coming from each side that lap over a beam or wall, the layout on one side will need to be offset from

the layout on the other perimeter sill or plate by 1½ in. to account for the lap.

After you have completed laying out the sills and midspan beams, check again to make sure that they jibe, and that the centers are accurate for the plywood. Now is the time to think; once you begin rolling joists, you just want to move.

Unlike rim joists, the end-blocking method requires cutting special-length blocks to correct for layout and accommodate double joists.

Getting ready to roll—There are two ways to secure the ends of the joists. The first is to run them the full width of the house, and block in between at their ends (photo above). This system has some structural advantages since the joist ends are locked in, but with all of those short, brittle pieces involved, it requires more fussing to get straight lines. The other way is to cut the joists 1½ in. short on each end, and run a 2x rim joist (also called a header joist, ribbon or band joist) perpendicular to the joists and face-nail it to them.

Whichever system you use, all the joists should be laid flat on the sills right next to their layout marks before you begin to frame. Stocking is basically a one-man job. As you pull each one up, sight it for crown. Many framers mark the crown with an arrow (make sure it points up), but it's not necessary if you start at one end of the building, and lay each joist down so that the crowned edge is leading. This way, when you return to the beginning point, you'll be able to reach for the leading edge of each one and tilt it up or roll it, leaving the crown up. Save the really straight lengths for rim joists. If any of the joists are badly crowned, cut them up for blocking.

Rim joists—The rim joist for one side of the building should be installed before any of the joists. Mark its top edge with the same layout as the sill it will sit on. If a large crown holds the rim up off the top plate or mudsill, cut the rim joist in half on a joist layout mark. Also, if this floor is over a crawl space, you will need to cut in vent openings. Place these every 8 ft., starting 2 ft. in from the corners, where air tends not to circulate. Typical vent openings are 5 in. high and 14½ in. long; these fit nicely between joists on 16-in. centers.

Install the rim joist with toenails from the outside face of the board down through the sill or top plate every 16 in. Once nailed, string the outside edge to make sure it is

straight and adjust it in and out if necessary with a few more toenails. Now snug up the joists that are resting flat on the plate or sill so they butt the inside face of the installed rim joist (this is easily done by catching the face of the joist that's up with the claw of your rip hammer and pulling toward the outside of the building, as shown below). If any joist ends are badly out of square, trim them.

Using the perimeter mudsills or plates, which have already been carefully plumbed and lined, as a kind of tape measure, mark

The author snugs a joist against the rim to check the end for square, using the building itself as a ruler for cutting the joists to length.

Kiln-dried framing makes better joists, but in the West green Douglas fir is much more common. Even so, the ideal moisture content of a joist is 15%. Many of the problems with floors—in particular squeaking and nail-pop—can be traced to wet joists shrinking away from nail shanks as the wood dries. Wet joists can permanently deflect once they begin to dry, causing a swale in the floor. Most of the so-called settling in a platform-framed house is due to the joists shrinking across the grain. Much of this can be eliminated by using dry lumber. Putting down a plastic moisture barrier over the earth in damp crawl spaces is a good way of keeping the floor above dry.

Cutting to length. The joists are squared up 1½ in. short of the building line (below) to leave room for the rim joist, and cut to length. With their crowns marked by an arrow, they are ready to be tipped up.

Rolling the joists. Joisting (right) is fast if a routine is established between partners. On the open side of the joists, an end toenail holds the joist on layout; toenailing the joist face to the plate will follow. On the other end, the rim joist is face-nailed to the joist with three 16d nails, then toenailed to the plate. The end joist in the foreground will be doubled later.

Installing the other rim. With joists nailed off, the open side gets its rim (bottom). This floor uses 2x14s; 2x10s are more typical.

Marking up the plans

The first step in laying out is to add to the floor plan any further information needed to detail the plates later on. You need to write in the header length for each door and tt if it will require double trimmers. Windows show two numbers—the header length and the height of the sill jacks. You also need to mark beam pockets and the point loads under them, rake walls, and the locations of non-standard stud and header heights. The last notation is the direction the stud layout will run.

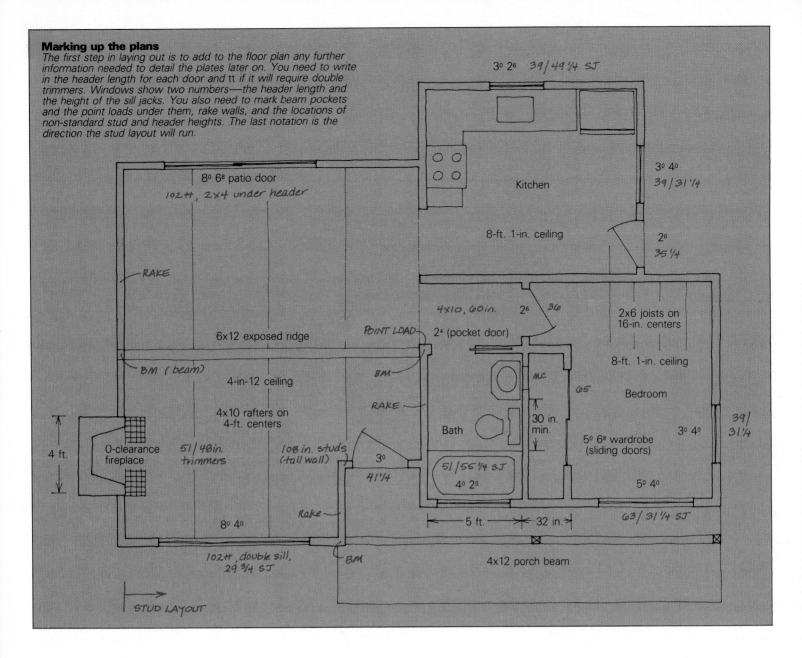

ments that should be part of the layout. I do this in the evening when I can slow down my pace a little bit. Thumb through the drawings just to review the general structure. Read the rough and finish carpentry specifications, and underline anything that isn't standard.

Next, go back to the first-floor plan in the blueprints. Write on the plans, next to the appropriate opening, the length of each door and window header. This will allow you to measure out and mark them on the plates later on without having to stop and figure, and will also allow whoever does the framing to make up a list and cut all of the header stock, sills and sill jacks at once. See the facing page for what these framing members are, and how to figure their lengths.

The drawing above shows a blueprint that contains many of the framing situations you will encounter. I'll use these same plans in explaining snapping out, plating and detailing. In this case, the blueprint is marked up with the necessary information for laying out. For doors, only the header length is written in; when nothing else is noted, this indicates a standard 6-ft. 8-in. door in an 8-ft. wall.

In the case of the aluminum patio door in

the living room, a 2x4 is needed to fur the header down to the right height. The *tt* next to the header length indicates double trimmers for this 8-ft. opening. The pocket door to the bathroom shows the length of the header, and its narrower width (4x10). Windows show two numbers: the first is the length of the header, and the second is the length of the sill jacks.

With the exception of the rake walls, which will be snapped out full scale on the deck, I have marked the only wall that doesn't use standard 8-ft. studs (the jog in the living room) with the notation *108-in. studs (tall wall)*. With a hodgepodge of elevations, it helps to mark up the plan view with colored pencil to differentiate stud heights.

The plans have also been marked for beam pockets *(BM)* on both ends of the ridge beam in the living room, and where the porch beam bears on the exterior wall. Resulting point loads (where significant loads have to be carried down to the ground) also have to be marked. The interior end of the ridge beam might not be picked up below without a notation. Also, studs should be doubled under joists that will be doubled on the floor above.

You'll also need to mark the dimensions of

any prefabricated items that will go into the house, like a medicine cabinet or roof trusses. (When using trusses, the building width can always err slightly on the narrow side, but don't ever make it too wide.)

One of the last things you'll want to note on the plans is at which end of the building you'll start the regular stud layout. The decision is yours. The side with the fewest jogs or offsets in the exterior walls is usually the place to start. When you reach a jog, compensate with the stud layout so that the studs, joists and rafters (or trusses) all line up, or *stack*. If the plans call for 2x6s on 2-ft. centers with trusses, code requires that they stack. It's a good idea anyway, both for increased structural strength, and for making multi-story mechanical runs like heat ducts easier.

In the drawing, I decided to pull my layout from the bottom left corner. By beginning there, the studs will pick up the rafter layout in the living room (exposed 4x10s on 4-ft. centers) and will work well with the ceiling joists and the porch overhang. You can also write down the cumulative dimensions of rooms, so that when you're snapping out, you'll be able to mark all of the intersecting

Laying out for Framing

A production method for translating the blueprints to the wall plates

by Jud Peake

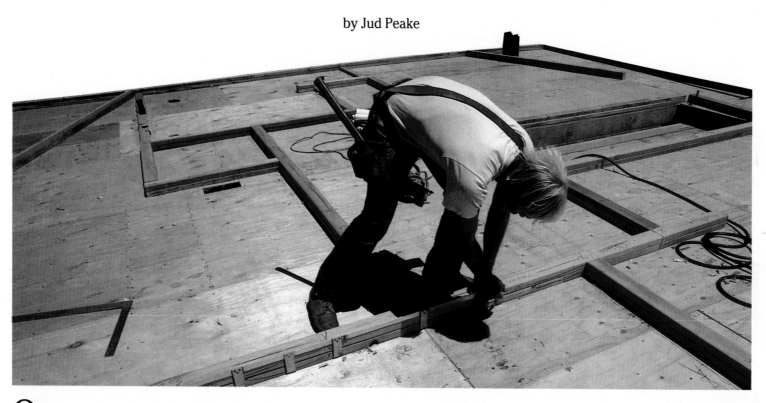

On big production projects where jobs are highly specialized, the carpenter assigned to layout uses a hammer very little. His tools are a lumber crayon (known in the trade as keel), a pencil, a layout stick, a channel marker and a tape measure. If he marks the top and bottom plates correctly, the carpenters may not even need a tape measure to frame up the walls. All of the figuring that involves the plans—and there are a lot of variables to anticipate—is done at the layout stage.

On large jobs, I don't necessarily frame the house I lay out. On houses I contract, I do both jobs. Either way, I treat layout as if there won't be anyone around to answer questions when the framing starts. Doing the layout as a separate operation increases the speed and accuracy of framing, whether you're building your own house by yourself or working as part of a big crew. The plans are an abstraction of the building to be constructed. Layout systematically translates your blueprints into a full-size set of templates—the top and bottom plate of each interior and exterior wall on a given level of the house. The pieces of the puzzle can then be cut and framed in sections.

There is little about a finished house that isn't determined by the layout. The actual procedure falls into four distinct steps. Within each step, you have to deal with wall heights

and the locations for windows, doors, corners, partitions, beams and point loads. You also have to deal with *specials* (which is anything else and usually means prefab components).

The first step in layout is to go over the blueprints and mark them up with the information you'll need, in the form it will be most useful. Step two is to measure out the slab or deck and establish chalklines representing every wall on that level. This is called *snapping out*. In step three you'll decide where walls begin and end and which ones will be framed first, and then cut top and bottom plates for each wall. This is called *plating*. Finally you *detail* the plates by marking them with all the information the framer will need to know to build the walls. Layout requires a thorough knowledge of framing. But even then, regional framing techniques vary widely.

Layout principles—Layout is based on parallel lines. If two lines are parallel and one is plumb, then the other will be plumb. Also, if a pair of lines meet at a right angle, then another pair of lines, each parallel to its counterpart in the first pair, will meet at a right angle. Stated less theoretically, put a 2x4 on top of another 2x4 and cut them to the same length. Using a square, draw a line across the edges of both every 16 in. Frame in between

these top and bottom plates with studs of equal length, and stand the wall up and plumb one end. Now all of the studs will be plumb. Doors and windows that have been laid out with these studs will be plumb. And if your foundation and deck are square and level, then any interior walls paralleled off the outside walls will also be square and plumb.

The second principle is to do all the layout at once. This way all the plates are fitted against each other, and you can be confident that the walls will work together before they grow vertically and become unmanageable.

The third principle is my own: avoid math. You can do most of your figuring in place. If you're looking for the plate and stud lengths of a rake wall (one whose top is built at the pitch of the roof it supports), snap it out full scale on the deck. Measure things as few times as possible. Once the walls are mapped out on the deck in chalk, don't measure them and then transfer this number to the plate stock sitting on sawhorses. Instead, lay the plate material right on the line and cut it in place. This saves time and reduces error—the kind that has to be fixed with a cat's paw.

Marking up the plans—Before you step out on the slab or deck, you need to go over the blueprints and systematically pick out the ele-

Rough framing isn't measured in thirty-seconds of an inch, and for good reason. Although a shabby frame can put finish carpenters in a murderous mood, it gets covered up and forgotten behind siding, drywall and paneling. But floors, if done poorly, will come back to haunt you. That squeak just outside the bedroom door is an annoyance for which there is no quick fix. But the problems can go beyond the floor itself. The blame for eaves that look like a roller coaster once the gutters are hung often rests squarely on a carelessly built floor two stories below.

Haste doesn't usually create these problems; using inappropriate materials and not knowing where to spend extra time does. In fact, by using production techniques and materials, rolling second-story joists (tipping them up and nailing them in place) and sheathing them with plywood on a modest-sized house is a day's work for my partner and me.

What a floor does—Most builders are very aware when they hang a door of the abuse it will get, but they don't think in the same way about the floor they are framing. This skin is the main horizontal plane the building relies on to transfer live loads (ones that are subject to change, like furniture and people) and dead loads (primarily the weight of the structure itself) to the bearing walls, beams and foundation. Critical here is the amount of deflection in both the sheathing and the joists. The subfloor must be stiff enough to handle both general and concentrated loading, and to support the finish floor that rests on it without too much movement. But at the same time it must be flexible enough for comfortable walking and standing, something a concrete slab can never be.

Floor systems—There are many different ways to build a floor, and the preference for one system over another has a strong regional flavor. Floor trusses—either metal or wood—are becoming popular because of their prefab economy, and because they can eliminate the need for bearing walls, beams and posts by free-spanning long distances. Girder systems are still used over crawl spaces in some areas. They are generally laid 4 ft. o. c. with their ends resting on the mudsill of the foundation wall, or in pockets cast into the wall itself, or on metal hangers attached to the foundation. The interior spans of the girders are typically posted down to concrete piers. Girders are decked with either 2x T&G or very heavy plywood designed for this purpose.

But most wood-frame houses still use joists. And whether a floor rests on foundation walls on the first story or on stud walls on a higher level, its elements are pretty much the same. Joists are typically 2x lumber, laid out on 12-in., 16-in. or 24-in. centers and nailed on edge. They are held in place on their ends by toenails to the sill or plate below, and then attached to a perimeter joist, or blocked in between. Blocking or bridging has also been used traditionally to stabilize a floor at unsup-

ported midspans. (For more on this, see the next page.)

The last element of any joist system is the skin—usually plywood sheathing. This decking not only forms the continuous horizontal surface, but is also the key to the integrity and structural continuity of any wood floor. I'll talk more about the choice of material and how to apply it later.

In the simplest floor system, joists rest on the mudsill of exterior foundation walls, or on the double top plate of the stud walls on higher stories. With the long spans found in most floor plans, joists have to be either very deep (say, 12 in. or 14 in.), or supported at or near the center of their length by beams or bearing walls, creating two shorter spans. In the case of even greater building widths, joists can come from either side of the building, lapping each other over these supports, with blocking nailed in between.

In much of the country where basements are used to get below the frost line, the joists bear on or lap over built-up beams of 2xs spiked together. These are usually supported by Lally columns—concrete-filled tubular steel posts with flanges top and bottom—or by hefty wood posts.

In the West, where I live, the first-story floor usually stretches over a crawl space. Perimeter support for the joists is provided either by stemwalls—mudsill-capped low foundation walls (most crawl spaces use the minimum allowable height of 18 in. from grade to the bottom of the joists)—or by cripple walls (sometimes called pony walls) that are framed up on hillside foundations to reach the first-floor level.

For supporting the joists in mid-length or for lapping them, floors over crawl spaces either use a carrying beam, or girder, on 4x posts that rest on concrete pier blocks, or use a crib wall—a low, framed wall studded up from an interior concrete footing that runs the length of the building.

Unlike girders, which are often crowned and leave a hump in the finish floor, crib walls can be plumbed and lined to make a perfectly straight, level surface. They will also shrink less than a large girder, and just feel a lot more solid to me. The posts under a girder are toenailed at the bottom to the wooden block that caps each concrete pier. This block inevitably splits, making the whole post-and-beam assembly feel a little cobbled together. I've also learned that concrete contractors get bored with straight lines at the point when they begin laying out piers, so supporting posts often get precious little bearing.

Bolting down the sill—Floors need to be level, square and precise in their dimensions. On a second floor, these conditions will depend on your accurately plumbing and lining the first-floor walls; but on the first floor, setting the mudsill is the critical step. First, check the perimeter foundation walls with a steel tape for the dimensions shown on the plans. Then pull diagonals to check for square. Run 3-4-5 checks on the corners for

square if the floor isn't a simple rectangle. Once you're sure of your measurements, snap chalklines that represent the mudsill's inside edge on the top of the foundation walls.

Be thinking ahead to the sheathing (called shear panel where I build) and siding if the foundation walls aren't arrow-straight. If these are one and the same, the plywood will need to hang down over the concrete foundation wall an inch or two. This means that you're better off having the mudsill overhang the foundation slightly, rather than the other way around. If finish siding is used over sheathing, this isn't as important, since the sheathing can butt the top of the foundation where it bows out from the mudsill.

Place the mudsill (either pressure-treated or foundation-grade redwood) around the perimeter of the foundation, and cut it to the lengths needed to fit on the stemwalls. You'll find foundation bolts placed on maximum 6-ft. centers; code requires each piece of sill to be held by at least two bolts. The cut sills should be laid in position on their edges along the top of the stemwalls so that they can be marked for where the anchor bolts fall along their length.

After carefully squaring these marks across the face of the sill, lay them flat so you can mark the bolt positions across their width. Measure the distance from the chalkline on the top of the stemwalls to the center of each bolt. Then transfer that measurement to the face of the mudsill, taking care to measure from the inside edge of the mudsill each time.

Production carpenters who do a lot of layout use a bolt marker. One of these can be made from an old combination-square blade, as shown in the drawing below. The end of the metal blade is notched so that it will index from the centerline of a ½-in. foundation bolt. At 5½ in. from this end for a 2x6 mudsill (or 3½ in. for a 2x4), a hole is drilled in the blade and fitted with a screw or nail.

To use this marking gauge, the mudsill has

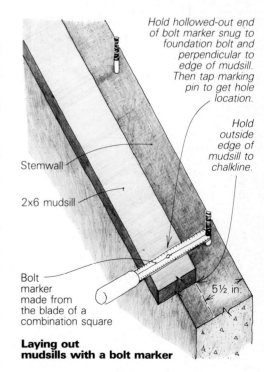

Hold hollowed-out end of bolt marker snug to foundation bolt and perpendicular to edge of mudsill. Then tap marking pin to get hole location.

Hold outside edge of mudsill to chalkline.

Stemwall

2x6 mudsill

Bolt marker made from the blade of a combination square

5½ in.

Laying out mudsills with a bolt marker

Bridging midspans

Although the days when all carpenters learned the trade through a formal apprenticeship are gone, construction knowledge is still passed down through the ranks, from those who know to those who want to. This kind of conservatism is often valuable, but it also means that outmoded methods die hard. Bridging between joists at midspan is a good example. Even on relatively short spans, a lot of knowledgeable builders are convinced that bridging guarantees a stiffer floor. But convincing studies show that it doesn't, and that information isn't new.

Bridging is a continuous line of bracing that runs perpendicular to the direction of the joists, and is installed between them. It can take the form of cross bridging (diagonal pairs of wood or metal braces that form an X in the joist space) or horizontal bridging (continuous, full-depth solid blocking).

A study done by the National Association of Home Builders Research Institute Laboratory in 1961 put most of the myths to rest. It proved with laboratory and field tests that bridging at midspan had little effect in stiffening a floor once it was sheathed. It was also of little help in transferring concentrated loads laterally to other joists, reducing floor vibration or preventing joists from warping. Bridging actually does all of those things as you install it,

but once the plywood subfloor is on, its net effect is negligible.

Still the myth persists. Building in the 1970s, I put in midspan solid blocking every 8 ft. on floors I built, to meet the code and because I was concerned with quality building. But there is only one way in which midspan bridging improves a floor substantially, and that is in providing lateral stability. Because joists are on edge, they tend to tip unless they are pinned in at their ends and at support points like girders and bearing walls. Plywood sheathing nailed and glued on top also helps. But over long spans, the joists can use a bit more help.

Of the three model codes in the U. S., only ICBO's Uniform Building Code requires using cross bridging or solid blocking with plywood sheathing under normal conditions, and that is only when the depth-to-thickness ratio is 6 or more, as with 2x12s. It is then required every 8 ft. along the span of the joists. BOCA requires bridging or blocking only if the joist depth exceeds 12 in. under normal loading conditions, and SBCC allows you to skip it entirely in single-family residential buildings.

If you do need to install midspan bridging or blocking on a floor you're building, a few simple tricks will make it go faster and eliminate some of the squeaks. First, you must decide between diagonal bridging, which can either be wood 1x3s or the newer sheet-metal bracing, and solid blocking.

I think that cutting and nailing the old wooden cross bridging is needlessly time-consuming, although all the standard construction texts cover it in detail. If I were going to use bridging, I would choose the metal variety that doesn't use nails. These straps have sharp prongs on their ends, and eliminate the squeaking you get from nails rubbing on the metal bridging when the joists flex underfoot. This kind of bridging can be installed from above or below the joists; a combination of the two is even better. Drive the upper ends of the bridging into the top of each joist face before sheathing, and return to drive the bottoms in from below when the joists have done most of their shrinking, and initial loading has flattened them some. Make sure to separate each member of the pairs to prevent contact squeaking.

Solid blocking has to be done just before sheathing, but it's straightforward. If you're going to insulate the floor yourself, it's the best choice. To begin, chalk a line across the joists where the blocking goes. This is typically at the center of any span under 16 ft., or in two rows equally spaced if the span is greater. Alternate the blocks on either side of this line so you can face-nail them through the joists.

Set a plank on top of the joists, parallel to the chalkline and about a foot away. Set it to the left of the line if you're right-handed. Spread out your blocks along this plank. Drive three 16d nails partway into the face of each joist down the line, alternating between the right and left side of the line. This allows you to position each block and then anchor it with a single blow of the hammer. Once you've traveled the entire length of the joists nailing the blocks on one end, you can turn around and travel back the way you came, nailing the other ends.

Construction adhesive recommended for solid blocking is supposed to cut down on squeaks, but it's messy and time-consuming. It's more practical to drive shims in any gaps that open up between the blocks and the joists. Don't use vinyl sinkers (vinyl-coated nails) or brights (uncoated steel nails). These will loosen up and squeak much sooner than ring-shanks, screw nails or hot-dipped galvanized. —*Paul Spring*

to be held flat with its outside edge on the chalkline. This is most easily done from inside the foundation. Make sure that the sill is exactly where you want it along its length. Then by holding the bolt marker perpendicular to the length of the sill, with the notch of the gauge pressing against the bolt, tap on the marking pin with your hammer. Whatever the location of the bolt, the correct bore center will be left on the sill.

After the sills are laid out for bolts and drilled with a slightly oversize bit (a 9/16-in. bit for ½-in. bolts is a good compromise between requirements for a snug fit and giving the carpenter a break), place them on the stemwall and hand-tighten the nuts. If termite shields or sill sealers (wide strips of compressible insulation that act as a gasket to keep air from passing between foundation and framing) are to be used, be sure to lay them in underneath the mudsill. Now is also the time to adjust the foundation for level by shimming under the mudsill where necessary with grout. Go back and cinch the nuts down tight once everything is set.

With a hillside foundation where the joists will rest on a cripple or pony wall, adjustments for level are made in the cripple studs themselves rather than by shimming the plate. If the foundation is stepped down the hill, these cripple studs will be in groups of various lengths. If a grade beam that parallels the slope is used, then the cripples will have to be cut like gable-end studs. In this case the beveled end will be nailed to the mudsill and blocked in between. In either case, getting the top plates precisely level (and straight) is worth the trouble. Do this by shooting the tops of the studs with a transit, or by setting up a string-and-batterboard system. Using a spirit level for long runs just isn't accurate enough, and trusting a hillside foundation to be true is a mistake you'll make only once.

Joist layout—The first thing I do to lay out the sills for rolling joists is to get out the plans again and find all the exceptions to the standard layout. These include stair openings that have to be headed off, lowered joists for a thick tile-on-mortar finished floor, plumbing runs that need to stub up on a joist layout, and the double joists used for extra support. Even a simple floor will require doublers to pick up the weight of the building at its most concentrated points.

The first place to double is at each end of the foundation, parallel to the direction the joists run. When joisting a second story, sandwich short 2x spacer blocks between two joists for a tripler (drawing, facing page). This will give you backing for the ceiling drywall below. Locate all interior walls running parallel to the joists and lay out the sills (and carrying beams if you have them) for doublers too. Since some plumbers feel that joists have just as much integrity cut in half as they do whole, it's a good idea to have a chat with your pipe bender before you complete the joist layout. This will save a lot of headaches later. For instance, the doublers under plumb-

The Pieces to the Puzzle
An introduction to the components of a modern frame and how to size them

If you've done some framing, it takes only a second of looking at the detailing on a set of plates to know what the wall will look like when it's framed and raised. After that it takes only a pencil and a 2x4 scrap to make up a cut list of headers, sill jacks, rough sills, trimmers, channels and corners. But if you're new to how all of this goes together, you'll need to understand the different framing styles and the many exceptions created by using different kinds of windows, doors and finish.

Below I've explained how the basics work. Since I learned my framing in the West, my explanation will focus on how it's done there using basic production techniques, but I'll also point out more traditional methods as I go along.

Stud walls—Studs (2x4s or 2x6s on 16-in. or 24-in. centers) hold up the roof or floor above, and provide nailing surfaces at regular intervals for the interior and exterior finish. Precut studs are 92¼ in. long and are used to build a standardized 8-ft. wall that works economically with standard plywood and drywall sizes. The actual height of this wall is 8 ft. ¾ in. once you add 1½ in. each for the *bottom plate (sole plate)* that sits under the studs, and the two *top plates* (the *top plate* and the second top plate, called the *double top plate* or *doubler*) that complete the wall. Architects sometimes add to the confusion by specifying this same wall as 8 ft. 1 in. Wall studs are seldom shorter, but where economy isn't as important, they are often longer. If walls are over 10 ft., they will require *fire stops*—horizontal blocking that slows down the upward spread of fire.

To find stud height, check the elevations and sections in the blueprints. What will be listed here are wall heights, usually shown as finished floor to finished floor *(F.F. to F.F.)*. In most cases, you can use the same dimension for rough floor to rough floor. To figure the stud length given this dimension, subtract 4½ in. (the thickness of the three plates), plus the subfloor thickness and joist depth.

Rake walls—These are also called gable-end walls, and they require the framing to fill in right up to the bottom of the pitched roof. This means that each stud will be a different length and will be cut at the roof pitch on top. Typically, a rafter will sit on the top of these walls. There are two ways to frame them. One is to build the lower part of the wall just as you do the walls under the eaves of the roof and then fill in the gables later. This is fine if there is a flat ceiling at the 8-ft. height. The other way is to build the wall in one unit with continuous studs. This is necessary for cathedral ceilings. See p. 73 for the specifics on figuring lengths and angles.

Wall intersections—When one wall intersects another, the framing has to provide a solid nailed connection between the two walls, and backing for the interior finish. The drawing above left shows how this is usually handled. Money can be saved by replacing backing studs with nail-on drywall corner clips, and by reducing corner units to a simple L, but there are disadvantages to these cutbacks.

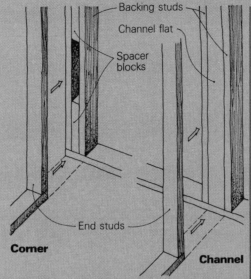

Corner

Channel

Backing studs
Channel flat
Spacer blocks
End studs

Doors—Door openings require *headers* to shift the weight of the roof in that area to both sides of the door. The vertical 2x support at each end of a header is called a *trimmer* (or cripple); the stud just outside of the trimmer that nails to the end of the header is called a *king stud*. The *rough opening* is the *rough-opening width*, measured between the trimmers, by the *rough-opening height*, which is measured from the floor to the bottom of the header.

Finding the height of the trimmers and the length of the header requires working backwards from the dimensions of the finished door. But you don't have to go through the full process each time—door headers should be 5 in. longer than the nominal width of the door. For example, a 37-in. header is needed for a 2-ft. 8-in. door. This 5-in. increment assumes that the trimmers will be framed very nearly plumb. Some carpenters use 5⅛ in. or even 5¼ in. to allow for sloppier framing.

Adding 5 in. to the header accommodates two 2x trimmers (1½ in. apiece, for a total of 3 in.) under the ends of the headers (drawing, above right). This leaves a rough opening 2 in. wider than the door. The remaining room is for two 1x side jambs (¾ in. apiece, for a total of 1½ in.), and ½ in. for shim space. (This leaves ¼ in. on each side.) Allow another ½ in. for exterior doors whose rabbeted jambs are closer to ⅞ in. thick. French doors will require closer to ½ in. of extra header length to account for the astragal (the vertical trim between the two doors that acts as a closure strip). If the door opening is wider than 8 ft., then the trimmers will need doubling, which requires another 3 in. of header length.

In the West, it's common to use 4x12 Douglas fir header stock in all 8-ft. walls. This system is fast, because all the framer has to do is cut the stock to length and nail it to the top plate—there are no *head jacks* to toenail, and you end up with a header at the right height that will span almost any opening. This system is admittedly wasteful, but gets around the cost of labor. Fir is still relatively inexpensive in large hunks, and the labor to cut and install the head jacks isn't.

If you aren't using 4x12s, check a span table for the correct header size (unless you are dealing with a non-bearing interior wall, where two flat 2x4s will do nicely). Typically, when

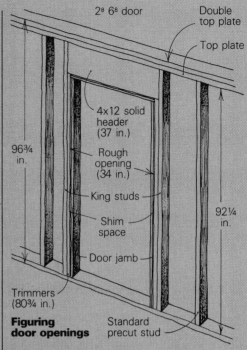

2⁸ 6⁸ door
Double top plate
Top plate
4x12 solid header (37 in.)
96¾ in.
Rough opening (34 in.)
King studs
Shim space
Door jamb
92¼ in.
Trimmers (80¾ in.)

Figuring door openings
Standard precut stud

solid headers aren't used in 2x4 bearing walls, the choice is a laminated header made on site from two lengths of 2x with ⅜-in. plywood sandwiched between. Using either system, if a wall exceeds 96¾ in. in height, head jacks (cripples) will be needed between the header and the top plate.

Using a standard 6-ft. 8-in. door, the trimmers should be cut 80¾ in., no matter how tall the wall is, or what kind of header you use. Once the bottom plate is cut out within the doorway, this will leave a rough-opening height of 6 ft. 10¼ in. This height will accommodate the door (6 ft. 8 in.), the *head jamb* (¾ in.), and enough play for the finish floor and door swing. An aluminum patio door requires 1½ in. of furring under the header. Pocket doors and some bifold doors require an extra 2 in. of rough-opening height for their overhead tracks. In an 8-ft. wall, a 4x10 held tight to the top plate works nicely. Even if you are not using solid header stock, you'll need trimmers that are 82¾ in. for these doors.

Windows—A rough window frame is like a door opening with the bottom filled in—that's just how you frame one. The rough width of a window is measured between the trimmers. The rough height is measured from the bottom of the header to the top of the *rough sill*. This is a flat 2x, doubled if the window is 8 ft. or wider, that runs between the inside faces of the trimmers. Unless it's otherwise noted on the plans, windows are framed with the same height trimmers as the doors.

In some areas of the country, the trimmers are installed in two pieces *(split trimmer, or split jack)* with the rough sill cut 3 in. longer and sandwiched between (see *FHB #15 p. 43*). If double trimmers are used, then the inside pair can be framed this way. Underneath the rough sill, the stud layout is kept by *sill jacks* (or cripples) which are in essence, short studs.

Finding the length of the rough sills and the sill jacks is fairly simple. Depending on how you deal with the trimmers, the length of the rough sill will be the same as the width of the

Figuring rough window openings

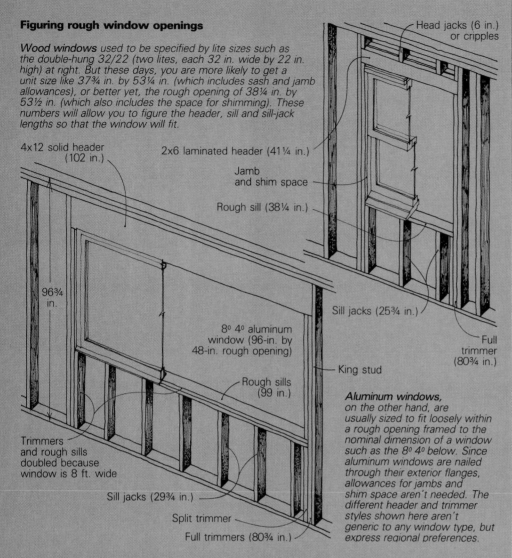

Wood windows used to be specified by lite sizes such as the double-hung 32/22 (two lites, each 32 in. wide by 22 in. high) at right. But these days, you are more likely to get a unit size like 37¾ in. by 53¼ in. (which includes sash and jamb allowances), or better yet, the rough opening of 38¼ in. by 53½ in. (which also includes the space for shimming). These numbers will allow you to figure the header, sill and sill-jack lengths so that the window will fit.

4x12 solid header (102 in.)

2x6 laminated header (41¼ in.)

Head jacks (6 in.) or cripples

Jamb and shim space

Rough sill (38¼ in.)

96¾ in.

8⁰ 4⁰ aluminum window (96-in. by 48-in. rough opening)

Sill jacks (25¾ in.)

King stud

Full trimmer (80¾ in.)

Rough sills (99 in.)

Trimmers and rough sills doubled because window is 8 ft. wide

Sill jacks (29¾ in.)

Split trimmer

Full trimmers (80¾ in.)

Aluminum windows, on the other hand, are usually sized to fit loosely within a rough opening framed to the nominal dimension of a window such as the 8⁰ 4⁰ below. Since aluminum windows are nailed through their exterior flanges, allowances for jambs and shim space aren't needed. The different header and trimmer styles shown here aren't generic to any window type, but express regional preferences.

rough opening or 3 in. longer. To get the length of the sill jacks, add the rough opening height to the thickness of the rough sill, and subtract this from the height of the trimmer.

The length of a window header has a lot to do with what kind of window will fill the hole—aluminum or wood. To get the header length on an aluminum window less than 8 ft. wide, just add 3 in. to the nominal window width to allow for a trimmer on each side (drawing, above). This is because aluminum windows are generally manufactured to fit loosely in a rough opening of their nominal size. Check with your supplier to be safe. The first number stated for windows (and for doors) is the width— a 3⁰ x 5⁰ aluminum window needs a rough opening 36 in. wide and 60 in. high. For windows wider than 8 ft., you'll need to use double trimmers, which will increase the length of your header 3 in.

Wood windows were once specified by the size of the lite, or glass, which didn't account for the sash or frame that surrounded it. The rules of thumb about how many inches to add for the sash to get the rough-opening size are quite general because the width of stiles and rails differs among manufacturers. The rules are also specific to window type (double-hung, single or double casement, sliders, awning and hopper) since the number and width of their stiles and rails vary.

Today, most manufacturers' literature gives a unit dimension (this includes sash and frame) and a rough opening dimension. Use the rough-opening measurements for laying out, and add 3 in. to get the correct header

length on windows under 8 ft.; 6 in. over 8 ft. If the rough-opening size isn't specified in the plans or by the manufacturer, you've got to measure the windows on site. If they are set in jambs, add 3½ in. (two trimmers plus ½ in. for play) to get the correct header length; if not, add 5 in., as you would for a door.

Specials—This catchall category includes any member or fixture in the frame that's big enough to worry about. Medicine cabinets and intersecting beams are *specials.* Toilet-paper holders aren't; they can be dealt with later.

Your main concern with most specials is to provide backing for the interior wall finish, and a break in the regular stud layout if an opening is required. Beam pockets are simple. Imagine a ridge beam that is supported on one end by a rake wall. There are a number of ways to frame the bearing pocket for the beam, but any of these schemes should include a post or double stud beneath the beam, and backing on either side of it for nailing wall finish or trim.

Tubs, showers and prefab medicine cabinets require both backing and blocking. Blocking is what it sounds like—short horizontal blocks or a 2x let into the studs for nailing the wall finish. Backing is usually an extra stud and does the same thing only vertically. In the case of a corner bathtub, then, the line of blocks that sits just above the lip of the tub where the drywall will be nailed is the blocking. At the end and side of the tub where the wallboard will butt, you'll find an extra stud for backing.   —*Paul Spring*

Snapping out. **Facing page: Peake uses his foot to anchor one end of the chalkline while snapping out a short interior partition. The red keel X by his right foot tells the framer on which side of the line to nail down the wall once it's raised. This project was unusual because the exterior walls were framed, sheathed and raised before the interior was laid out.**

interior walls along one side of the building by pulling the tape from just one point rather than having to do this room by room.

Snapping out—Snapping out isn't much more difficult than redrawing the architect's floor plan full size on the deck. Using a chalkbox, you need to snap only one side of each wall and draw a big X with keel every few feet on the side of this line that the wall will sit.

Measure for the outside walls first. Come in 3½ in. (for a 2x4 wall) or 5½ in. (for a 2x6 wall) from the edge of the deck. Be sure to measure from the building line, not from the edge of the plywood, in case it's been cut short or long. Don't even trust the rim joists without checking them with a level for plumb, since they may be rolled in or out.

Snapping lines should go quickly. To hold the end of your string on a wood deck, just hook it over the edge of the plywood, or use a nail or scratch awl driven into the deck. If the slab is very green (poured less than a week before), a drywall nail will usually penetrate the concrete; if not, use a concrete nail. You can even hold the string with one foot if the wall that you're snapping is short (photo facing page).

On very long walls, especially on windy days, have someone put a finger or foot near the middle of the line and snap each side separately. This will keep the chalkline truer. If you're working alone, close the return crank on the chalkbox to lock it and hook it over the edge of the deck so it's secure on both ends.

I use red chalk for layout. Lampblack also shows up well, but blue isn't a good choice in my area because it's the favorite of plywood crews. When I expect rain or even heavy dew, I use concrete pigment (Dowmans Cement and Mortar Colors, Box 2857, Long Beach, Calif. 90801) instead of chalk. If you're doing a lot of layout, get yourself a couple of chalkboxes with gear-driven rewind; if you've made a lot of mistakes, correct them with a different color chalk. When it comes to keel, I use red mostly; blue has a way of disappearing.

In snapping out the exterior walls, don't be concerned about intersecting lines where the walls come together. This problem will be solved when you do the plating. Just concentrate on getting the lines down accurately. Once you've done that, check the lines for square by measuring diagonals, and check the dimensions again carefully. On slabs and first floors, check to see that where you have put the exterior walls will allow the siding to lap over the concrete for weathertightness.

I usually snap out rake walls on the deck (sidebar, facing page). This will let the framer cut the studs in place between the chalklines, once again avoiding math. Make sure that you

Rake walls

Rake walls are fairly simple to plate and detail if you snap them out full scale on the deck, and keep in mind how they will relate to the rafters that will eventually sit on them. The bottom plate of the wall will look like any other. The trick is locating the top plate so the framer can fill in the rake-wall studs without doing any calculations.

In this case the rake wall will intersect a standard 96¾-in. wall. The first real complication is dealing with the bird's mouth on the rafter. With 2x4 walls, I use a 3½-in. level cut so that the rake wall dies into the 8-ft. wall at the top inside edge of its double plate. This also allows me to measure the run of the rake wall from the inside of the 8-ft. wall to the near face of the ridge beam. In this case, that's 10 ft. Since the pitch is 4-in-12, the rise between those points is 40 in.

Now back to the deck. Measure along the 8-ft. wall, 96¾ in. from the chalked *baseline* of the rake wall. This baseline was snapped out like the other wall lines as a guide for positioning the inside edge of the bottom plate once the wall is framed and raised. But it is also a convenient starting point for the full-scale elevation of the rake wall that you are going to snap out on the deck. In this case, it will be used to represent the bottom of the bottom plate.

The next step is to lay out the

bird's mouth of the rafter above the 96¾-in. mark you just made along the 8-ft. wall. Then take two pieces of 2x scrap and lay them inside the line at approximately a 4-in-12 pitch to represent the rake top plates. The reason you use two pieces of 2x scrap instead of just subtracting 3 in. for the rake top plates is that the vertical thickness of these plates when they are at a pitch will be greater than when they are horizontal. In the case of a 4-in-12, two plates add up to about 3¼ in.

Now make a mark on the chalkline of the 8-ft. wall just below the two scraps. This point represents the top (short point) of the shortest stud in the rake wall, and will be used to establish the line of the rake top plate. To complete this line across the deck, you need to create a large 4-in-12 triangle. In this case, a 4-ft. leg and a 12-ft. leg work out nicely. Actually, this triangle can be any size as long as its proportions are correct, it is close to the length of the rake wall, and it is positioned so that its hypotenuse represents the bottom of the rake top plate.

To lay out this triangle, first make a mark on the rake baseline 12 ft. out from the inside of the 8-ft. wall. From this point, measure up 93½ in. (the height of the 8-ft. wall less the thickness of the two top plates) parallel to the 8-ft. wall, plus another 4 ft. to take care of the rise. Now use a snapline to connect

this point with the one on the 8-ft. wall that you established below the scrap top plates. This is the hypotenuse of the 4-in-12 triangle, and as the bottom edge of the rake top plate it will be used by the framers to cut the tops of the rake studs to length.

To establish the top end of the rake wall, find the point along the rake baseline that represents the near face of the ridge beam. When the rake wall is framed, the top plates will die at the top inside face of this beam. Now snap a line that is parallel to the 8-ft. wall that starts at this point on the rake baseline and ends by intersecting the rake top plate line you just snapped in the center of the deck.

To check your layout, you can use your tape measure to get the height of the longest stud in the rake wall at its long point. Do this by measuring along the chalkline you just established at the inside face of the ridge beam. Stretch your tape from the baseline of the rake wall to the bottom of the top plates, and then subtract 1½ in. for the bottom plate. This measurement should be the same as the shortest stud plus the rise of 40 in. that was figured earlier.

To finish up, determine the position of the ridge beam by laying out the bird's mouth at the top. By subtracting the depth of the ridge beam, you can also determine the length of the posts beneath it. —*J. P.*

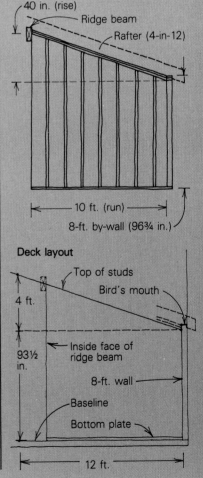

Standing rake wall

40 in. (rise)

Ridge beam

Rafter (4-in-12)

10 ft. (run)

8-ft. by-wall (96¾ in.)

Deck layout

Top of studs

Bird's mouth

4 ft.

Inside face of ridge beam

93½ in.

8-ft. wall

Baseline

Bottom plate

12 ft.

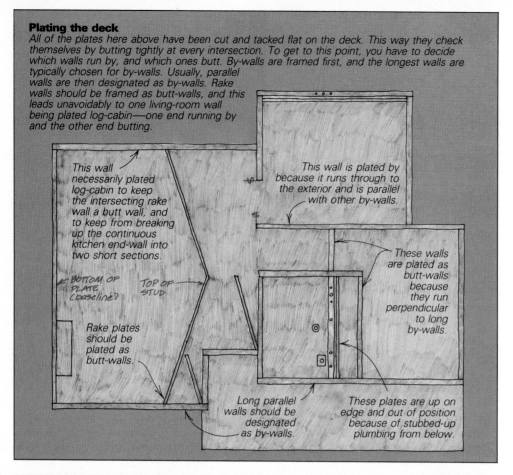

Plating the deck
All of the plates here above have been cut and tacked flat on the deck. This way they check themselves by butting tightly at every intersection. To get to this point, you have to decide which walls run by, and which ones butt. By-walls are framed first, and the longest walls are typically chosen for by-walls. Usually, parallel walls are then designated as by-walls. Rake walls should be framed as butt-walls, and this leads unavoidably to one living-room wall being plated log-cabin—one end running by and the other end butting.

This wall necessarily plated log-cabin to keep the intersecting rake wall a butt wall, and to keep from breaking up the continuous kitchen end-wall into two short sections.

BOTTOM OF PLATE (baseline) *TOP OF STUD*

Rake plates should be plated as butt-walls.

This wall is plated by because it runs through to the exterior and is parallel with other by-walls.

These walls are plated as butt-walls because they run perpendicular to long by-walls.

Long parallel walls should be designated as by-walls.

These plates are up on edge and out of position because of stubbed-up plumbing from below.

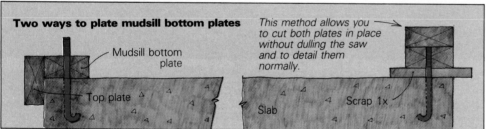

Two ways to plate mudsill bottom plates

Mudsill bottom plate

Top plate

This method allows you to cut both plates in place without dulling the saw and to detail them normally.

Slab

Scrap 1x

Plating. **Peake cuts in the top and bottom plate of a small raceway (left), and tacks each of the two sets of plates together with 8d nails (right) so they can be detailed. One of the first decisions in plating is which walls will run by, and which will butt. Here the long interior wall is a by-wall, and the parallel wall of the raceway is plated the same way.**

let the framers know with directions in keel whether your lines represent the top and bottom of studs, or the plates themselves.

Once the exterior walls have been chalked out, snap the interior walls by paralleling them (measuring out the same distance from several points and connecting them). When a wall ends without butting another wall, indicate where with a symbol that looks like a dollar sign with a single vertical bar. Follow the written dimensions given on the blueprints. Before you use an architect's scale to get a missing dimension, make sure that you can't find it by adding and subtracting others. If a mistake is made—whether it's yours or the draftsman's—you're the one who's going to fix it. Yet it's always better if what you've done reflects the approved plans.

Occasionally, though, you will have to adjust room sizes to accommodate some unanticipated condition. The rule of thumb here is to adjust in the largest rooms. The sizes of smaller spaces are usually dictated by the building code or some prefabricated item.

Plating—Plating involves cutting a top and bottom plate for each wall, tacking them together and laying them in place on the deck. To do this, you must decide at each intersection of walls which ones run through (*by-walls*) and which ones stop short (*butt-walls*). Walls that should be framed first—usually the longest exterior walls—are designated by-walls. Usually, the walls parallel to these will also be plated as by-walls. The framing and plating is simpler if you plate rake walls as butt-walls since the top plates are detailed in place (on their full-scale layout lines) near the middle of the deck. If plated as a by-wall, the top rake plate would lap over the plates of exterior walls that run perpendicular.

What you want to try to avoid in plating is *log-cabining*—building walls that run by at one corner and butt at the other. Such walls will probably have to be slid into position after they're framed and raised, which isn't easy, especially with a heavy wall.

The drawing top left shows the plating for this floor plan, and lists reasons why some walls have been designated butt-walls and others by-walls. Notice that the living-room wall at the top of the drawing has been plated log-cabin. So much for hard-and-fast rules. In this case, the rake wall at the end of the living room had to be a butt wall, and the kitchen wall on the other end had to run by so it wouldn't end up in two extremely short sections. The result is necessary log-cabining.

To do the plating, spread the top and bottom plate stock near the snapped lines. Use long, straight pieces. A crooked top plate can drive a framer crazy when it's time to straighten the raised wall with braces. To make the length on long walls, top plates will have to butt together at the center of a stud. The middle of a solid header is an even better spot. Breaks in the top plate and the double plate (the framer will be supplying this permanent tie between walls) have to be at least 4 ft. apart. That means to stay at least 4 ft. away from intersecting walls when laying out a break in the top plate, since the double plate of this wall will have to end there.

Very long walls will have to be framed in sections. With an average number of headers in an 8-ft. high 2x4 wall, each carpenter on the site should be able to handle at least 10 lineal feet of wall when it comes time to raise it. If you're going to break a long wall into separate sections, you'll need to end the top plate at the center of a stud. The bottom plate should be broken at the same place.

The plate stock for the bottom plate should be tacked flat to a wood deck along the snapped line with an 8d nail near each end. It's fine if it laps over nearby wall lines, because the next step is to crosscut it in place by eyeballing the chalkline of the intersecting wall. Now lay down the top plate in the same way, cut it and tack it to the bottom plate with two more 8d nails (photos facing page). The only exception to nailing the plates together is a rake wall where the top plate will be left on the deck on its angled layout line.

In the case of a slab, the bottom plate will be a mudsill. I use a bolt marker like Don Dunkley's (see p. 7). I usually make mine out of a piece of 1x2 with a joist-hanger nail at 3½ in. and at 5½ in. Once the mudsill is drilled out and set on the bolts, you have to deal with the top plate, which won't tack down to the mudsill because of the bolts.

Some carpenters hang the top plate off the edge of the mudsill (drawing facing page, center left), and then detail the layout across the edge of the top plate and the flat of the mudsill. But the framers will like it better if you shim up the mudsill with 1x scraps (drawing facing page, center right). This allows you to cut both the mudsill and the top plate in place without dulling your sawblade on the concrete, and to detail them normally.

Detailing—After all the plating is complete, detailing can begin. It is done in three stages: recording the information that you've added to the blueprints on the plates, marking out the precise measurements for headers, corners and intersecting walls on the face and edges of the plates, and then adding the stud layout. The drawing on the next page shows the floor plan with the plates fully detailed.

Layout style varies widely from region to region. One difference is in detailing shorthand. Layouts often contain more detailing than is really necessary. For instance, if you indicate where the end of a header falls with a line, and then make an X for the king stud beyond the line, it will be evident to the framer that the trimmer goes on the other side of the line. Writing T for trimmer (or C for cripple, de-

pending on the terminology you use) takes time, and doesn't add any information.

Another difference is the orientation of the plates. Some production carpenters tack the plates together and then toenail them along the chalkline with their edges up, rather than flat. But there are several advantages to running them flat. The first is that they check themselves. They can't be too long or too short because they are laid in the precise positions that they will occupy once they have been framed. Second, the location of headers and wall intersections is easy to see when it's detailed on the top of a flat plate, and won't get overlooked or misframed. Last, all the information necessary for the framers to cut the double plates that interconnect the walls is

marked on the surface to which they will be nailed—the top face of the top plate.

Almost all the marks that you'll put on the plates will be on the top of the top plate, and on one set of edges. The way to determine which set of edges is to approach a pair of tacked-down plates as if you are going to frame the wall. The trick is to figure out from which direction the top plate will be separated, the studs added and the wall raised. With an exterior wall this is easy. The top plate will be walked to the interior of the deck with its top markings facing the opposite side and its stud layout (which is on the edge) up. This way, once the vertical members have been filled in, the wall can merely be tilted up into place without having to reorient it. Exteri-

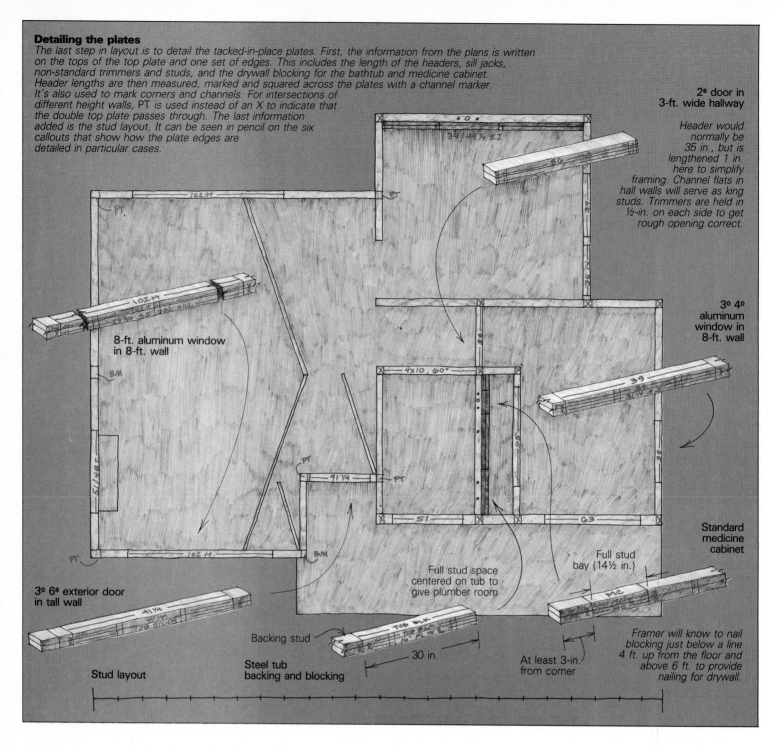

Detailing the plates

The last step in layout is to detail the tacked-in-place plates. First, the information from the plans is written on the tops of the top plate and one set of edges. This includes the length of the headers, sill jacks, non-standard trimmers and studs, and the drywall blocking for the bathtub and medicine cabinet. Header lengths are then measured, marked and squared across the plates with a channel marker. It's also used to mark corners and channels. For intersections of different height walls, PT is used instead of an X to indicate that the double top plate passes through. The last information added is the stud layout. It can be seen in pencil on the six callouts that show how the plate edges are detailed in particular cases.

2⁶ door in 3-ft. wide hallway

Header would normally be 35 in., but is lengthened 1 in. here to simplify framing. Channel flats in hall walls will serve as king studs. Trimmers are held in ½-in. on each side to get rough opening correct.

8-ft. aluminum window in 8-ft. wall

3⁰ 4⁰ aluminum window in 8-ft. wall

Standard medicine cabinet

3⁰ 6⁸ exterior door in tall wall

Stud layout

Backing stud

Steel tub backing and blocking

30 in.

Full stud space centered on tub to give plumber room

Full stud bay (14½ in.)

At least 3-in. from corner

Framer will know to nail blocking just below a line 4 ft. up from the floor and above 6 ft. to provide nailing for drywall.

---

or plates, then, are typically detailed on their outside edges. Interior partitions, since they can often be tilted up from either direction, require the layout carpenter to guess how and where the framers will set up.

You're now ready to do the detailing. Forget your tape measure for the moment, take the plans in hand and walk around the deck copying the information you've written there onto the plates. For each window and door, write its header length on the top of the top plate near where the header will be nailed, and *tt* if it requires double trimmers. Now lean over the plates and mark the outside edge of the top plate with the same information.

As long as you detail the rough-opening dimensions of the windows on the plates when you lay out, you don't have to figure the height of the rough sill or the length of the sill jacks. The framer can do this by measuring the rough-opening height down from the header, installing the rough sill, and then filling in with the sill jacks. However, if you figure the length of the sill jacks at the layout stage, they can be precut for the framing. Write this length in keel on the outside edge of the bottom plate and follow it with the letters *SJ*. For any walls with studs longer than the standard 92¼ in., write their lengths on the face of the top plate and on the outside edge of one of the plates. Note the location of prefab items, like medicine cabinets (*MC*), and drywall blocking around fixtures like bathtubs (*TUB BLK*). Label rake walls *RAKE*.

At this point you can begin to mark corners and channels on the top and edges of the plates. Both of these are just wall intersections—one at the end of a wall, one in the middle—and are drawn the same width as the plate stock. When these intersections are

framed, they'll need backing studs for wall finish, but since corners and channels are usually made up as units that include their backing studs, you need only to show the intersection of the plates for the framer to get it right.

To detail the corners, use a pencil to scribe a line onto the inside edge of the by-wall plates along the side of the butt-wall plates. Then use a channel marker (see the box on the facing page) to continue that line across the face of the top plate and down the other edges. These lines will show exactly where to nail the walls together once they are framed and raised. Channels aren't much different from corners, except you have two lines to scribe—one on each side of the intersecting plates. Put a big keel *X* between the lines on the face of the top plate, and on each of the outside edges of the plates to show where the double plates of the intersecting wall will create a half-lap. If the walls connect at different heights, the double plates of the by-wall shouldn't be broken for the double plate of the butt wall. The framer should be warned by marking the corner with the letters *PT*, which tell him to *plate through*.

A common framing mistake, usually discovered after the wall is raised, is putting the flat stud of the channel on the wrong side of the wall from the intersecting partition. As long as the framer knows the flat-plating method used here and doesn't reverse the top and bottom plates, where to locate the flat will be obvious. The key is the *X* that marks the channel. Because you are prevented from making any marks on the inside edges of the by-wall plate when you scribe the butt-wall plate to it by the plates themselves, the *X* will get marked only on the opposite set of edges from where the intersection will actually happen. So the framer should nail the channel flush with the edge of the plate that doesn't have an *X*.

Now detail the window and door openings. Following the blueprints, measure accurately to each end of the headers and use your channel marker to square the lines across the top plate and down the outside edges. Make an *X* on the outside of each of these lines to indicate the king stud.

When making an *X* over the edges of the plates, you can save yourself an extra motion by making two intersecting half-circles. This will leave an *X* on each plate when they are separated. The only time I show the location of trimmers is when they are doubled, which I indicate with *tt* (see the 8^0 4^0 living-room window callout in the drawing, facing page).

Interior doors are often placed near the corner of the room they serve. The standard way to frame them is to let the king stud act as one of the backing studs in the channel, compressing space. Once the drywall is hung, this leaves a little less than 3 in. for casing. If the casing is wider than this, it will have to be scribed to the wall. If the space is even narrower and the door is in a butt-wall such as the 2-ft. 6-in. door in the 3-ft. hallway in the plan, you'll have to use the channel flat as a king stud. A useful rule of thumb is that the space left for the trim is about the same as the distance from the studs of the intersecting wall to the inside face of the trimmer.

There are a few special items in bathrooms that need detailing. The medicine cabinet fits between studs, but it will need blocking above at 6 ft. off the floor and below at 4 ft. If the medicine cabinet is near a corner, double the end stud that nails to the channel to give the necessary room for the swing of the cabinet door. Bathtubs and showers should be blocked along their top edges. Detail this by specifying a height on the plate that runs from the floor to the centerline of the blocking. A double stud or flat stud should be laid out to pick up the side and end of the tub or shower. In the drawing, you can see that I've also centered a standard stud space on the plumbing end of the tub to make it easier for the plumber to run supply lines and a drain.

Beam pockets can be detailed with your channel marker, but label them *BM* so they are not confused with channels. The posts under these beams are detailed with their actual width marked on the edge of the plates with keel, and their nominal size written in between these lines. Also give a length for the post if it is different from stud height.

Stud layout—The regular stud layout comes last and is done on the outside edges of the plates. Pull all the outside and inside walls that run perpendicular to the joists and rafters from the same end of the building. Do not break your layout and start again at partitions, but continue the full length of the wall. The standard 16-in. and 24-in. centers are meant to work modularly with 4-ft., 8-ft. and 12-ft. sheet materials. Remember that 16 in. o. c. means from the end of the building to the center of the first stud, so reduce your layout by ¾ in. each time when pulling from the corner (15¼ in. to the first stud, 31¼ in. to the second stud, and so on). This way the plywood sheathing and subfloor will work out with a minimum of cutting and waste.

It's a common mistake to have drywall on your mind when laying out studs. Drywall is relatively cheap, easy to cut, and can be bought in 12-ft. lengths. It should be at the bottom of your list of worries when laying out.

Stud layout should be done in pencil. If you are using a layout stick, you can put your tape measure back in your nail bag once you get started. Scribe along both sides of each finger of the layout stick to mark for the studs, then reposition it farther down the plate and repeat. If you aren't using a layout stick, stretch your tape the length of the wall and make marks at 15¼, 31¼, etc. Then come back with a combination square set at a depth of 3 in., square these marks down the outside edges of the plates, and make an *X* on the leading side of the line. Don't bother to draw a line for both sides of the stud, but don't lose your concentration either when making the Xs. Putting them on the wrong side of the line will cause big headaches later.

Rake top plates can be laid out by stretching a tape from the by-walls they butt, but you'll have to hold the tape perpendicular to the by-wall and keep moving the end of it farther down the by-wall so that the stud centers on the tape will intersect the angled rake plate. A better method is to measure the distance between studs along the angled plate after marking the first few, and then use this increment to mark the studs thereafter.

The last thing to do before you leave the deck is to look over the tops of the plates and make sure that every room has a door in it (a common but embarrassing mistake), and that all the channels are marked. These marks are easy to spot because they are on the top of the plates. Finally, to get the framing off to a good start you can cut all the headers, sills, sill jacks, and specials, as you already have a cut list on the marked-up plans. □

Jud Peake is a contractor and a member of Carpenters Local 36 in Oakland, Calif.

Tools of the trade

Most production layout tools were born of necessity on the site and made with available materials on a rainy day. One such tool is the channel marker (middle photo), a simple square made out of short pieces of plate stock and used for outlining corners and channels. It should have a leg 3 in. long (the depth of two 2x plates) and another leg 3½ in. long (the width of a 2x4). Both legs are 3½ in. wide. I make a more durable version with aluminum flat stock that includes a 1½-in. flange at the top. By turning the square over, you can lay out the thickness of a stud with this flange.

Two more tools that will speed things up are a layout stick (top photo) and a keel/pencil holder (bottom photo). This last item is just a short piece of ½-in. clear plastic tubing that will take a carpenter's pencil in one end, and your keel in the other. Layout sticks can be made out of standard aluminum extrusions riveted together. The 1½-in. wide and 3-in. long fingers on mine are laid out for 16-in. centers and 24-in. centers. I even threw a hinge into my stick so that it could fold up to fit in a standard carpenter's toolbox. —*J. P.*

Timber-Frame Layout

Systems for labeling the timbers and adjusting the joinery keep the frame plumb and true despite variations in dimensions

by Tedd Benson

The mystique that surrounds the craft of timber framing often clouds a full understanding of the kind of work that goes into cutting, assembling and raising a frame. It's easy to imagine yourself paring off fine shavings with a razor-edged chisel and raising timbers in communal euphoria. There's real satisfaction in pushing a tenon home into its perfectly mated mortise, or in driving the pegs to lock the joint. All these things contribute to the pleasures of framing with timbers.

But precise joinery is only a small part of

Careful planning, hard work and strict adherence to labeling, layout and mapping rules made it possible for five workers and a crane to raise this frame in one day. It has 203 individual timbers and 382 joints.

the process. Many of the frames we build contain well over 200 timbers and 350 connected joints. The frame shown in the photo above was raised in one dramatic day by a crew of five and a crane. But this day was merely the culmination of all the work that preceded it. There were five days of sanding timbers and assembling bents, and before that, many hours of work in the shop.

In order to work with speed and efficiency, there can't be any mystery whatsoever about what timber goes where. And all the timbers must fit exactly—a single misaligned joint stops a raising dead in its tracks.

We lay out, cut, and finally truck a completed frame to the site without test-fitting the joints. We're able to do this only because every frame evolves with a great deal of plan-

ning, some applied geometry, and an organized approach to layout and cutting.

With an organized system and good preparation, a beginner can get through a difficult project. Without them, even a master joiner couldn't possibly succeed.

We apply three systems to the timbers and joints. *Labeling* is a system for identifying each member so we know where it fits within the overall frame. *Layout* is a method for locating and marking each joint so that all timbers will align correctly. *Mapping* is a system of accounting for variation in timber dimension and adjusting the length of adjoining timbers accordingly, so the completed frame will be true to its planned measurements.

To demonstrate how these planning principles work, I'll use the frame shown above as

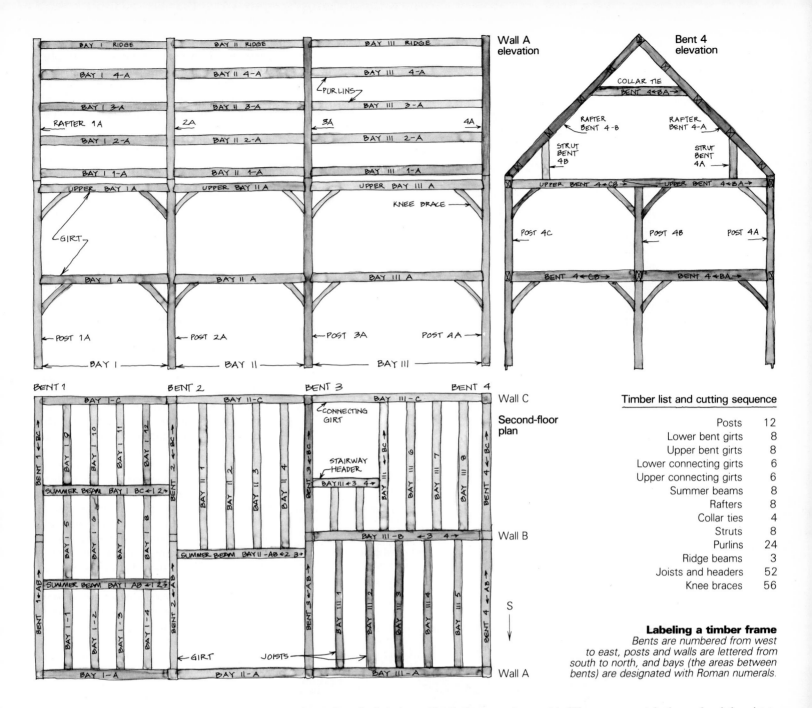

Wall A elevation

BAY I RIDGE BAY II RIDGE BAY III RIDGE

BAY I 4-A BAY II 4-A BAY III 4-A

PURLINS

BAY I 3-A BAY II 3-A BAY III 3-A

RAFTER 1A 2A 3A 4A

BAY I 2-A BAY II 2-A BAY III 2-A

BAY I 1-A BAY II 1-A BAY III 1-A

UPPER BAY I A UPPER BAY II A UPPER BAY III A

KNEE BRACE

GIRT

BAY I A BAY II A BAY III A

POST 1A POST 2A POST 3A POST AA

BAY I BAY II BAY III

Bent 4 elevation

COLLAR TIE

BENT 4←BA→

RAFTER BENT 4-B RAFTER BENT 4-A

STRUT BENT 4B STRUT BENT 4A

UPPER BENT 4←CB→ UPPER BENT 4←BA→

POST 4C POST 4B POST 4A

BENT 4←CB→ BENT 4←BA→

BENT 1 BENT 2 BENT 3 BENT 4

Wall C

Second-floor plan

BAY I-C BAY II-C BAY III-C

CONNECTING GIRT

STAIRWAY HEADER

BAY III←3 4→

SUMMER BEAM BAY I BC←1 2→

BAY III-B ←3 4→

SUMMER BEAM BAY II-AB←2 3→

Wall B

SUMMER BEAM BAY I AB←1 2→

S

GIRT JOISTS

Wall A

BAY I-A BAY II-A BAY III-A

Timber list and cutting sequence

Posts	12
Lower bent girts	8
Upper bent girts	8
Lower connecting girts	6
Upper connecting girts	6
Summer beams	8
Rafters	8
Collar ties	4
Struts	8
Purlins	24
Ridge beams	3
Joists and headers	52
Knee braces	56

Labeling a timber frame
Bents are numbered from west to east, posts and walls are lettered from south to north, and bays (the areas between bents) are designated with Roman numerals.

---

an example. These systems apply to every frame we construct, but we have to modify them somewhat to meet the specific demands of each new frame.

Labeling—After we finish the blueprints for a house, we draft a complete set of shop plans for the frame. This set of working drawings typically includes elevations for each bent and wall, a plan view of each floor to show joist and beam locations, and large-scale sections or blow-ups of any unusual joinery details. At this stage, all joinery decisions have been made, and every timber has been sized in dimension to support its respective load.

Labeling begins when the plans are drawn. A consistent identification system allows us to distinguish between the many timbers in any frame that look alike but may not be dimensionally identical.

Unless your label tells you the exact location and orientation of every piece in the frame, you'll be plagued by constant remeasuring, and you run the risk of putting timbers

where they don't belong. Here's how our labeling system works:

We draw plan views of the frame with south at the bottom of the page. Moving from west to east and from south to north, we number the bents, letter the posts, and assign Roman numerals to the bays (the spaces between the bents). For example, bay I is between bent 1 and bent 2; bay II is between bent 2 and bent 3; and so on. The posts in the southernmost row are A posts, the posts in the next row are B posts; those on the north side of the house are C posts and could be D or E posts on a very large frame. Posts also carry the number of the bent in which they stand. The southwest post will be 1A, for instance, and the post on the northeast corner will be 4C. The drawings above show how the various parts of the frame are labeled.

Bent girts are the horizontal timbers that join posts together to form a bent. Since the bent for a two-story frame will have more than one girt, a label like upper bent 1 ←B A→ would identify an upper girt that joins posts 1B and

1A. The arrows match the ends of the girt to the posts they will join.

Connecting girts, which we sometimes call bent connectors, are bay timbers that span between posts. They join one bent to another, and carry the braces that keep the frame rigid in the direction of the wall. They're labeled according to the bay and wall they fit in. For example, I-B would indicate a girt that falls in the first bay, in the A wall.

Summer beams, purlins and floor joists all fall within bays of the frame. Summer beams are connected to adjacent bent girts and hold the floor joists. They're labeled according to the bay they're in and by the posts and bents they fall between, for example, summer beam I AB ←1 2→ (bay I, between A and B posts and between bents 1 and 2).

Purlins are horizontal timbers that connect adjacent rafters, and are among the pieces that often appear identical but aren't, as a result of variations in rafter dimensions. A purlin would get a label like purlin III 4-A (bay III, A wall, fourth purlin up from the eave). Floor

Enclosing the Timber Frame

Covering the frame with stress-skin panels provides insulation, interior finish and exterior sheathing in a single step

by Jeff Arvin

In the last fifteen years, there's been a revival of interest in timber-frame construction. But a lot of builders (both novice and professional) who otherwise might have been interested have hesitated because of the problems of enclosing the completed frame. The look and feel of a finely joined skeleton of big timbers has always been appealing, but the need to build separate wall and roof systems to hold insulation, interior finish and exterior

siding has made timber framing expensive and frustrating. Fortunately those days may be behind us, thanks to the development of stress-skin panels—rigid sheets of foam with exterior sheathing (or siding) bonded to one face and drywall bonded to the other. Attaching these laminated panels to the outside face of the frame effectively insulates the building and provides exterior sheathing and interior drywall in one step. And the quality of the in-

sulation is better than that of a stud wall because the stress-skin wall has a core of solid foam insulation uninterrupted by framing members. Using the installation techniques discussed here, a crew of four can enclose an average-size timber frame in three to six days.

A brief history—Early European timber framers used wattle-and-daub infill between posts. A grid of small saplings (wattle) was woven between posts and plastered on both sides with a mixture of mud and straw or (later) lime mortar. Several applications were required. Another traditional method, brick nogging, called for brick walls to be built between posts. Neither of these methods offers much in the way of insulation, and the fact that the walls were built between and not outside the posts was also a shortcoming. With the outside face of the frame exposed to the weather, seasonal expansion and contraction of the timbers opened up gaps between framing and walls that required constant chinking. Since energy efficiency is such a major concern today, these older methods of enclosure don't have much of a place in modern timber-frame construction.

Most timber framers today have had some experience in enclosing a frame with standard 2x framing lumber. With this more recently developed method, the 2x frame is built against the outside face of the timber frame, creating insulation space and nailing surfaces for interior drywall and exterior siding. Here again, there are some major drawbacks. Apart from the time and expense of building walls around walls (so to speak), finishing a frame enclosed this way calls for scribe-fitting gypboard or paneling to the often irregular surfaces of timbers and braces. And as the timbers move slightly with changing temperature or humidity, what was once a good scribed fit shows a gap. It's not difficult to understand why timber-frame houses developed the reputation of being expensive and tricky to enclose and finish.

Stress-skin advantages—Sometimes known as laminated building panels, curtain-wall panels or preformed wall and roof panels, these rigid sandwiches of foam (either urethane or expanded polystyrene) and sheet materials (gypsum board, plywood, waferboard or even steel or aluminum) have been around for over 30 years. In various forms, they've been used to build recreational vehicles, refrigerated warehouses and commercial and industrial structures. Timber framers discovered stress-skins only about nine years ago. The name "stress-skin" refers to the panel's skin of sheathing materials, which is held rigid by the foam core. Stabilized in this way, the panel performs like a torsion box, offering excellent resistance to twisting and racking.

In addition to a high strength-to-weight ratio, stress-skin panels are very good insulators. The 5½-in. thick expanded polystyrene (EPS) foam that we use at Riverbend Timber Framing in wall panels tests out to R-24. Fastened to the outside of a timber frame with all panel joints and shim spaces (for window and door frames) sealed with site-applied foam adhesive, stress-skin panels let us create a tight blanket of insulation that's uninterrupted by studs and other framing members. The oak in a timber frame can easily weigh 30,000 lb., and enclosing this mass within insulated walls definitely provides some thermal storage in each room of the finished house. Air infiltration is reduced to a negligible level, and if

you're using urethane-core panels, the foam itself can act as the vapor barrier (the Perm rating for urethane is 1). Expanded polystyrene, or EPS, has a Perm rating of 4, so vapor-barrier paint should be applied on the drywall if you want a lower Perm rating. We've found, however, that conventionally painted EPS-core panels do just fine under normal circumstances. EPS absorbs a negligible amount of water, and won't support mildew or bacteria at all, unlike the wood in a moist stud wall.

Another way that stress-skin panels complement timber frames is in locating and installing windows and doors. Because the panels are so strong, you don't have to frame the rough opening with king studs, trimmers, headers and rough sills, as in stick-frame construction. This gives you great flexibility in planning window and door locations. You can cut the rough openings in the panels as they are being installed, or leave the walls solid after the frame is enclosed, installing windows as time and budget allow.

No matter where you get your panels (see p. 31 for a list of manufacturers), they're fairly expensive. Prices range from about $1.50 to $5 per square foot, depending on the type and thickness of the foam and the laminating materials that are used. Panel quality also varies, and poorly made panels have been known to delaminate and deform over time.

The panels we use have an EPS foam core 5½ in. thick for walls and 7¼ in. thick (R-35) for roofs. You can specify other thicknesses if you want more or less insulation. Like many stress-skin panels, the ones we use are manufactured with a tongue-and-groove joint. The sheathing materials are offset ¾ in., creating a foam tongue along one long side and a drywall and sheathing-faced groove along the other. Unassembled, these edges are delicate and must be handled carefully. A few crumbled edges are inevitable, and can be repaired after enclosure with either spray-on foam sealant or drywall compound. But you can't afford to throw away many damaged panels when they cost $60 to $90 apiece. Before you buy panels, it pays to check out a few panel suppliers for price, quality and handling policy.

Planning and layout—At the layout stage we take a close look at the timber frame to determine what panel size or sizes will work best. The standard size is 4x8, but it's possible to get 4x9s, 4x10s and 4x12s from some suppliers. On walls, we hang panels vertically. The lower edge of the panel rests on an extended sill made by bolting a pressure-treated 2x10 or 2x12 atop the foundation. The upper end of each panel is spiked to a girt (drawing, p. 29). It's not necessary for vertical panel edges to land on posts so long as the connections at top and bottom are solid. Bonded together with gap-filling foam sealant, the tongue-and-groove vertical joints between panels are extremely strong.

The location of large windows or doors can sometimes affect panel layout. Window and door openings are usually cut after the panels are up, but if a large opening is needed, we

sometimes skip a panel and piece together the opening with panel offcuts.

Sheathing materials are also specified when you order panels. We use type X, ½-in. fire-rated drywall against the interior face of the foam. Where an interior nailing base is required (such as a kitchen with wall-hung cabinets) we have waferboard and then drywall laminated to the inside face of a foam core. The core thickness has to be adjusted (by the thickness of the waferboard) so that overall panel thickness remains consistent. We can also have 1x4 nailers for wainscoting or baseboards let into the foam. Some stress-skin manufacturers are more flexible than others when it comes to varying panel specifications.

On the exterior face of the panel, you can usually choose between plywood or waferboard sheathing and plywood siding. Many of the clients we work with opt for T-111 siding on sidewall panels because it gives a finished exterior surface. The vertical-channel, rough-sawn siding shown here isn't the best-looking siding around, but there's no reason why it can't act as a nailing base for shingles or clapboards a year or two down the road, when the owner can afford the extra material.

Installing panels—As with any new material, stress-skin panels have led to the development of new methods and procedures. The techniques we use for cutting panels, trimming foam, building corners, and attaching panels work well for us, but we know other timber framers who do things slightly differently. For example, panels can be joined together using wood splines rather than tongue-and-groove joints. We've had good results with the T&G system, so we stick with it.

Skinning out a frame isn't a complex job, but it is hard work. The panels weigh about 130 lb. apiece, and their size makes them unwieldy. Three people are adequate for the job, four are better. Even with four workers, you'll probably need some mechanical assistance to get panels up to the second-story and roof. We've done jobs using a block-and-tackle, scaffolding and muscle, but that isn't the safest or the fastest way. A high-extension forklift makes things a lot easier if your building site isn't too rugged. On some jobs, a crane makes quick work of setting roof panels.

Panels are attached to the frame with spikes that penetrate at least 2 in. into the wood. If you're spiking into oak, you've got to use hot-dipped galvanized or stainless-steel spikes to avoid corrosion problems. For a softwood frame, ringshank spikes are needed for extra holding power. The spikes should be driven every 12 in.

The techniques for cutting panels are fairly straightforward. We use a 16-in. circular saw for most cuts, setting the blade depth so that it's only about ⅛ in. shy of cutting through the full thickness of the panel. We cut from the sheathing side, and the blade cuts far enough into the drywall so that we can simply flip the panel, fold it over and slice the paper to separate the two pieces. The 8½-in. thick roof panels we use have to be cut from both sides. We

snap chalklines to lay out the cuts, and use a combination blade (photo top left).

Cutting openings for windows and doors works the same way. You cut to the outline of the rough opening on one side of the panel and drive spikes through it at the corners. Then flip the panel (or go to the opposite side if the panel has already been installed), snap lines between protruding spikes and cut through to the saw kerfs made from the other side. Alternatively, you can plunge-cut from one side with a 16-in. saw, and then through-cut at corners with a handsaw. I've also heard of carpenters using an electric chainsaw to through-cut from one side.

We usually begin installing panels at the corners, using an overlapping edge detail like the one in the drawing on the facing page. Here the advantage of using 5½-in. thick foam-core sheets becomes apparent. The 2x6 nailers that are let into the panel edges at corners are 5½ in. wide, so the fit is quick and snug, once you've hollowed out a channel along the panel edge. To create this channel, we use a hot-wire bent to the 1½-in. by 5½-in. section of a 2x6. Our hot-wire is nothing more than a thin steel wire with a positive lead connected to one end and a negative lead connected to the other. It is mounted on a short wood scrap that rides on the edges of the sheetrock and sheathing, preventing the wire from digging too deeply into the foam. You need a transformer to moderate current flow. Once the juice is turned on and the wire heats up, you pull it along the edge of the panel, melting away the foam (photo bottom right). We use a straight hot wire if we need to trim the foam tongue off a panel. Sometimes you have to touch up a hot-wired edge with a utility knife or a chisel to get a good-fitting edge. And if your stress-skin panels have urethane cores, the hot-wire technique won't work, so you have to use a router or a grinder with an extra-wide bit.

Before pressing the 2x6 into the channel and nailing it fast along both drywall and sheathing edges, we spray some Fomofill sealant (Fomo Products, P.O. Box 4261, Akron, Ohio 44321) along the joint. This spray-can polyurethane foam looks a lot like whipped cream and expands slightly before hardening. It serves the double function of filling gaps and bonding adjoining surfaces together.

Once a corner is complete, we proceed down the walls, filling each tongue-and-groove joint with aerosol foam (photo bottom left) and sliding adjoining panels together. Sometimes you have to chamfer the foam tongues slightly with a utility knife to help start the joint. If more persuasion is needed, we hold a beater board across the free edge of the panel and tap it into place with a sledge-hammer. It's important to keep checking for plumb as each panel is hung.

On two-story structures, you'll need to install a second course of panels to reach the eaves. The horizontal joints between panel courses can simply be butted and glued shut with foam, but it's better to install a horizontal 2x spline between first and second-story pan-

Stress-skin panel joinery. These EPS-core panels are faced with T-111 exterior siding and ½-in. drywall. Top, cuts are made with a 16-in. circular saw from the sheathing side. Then the panel is flipped, folded over along the cutline, and razored apart with a utility knife. At left, all panel-to-panel and panel-to-wood joints are sealed with site-applied foam. The foam looks like whipped cream when it leaves the can but will harden into a gap-filling adhesive that keeps the insulative blanket continuous around the frame. Below, a hot wire is used to melt out a thin channel in the foam along a panel edge for joining ripped panels or joints that require a 2x spline. This channeled edge is identical to half of the manufactured tongue-and-groove joint used to connect panel edges that don't land on timbers.

Illustration: Christopher Clapp

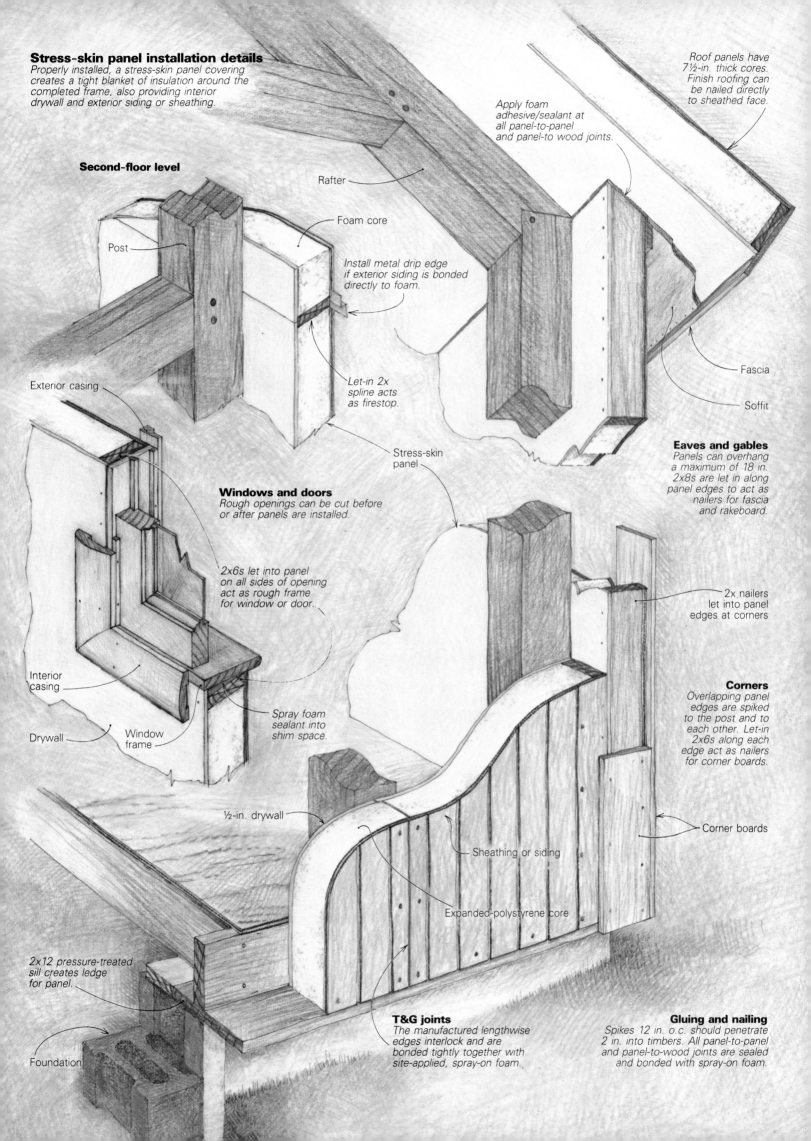

Stress-skin panel installation details
Properly installed, a stress-skin panel covering creates a tight blanket of insulation around the completed frame, also providing interior drywall and exterior siding or sheathing.

Second-floor level

Post

Foam core

Exterior casing

Install metal drip edge if exterior siding is bonded directly to foam.

Let-in 2x spline acts as firestop.

Apply foam adhesive/sealant at all panel-to-panel and panel-to wood joints.

Rafter

Roof panels have 7½-in. thick cores. Finish roofing can be nailed directly to sheathed face.

Fascia

Soffit

Stress-skin panel

Windows and doors
Rough openings can be cut before or after panels are installed.

2x6s let into panel on all sides of opening act as rough frame for window or door.

Interior casing

Drywall

Window frame

Spray foam sealant into shim space.

Eaves and gables
Panels can overhang a maximum of 18 in. 2x8s are let in along panel edges to act as nailers for fascia and rakeboard.

2x nailers let into panel edges at corners

Corners
Overlapping panel edges are spiked to the post and to each other. Let-in 2x6s along each edge act as nailers for corner boards.

Corner boards

½-in. drywall

Sheathing or siding

Expanded-polystyrene core

2x12 pressure-treated sill creates ledge for panel.

Foundation

T&G joints
The manufactured lengthwise edges interlock and are bonded tightly together with site-applied, spray-on foam.

Gluing and nailing
Spikes 12 in. o.c. should penetrate 2 in. into timbers. All panel-to-panel and panel-to-wood joints are sealed and bonded with spray-on foam.

Panels are fastened to the frame with spikes that penetrate at least 2 in. into the timber and are spaced 12 in. apart.

be measured up from the rafter tails a distance that equals the length of the panel less the desired overhang. The rake overhang can be snapped out on the underside of rake panels and lined up with the gable rafter as rake panels are installed.

Since the foam cores of our roof panels are 7¼ in. thick, we let in 2x8 nailers along all overhanging edges. At the eaves, these nailers get covered with fascia boards; at gable ends, we nail on rake boards. Then the drywall undersides of all overhangs are covered with finish soffits. The soffits are supported at the wall juncture by a frieze board.

Panel installation on an average-size house (1,600 sq. ft.) rarely takes more than five working days with an experienced crew. To speed the operation, we've recently experimented with pre-assembling several panels on the ground and setting large roof sections in one shot with a crane. We cut and assemble sections at least eight hours ahead of the crane to give the foam sealant time to cure.

Windows and doors—The stress-skin panel itself takes the place of a rough opening's usual structural framing, so windows and doors are surprisingly easy to install. After cutting the specified opening (use the rough-opening size specified by the window manufacturer, or simply add ¼ in. to the outside dimensions of the unit), you remove 1½ in. of foam from around its perimeter, using either a hot wire or a router and utility knife. The channel doesn't have to be perfect because voids can be filled with aerosol foam as you let in 2x nailers on all four sides. The face of the 2x should be flush with the drywall and sheathing edges. Nail through both drywall and sheathing into the 2x to complete the rough opening (see the drawing on p. 29).

The next step is to install the window or door, shimming it plumb and level just as you would in a standard stud wall. Nail the window or door frame in several places to the 2x frame, and then spray foam sealant into the shim space. One note of caution here: Make sure that you use a foam sealant with a low expansion ratio—foams that expand 300% or more can create enough pressure to foul up some window mechanisms.

In addition to using stress-skins for enclosure, we've also found that they work well inside the house as floor panels. Here the foam core needn't be quite so thick— we use 3½-in. cores. Installed over summer beams and ceiling joists, the drywall face of the panel is the finished ceiling and the plywood or waferboard face is underlayment for the second floor. The foam insulation is ideal for zone heating schemes. If you use this type of floor system, however, the panels must be rigidly connected. Where there's no tongue-and-groove joint, we join panels with a 2x4 spline. Panels should be laid with their long edges perpendicular to the floor joists so that each panel rests on at least three joists. □

Jeff Arvin works with Riverbend Timber Framing, Inc., in Blissfield, Mich.

els as a fire stop. At the eave, wall panels are beveled to match the roof pitch, and we cut the bevel back slightly to accept ¼ in. to ½ in. of foam sealant. Gable-end panels are cut to size on the ground and then lifted into position and nailed up (photo above). All joints are sealed with foam.

The joints between roof panels should land on purlins and rafters wherever possible. It's also especially important for the spray-on foam adhesive in panel joints to form a strong, solid bond. Otherwise, water vapor that collects near the ceiling can find its way through voids in the joints, freezing at or below the shingle line.

When positioning roof panels, extra care must be taken to create straight overhang lines at eaves and rakes. The maximum unsupported overhang we recommend is 18 in. To keep the eaves even, snap a line along the rafters and purlins where the top edge of the first panel course will land. This line should

How stress-skin panels are made

The techniques for making stress-skin panels can be surprisingly basic or very sophisticated, depending on what type of foam is used and how the sheathing materials are bonded to the foam core. At Neilsen-Winter, Inc. (see Sources of Supply, facing page, for the address), urethane-core panels are foamed in place, and the foam itself acts as the adhesive for the various facings available. Sheathing materials move on parallel conveyor belts, and the foam is injected between them. It gradually expands to fill the insulation space and bonds to drywall and waferboard. According to Amos Winter, who developed the process, this machinery can turn out panels up to 4 ft. wide and 28 ft. long.

Foam Laminates of Vermont, on the other hand, uses expanded polystyrene foam (widely known as EPS) as core material and laminates each stress-skin panel by hand. Their panels are made by gluing drywall and waferboard to opposite faces of the foam. PVA adhesive (ordinary white glue) is troweled generously over the inner faces of drywall and waferboard, and then these sheets are bonded to the core. By using a pair of jigs, workers offset the sheathing materials, creating the tongue-and-groove edges that facilitate panel joinery (photos, facing page). Laminated panels are stacked one on top of the other in a press made from I-beams, chain and threaded rod. Tightening nuts on the threaded rod causes a pair of I-beams to compress the stack. Several hours after tightening, the panels can be removed and stored under cover.

Even though competition among manufacturers continues to drive prices down, you can still save money by laminating your own panels. Most on-site laminating is done with EPS cores because they're more widely available than urethane and because they hold glue better. Whether you're buying or building panels, however, read on. There are important differences between urethane and EPS foam that might influence your choice of materials, and there are also several types of adhesives to choose from if you decide to laminate your own.

EPS or urethane?—The builders and manufacturers who favor urethane-core panels cite higher insulation value and greater high-temperature stability as the main advantages over EPS-core panels. A Neilsen-Winter panel with a 3½-in. thick urethane foam core provides about the same insulation value (R-24) as an EPS panel with a 5½-in. thick core. The thinner panels enable you to use standard window and door jambs; extended jambs are necessary with thicker panels.

Another advantage to urethane foam is that it provides a better vapor barrier. The Perm rating for 3½ in. of urethane is around

1, while 5½ in. of EPS has a Perm of 4, which disqualifies it as an effective vapor barrier. Remember, though, that the necessity of minimizing vapor movement through the wall isn't as crucial with EPS as it is with 2x studs. EPS doesn't hold as much moisture as wood, and won't rot or mildew either.

The benefits of urethane foam carry a higher price tag. These panels can cost twice as much as their EPS counterparts. Another problem with urethane is that it's hard to glue. The foam-in-place method of making urethane stress-skins creates an excellent bond between the foam and the sheathing material. But once urethane foam solidifies and cures, its brittle, porous surface tends to hold a fine dust that interferes with good adhesion. You really need to use a heavy-bodied mastic-type adhesive and clean the foam surface thoroughly before applying it. This explains why most on-site stress-skin panel fabrication is done with expanded polystyrene.

Both urethane and expanded polystyrene are U.L. (Underwriters Laboratories) listed with approved ratings for smoke and flame spread. But EPS tends to deform at sustained temperatures above 200°F. This low transition temperature, as polymer manufacturers call it, could cause extensive delamination during a fire. In hot climates, roof panels with EPS cores should be covered with light-colored roofing to reflect the sun. An alternate method to prevent EPS overheating is to install furring strips beneath the roofing, so that there's an insulative airspace.

Several adhesives can be used to laminate EPS-core stress-skin panels. Peter McNaull, of Foam Laminates of Vermont, has had good results using PVA glue. An adhesive called Mastic Adhesive 4289 was developed by 3M (Adhesives, Coatings and Sealers, 3M Center, St. Paul, Minn. 55144) especially for stress-skin lamination. HC6478 is the name of the adhesive developed by Hugh Products (PPG Industries, 1400 East Avis Dr., Madison Heights, Mich. 48071) for laminating panels. Isoset (Ashland Chemical Co., Box 2219 Columbus, Ohio 43218) and Mor-Ad 336 (Morton Chemical Co., 110 N. Wacker Dr., Chicago, Ill. 60606) are more expensive adhesives that are used by manufacturers to make structural panels. No matter what adhesive you use, the panel should be compressed following glue-up until the adhesive sets.

The electrical question—Some consideration should be given to running electrical wiring in or along your stress-skin panels, whether you're laminating your own panels or ordering manufactured ones. Many panels have a void cut into the foam just beneath the drywall that you can fish wires through. Other companies incorporate thinwall conduit in the panel, or cut a thin channel along the tongue-and-groove edge. Alternatively, you can leave your panels as they are and run most of your wiring in a hollow baseboard or in the interior walls. —*Tim Snyder*

Gluing and pressing
At left, PVA adhesive is troweled generously over a 4x8 sheet of drywall at Foam Laminates of Vermont. The waferboard sheet that will be glued to the opposite face of the foam gets the same treatment. Below, jigs are used to offset the sheathing materials as they're bonded to the foam. A chain and I-beam press compresses a stack of nine panels until the glue sets.

Sources of supply

Many of the stress-skin panel manufacturers listed below also supply spikes for fastening panels to a structural frame, aerosol foam sealant, and installation information.

Atlas Industries, 6 Willows Rd., Ayer, Mass. 01432 (Atlas will also sell urethane foam cores).

Branch River Foam Plastics, Inc., 167 Mill St., Cranston, R.I. 02905.

Chase Panels, 2755 S. 160th St., New Berlin, Wis. 53151.

Homasote Co., P. O. Box 7240, West Trenton, N.J. 08628.

Foam Products Corp., 2525 Adie Rd., St. Louis, Mo. 63043.

Foam Laminates of Vermont, Box 102, Hinesburg, Vt. 05461 (unfaced EPS cores also available).

Foam Plastics of New England, P. O. Box 7075, Prospect, Conn. 06712 (EPS cores available).

J-Deck, Inc., 2587 Harrison Rd., Columbus, Ohio 43204.

Neilsen-Winter, Inc., Main St., West Groton, Mass. 01474.

NRG Barriers, Inc., 61 Emory St., Sanford, Maine 04073.

Riverbend Timber Framing, Inc., P. O. Box 26, Blissfield, Mich. 49228.

Enertex Systems, Inc., 1425B North Park Drive, Fort Worth, Tex. 76102.

Building with Stress-Skin

Laminated, insulated panels offer new ways to build economical, energy-efficient houses

by Alex Wade

Many of the clients I work with are owner-builders who want livable, energy-efficient houses for as little money as possible. The post-and-beam designs I've come to specialize in are meant to be built quickly and economically, with special consideration given to the local climate and available materials, as well as to the skill of the builders. Over the last 15 years, I've been able to improve the speed, economy and quality of this kind of construction by using stress-skin panels to enclose simplified post-and-beam frames.

Stress-skin panels are rigid sheets (usually 4 ft. by 8 ft.) made from foam insulation and various sheathing materials such as plywood, wafer-board and drywall. The sheathing is bonded to both faces of the foam (see pp. 30-31 for information on how stress-skin panels are made), producing a laminated panel that has unusual shear strength and insulative value.

Laminated building panels had been used for years in commercial construction before a handful of timber framers recognized their suitability for enclosing finely joined frames some 10 years ago. Stress-skin panels gave these traditional-style framers a fast, effective way to fill the space between timbers with insulation, exterior sheathing and drywall (see pp. 26-30). It hasn't taken very long for stress-skins to catch on, and they have uses that reach far beyond timber frames. In fact, stress-skin panels and low-cost construction go hand in hand.

Stress-skin construction—To understand how stress-skin panels affect the economy and strength of a building, let's consider an analogy based on the evolution of auto designs. The traditional oak timber frame with its pegged, tight-fitting mortise-and-tenon joinery can be likened to the heavyweight chassis of an old-fashioned, full-size automobile designed with a separate body and frame. The often ornate body parts fastened to this gas-guzzling structural system weren't expected to make it stronger. In similar fashion, the traditional oak timber frame is massively sized, and then strengthened further with knee braces. Joined together properly, the frame can stand on its own perfectly well, so the

Builder Kevin Berry and crew tilt a stress-skin panel into place against a frame designed by the author. Revising construction details to take advantage of stress-skin's strength, size and insulative qualities will lead to a new generation of economical, energy-efficient houses.

Types of stress-skin construction

Enclosed-frame designs call for nonstructural stress-skin panels (one face of the foam core is clad with drywall) to be fastened to the outside face of a timber frame with spikes and adhesive caulk. Even nonstructural panels have enough shear resistance to replace the many knee braces used in most traditional timber-frame buildings. As a result, frame joinery can be simplified, cutting lumber costs and speeding erection time. In the frame shown above, major posts are 4x4s, and simple lap joints are used where girts meet over posts. Temporary braces hold the frame plumb and square until the stress-skin panels are applied.

Structural stress-skin panels must have plywood or waferboard bonded to both faces of the foam core with construction adhesive rated for structural use. Structural panels also have to be joined together edge to edge more solidly than non-structural panels, usually with wood splines. Once trucked to the site, these panels are simply tilted up onto a footing, subfloor or slab and joined with splines and spray-on adhesive to form the walls of the building (photo above right). Rough openings for doors and windows can be cut before or after wall raising. With cross-bracing (usually floor or ceiling joists) to keep wall panels from bowing out, structural stress-skins can also be used for the roof.

Hybrid stress-skin panel designs borrow construction details from structural-panel systems and enclosed-frame designs. In the design shown in the drawing, less expensive non-structural panels can be used in 4x8 size. Horizontal and vertical 2x6 splines tie panels together, supporting the stress-skin walls and also the ledger board to which the floor joists are fastened. Interior loads are carried by the ledgers and a single girder that is supported by 4x4 posts. The simple girder-and-post arrangement is the subframe in the hybrid design. It runs parallel with the ridge, supporting the floor joists and a second beveled girder at the roofline. —A. W.

Photo top left: Alex Wade; Photo right: Neilsen-Winter

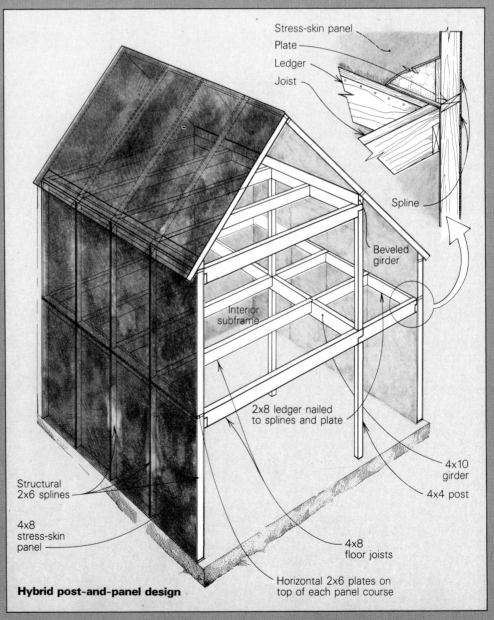

Stress-skin panel
Plate
Ledger
Joist
Spline

Beveled girder

Interior subframe

4x10 girder
4x4 post

2x8 ledger nailed to splines and plate

4x8 floor joists

Horizontal 2x6 plates on top of each panel course

Structural 2x6 splines

4x8 stress-skin panel

Hybrid post-and-panel design

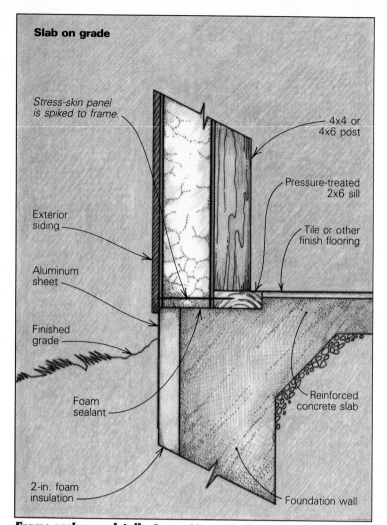

Slab on grade

Stress-skin panel is spiked to frame.

Exterior siding

Aluminum sheet

Finished grade

Foam sealant

2-in. foam insulation

4x4 or 4x6 post

Pressure-treated 2x6 sill

Tile or other finish flooring

Reinforced concrete slab

Foundation wall

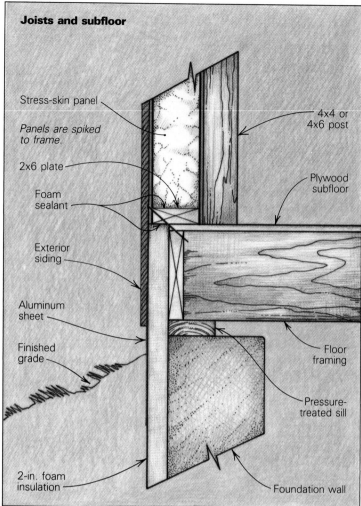

Joists and subfloor

Stress-skin panel

Panels are spiked to frame.

2x6 plate

Foam sealant

Exterior siding

Aluminum sheet

Finished grade

2-in. foam insulation

4x4 or 4x6 post

Plywood subfloor

Floor framing

Pressure-treated sill

Foundation wall

Frame enclosure details. **Stress-skin panel dimensions should be a major factor in the frame design. Above, frame and foundation detailing are integrated with panel thickness to create a flush finish at the base of the exterior wall. Both designs shown above provide a solid ledge for the panels' bottom edges. The width of the ledge aligns the sheathed exterior face of each panel with the foundation insulation.**

rigidity and shear resistance provided by the stress-skin panels are quite superfluous.

In their quest for better fuel economy, auto manufacturers have since learned that by combining the structural qualities of body parts and the chassis (called *unibody* or *monocoque* construction) they can reduce curb weight (and also manufacturing costs) without adversely affecting the overall strength of the car. Steel ribs and stiffeners are designed into floor pans, fender walls and other body parts, integrating the car's structure with its interior and exterior skin. Similarly, stress-skin panels can be used as stand-alone structural members.

Many mobile-home manufacturers use structural panels, but so far only a few companies use structural stress-skins for site-built houses. Delta Industries (1951 Galaxie St., Columbus, Ohio 43207) and J-Deck Inc. (2587 Harrison Rd., Columbus, Ohio 43204) are the two most successful structural stress-skin builders that I know of, and Neilsen-Winter Corp. (Main St., West Groton, Mass. 01472) has recently developed a structural panel system. Delta Industries has even had success in using structural stress-skin panels clad with pressure-treated plywood for below-grade foundation walls.

Hybrid stress-skin designs can be even less expensive to build than either a structural panel house or a lightweight frame design that's enclosed with non-structural panels. The use of structural splines between inexpensive non-structural panels and a small subframe for interior loads give this type of system a great potential for simplicity and economy in a small house design. Hybrid panel systems could easily compete in cost with factory-built prefab houses, and they offer more in the way of energy efficiency and aesthetics. In order for this to happen, though, building codes and especially building inspectors will have to change.

Stress-skins and the unbraced frame—At this stage, small-scale contractors and owner-builders generally favor using some kind of exposed frame with stress-skin panels. Building officials are likely to put more faith in structural systems that they can see, and the exposed timbers in a frame house have a definite aesthetic appeal. In addition, non-structural 4x8 panels have become widely available, and competition among panel manufacturers continues to drive prices down.

Even though panels faced with drywall on one side aren't rated to stand alone, they still perform like torsion boxes when spiked and glued to the outside of a frame. The foam core and the adhesive bond between core and sheathing materials provide the racking resistance. This means that the post-and-beam frame doesn't require corner braces. In fact, I've slimmed down my post-and-beam designs spe-

cifically to take advantage of the racking resistance that stress-skin panels provide.

Apart from saving time and timber, the unbraced frame (especially in a small house) looks cleaner and offers far more freedom in locating windows and doors. There's no need for complex timber joinery either. Where traditional timber framers use housed dovetails and pegged tenons, I usually specify nailed lap joints reinforced with metal truss plates that are eventually hidden beneath panels or flooring. Kevin Berry and Jeff Seeley, two builders I work with frequently, are often able to put together one of these simplified frames in just two days. The cost for a completed shell, with windows and doors installed but without finish siding or roofing, is usually around $15 per square foot.

In most cases, all the joinery for a simplified frame can be cut on site. Whenever possible, we try to get a good price from a nearby mill on roughsawn lumber (usually hemlock), and I try to hold the thickness of all framing members to no more than 3½ in. so that most of the joints can be cut with a skillsaw.

In houses that use a light post-and-beam structure, panels are spiked to the frame with either ringshank or hot-dipped galvanized spikes that penetrate 2 in. or more into the timber. The interior face of the panel should be seated in a thick bead of caulk where it rests against the frame. Where panels join vertically, the joint is

Illustrations: Elizabeth Eaton

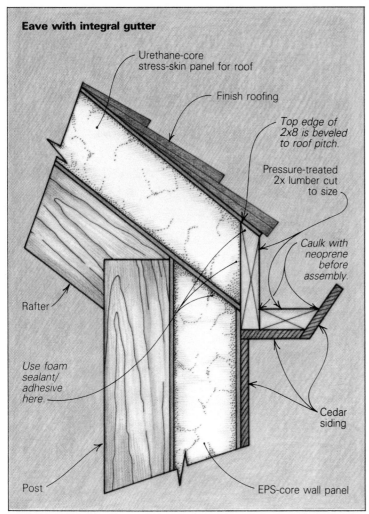

Eave with integral gutter

Urethane-core stress-skin panel for roof

Finish roofing

Top edge of 2x8 is beveled to roof pitch.

Pressure-treated 2x lumber cut to size

Caulk with neoprene before assembly.

Rafter

Use foam sealant/ adhesive here.

Cedar siding

Post

EPS-core wall panel

Above, eliminating eave overhang and incorporating a wood gutter are economical roof-construction details that also look good on the finished house. Above right, a scrap 2x6 pad and a hand sledge are used to close the tongue-and-groove joint between two wall panels. The posts and girts in this frame are located to provide nailing surfaces along panel edges.

usually a manufactured tongue-and-groove that is bonded with site-applied aerosol foam. Horizontal joints between panels are usually butt joints if they fall over solid backing. Otherwise, a 2x spline is necessary. In either case, the joint is sealed with spray-on foam.

In order to save time and money building a post-and-panel house, you have to plan post and girt locations carefully so that the panels are easy to install and waste from panel offcuts is minimal. I locate posts on 8-ft. centers. This way, wall panels can be installed vertically, and every other vertical panel joint lands on a post. Girts between perimeter posts should be located on vertical 8-ft. centers to provide a nailing surface for panel edges.

Because the stress-skin panels provide the wall's shear resistance in this type of design, the unbraced frame needs temporary braces to hold it plumb and level until the panels are nailed up. We temporarily brace posts to girts and to the ground as the frame is built. These braces stay on until the stress-skin wall is applied.

I've seen all sorts of arrangements tried at the base of the wall, where the frame, panel and foundation all converge. Many of them aren't very attractive, and not all of them are as easy or as economical to build as they could be. I prefer the details shown in the drawings on the facing page because they're uncomplicated and because the finished appearance unifies the insu-

lated foundation with the insulated wall. The frame shown in the left-hand drawing has a pressure-treated 2x6 sill bolted atop a 10-in. block wall that extends just a few inches above finished grade. This sill location leaves the outer 4½ in. of the foundation's top edge exposed, and this is where the 6½-in. thick stress-skin panels are set. The panel overhangs the foundation by 2 in.—just the thickness of the foam foundation insulation coming up from below. The outer face of the panel is flush with the foundation insulation. Kevin Berry likes to flash this grade-level joint with low-cost aluminum roll-stock before nailing up the exterior siding (photo below left, next page). Alternatively, you could parge on stucco finish over wire mesh. The cheapest designs have poured-concrete ground floors. But with some modifications, you could achieve the same flush-fit appearance with a conventional subfloor (drawing, facing page, right).

The roof—I recommend urethane-foam insulation for the roof—either in the form of stress-skin panels or rigid foam board installed over decking. Urethane provides more insulative value per inch than expanded polystyrene (EPS), and it's also a more effective vapor barrier. If EPS panels are used on the roof, they usually have to have 7½-in. thick cores, and the inner drywall face of the panel has to be sealed with

vapor-barrier paint (in the Northeast, where I do most of my work).

The nicest eave detail that I've seen on a stress-skin house incorporates a site-made gutter and fascia board in the first course of roof panels (drawing, above). To build this type of eave, you have to bevel-cut both the roof panel and the fascia. I use only pressure-treated 2x lumber for the back and bottom of the gutter, and cedar for the front side, since it's a better matching wood for roughsawn siding.

It's possible to let panels overhang by about 1 ft. at eaves and gables, but remember that neither the drywall nor the foam edge can be left exposed to the weather. For the sake of economy and ease of construction, eliminating overhangs altogether makes the most sense.

Windows and doors—Windows are easy to install in stress-skin houses if you've got a small chainsaw or a 16-in. circular saw. Either one of these tools has sufficient depth of cut to saw the rough opening from inside the house. Otherwise, the opening has to be cut from both sides of the panel. Either way, it should be sized about ¼ in. larger than the window frame.

Since the wall panels we use have 5½-in. thick cores, window frames have to have extension jambs made for 2x6 stud-framed walls. Though some builders recommend letting in 2x6s around the rough opening, we've found

Corner construction. **Leaving the corner open when panels are installed provides access to electrical raceways routed in the interior face of the panel's foam core. As shown at top left, a 2x6 panel edge has to be notched in line with the raceway before it is let into the panel and secured with caulk and drywall nails. Once both corner panels have been spiked to the frame, above right, plastic-sheathed cable can be run through the sill from the service panel. Then the cable is snaked through panel raceways down the wall to outlet-box locations cut at raceway height inside the house. Above left, the corner is enclosed when wiring is complete. Proper sill and post location allows the sheathed face of the panel to align flush with the foundation insulation. Roll aluminum, installed beneath the siding, extends below grade as a low-cost but nice-looking finish detail.**

that the rigidity of the panels eliminates the need for this extra framing unless you're installing a large window or a door.

To install the window or door, you shim it plumb and square in the opening, nail through the exterior trim and into the sheathing, and then fill the shim space with aerosol foam. This foam spray expands and bonds wood to foam, securing the frame. You'll have to trim off some foam squeeze-out before installing interior trim.

The open corner—Most builders who use stress-skin panels to enclose frames overlap panel edges at the corners. The open corner design that I use (photo above right) saves the 6½ in. of panel width lost to the overlap and also provides a very workable raceway for electrical wiring. Electrical wiring has been the bane of stress-skin systems since their development. Some wiring schemes call for custom-made hollow baseboards, while others demand that the electrician be on site as panels go up in order to snake wiring between panel joints. Both of these approaches can be costly and troublesome.

With open corners, neither corner panel ex-

tends beyond the face of the corner post that it's nailed to. This creates a 6½-in. by 6½-in. open area that is later boxed in and insulated. Before this happens, though, the open corner provides access to the wiring raceways that run the full lengths of both walls once all panels have been installed. The raceways are small channels that were routed along the inside face of the foam core before the drywall was glued on. The panels shown here have raceways cut 18 in. from their top and bottom (4-ft. wide) edges. This is a convenient feature, because panels can be flipped without misaligning raceways and because each wall can have outlet locations near ceiling and floor. Not all panel manufacturers are set up to incorporate raceways in their panels, so this could be an important consideration when deciding which supplier to use.

Using this open-corner detailing requires that you let in a 2x6 along each 8-ft. panel edge that will face the corner. This strengthens the panel edge and provides a nailing surface for filling in the exterior corner once the wiring has been run. As shown in the photo top left, the 2x6

edge board has to be notched out to give access to the panel's two raceways. Then it's let into the panel edge and secured with sealant and drywall nails.

Once all the panels are up, plastic-sheathed cable (Romex) is run into the corner from the service panel through a slot or hole made in the 2x6 sill. Using a double sill lets you drill out this hole with less risk of running your bit against masonry. Inside the house, the electrician can locate the raceway simply by measuring 18 in. up from the base of the panel. Holes for outlet boxes are usually cut with a utility knife right at raceway height.

Electrical cable can be run toward the center of the wall from one or both corners, depending on how your circuits are mapped. And by continuing the open corner above the first course of panels, upper floors can be wired the same way. Stud-frame partition walls inside the house are wired conventionally. □

Architect Alex Wade's Guide to Affordable Houses *is available from Rodale Press (33 E. Minor St., Emmaus, Pa. 18049).*

Stud-Wall Framing

With most of the thinking already done, nailing together a tight frame requires equal parts of accuracy and speed

by Paul Spring

I spent much of my former career as a carpenter building a reputation for demanding finishwork, but some of my best memories center around the sweaty satisfaction of slugging 16d sinkers into 2x4 plates as fast as I could feed the nails to my hammer.

The emphasis in framing is on speed. A lot has to happen in a short time. Accuracy, however, is no less important. The problems created by sloppy framing—studs that bow in and out, walls that won't plumb up and rooms that are out of square—have to be dealt with each time a new layer of material is added.

The fastest framing is done using a production system. But these techniques have long been the domain of the tract carpenter, and bring to mind legendary speed coupled with a legendary disregard for quality. However, production methods don't have to dictate a certain level of care. Instead, they teach how to break down a process into its basic components and how to economize on motion.

Done well, production framing is a collection of planned movements that concentrates on rhythmic physical output. It requires little problem-solving since most of the head-scratching has been done at the layout stage. As long as the layout has been done with care, a good framer can nail together and raise the walls of a small home in a few days, and still produce a house in which it's a pleasure to hang doors and scribe-fit cabinets. And this pace will give both the novice and the professional builder more time on the finish end of things to add the finely crafted touches that are rare in these days of rising costs.

If you know what the basic components of a frame are, nailing the walls together is simple. If not, you'll need to read the article on layout on pp. 13-21. After figuring out which walls get built first, you will separate the bottom and top plates (which were temporarily nailed together so that identical layout marks could be made on them); fill in between with studs, corners, channels, headers, sills, jacks and trimmers; and then nail them all together while everything is still flat on the deck or slab. The next step is to add the double top plate and let-in bracing. Finally you'll be able to raise the wall and either brace it temporarily or nail it to neighboring wall sections at corners and channels. Before joists or rafters are added, everything has to be *plumbed* and *lined*—this means racking and straightening the walls so that they are plumb and their top

Walls can be framed with surprising speed if they're laid out well. The pace isn't frantic—it's rhythmic, and based on coordination, economy of motion and anticipation.

plates exactly mimic the layout that was snapped on the deck—but that subject will be covered in another article (see pp. 46-49).

For the sake of simplicity, I have stuck largely to giving directions for nailing together a single exterior 2x4 wall with most of the usual components. I have tried to mention how this process would be different under different circumstances, and how each section of the wall is part of a larger whole. If you are using 2x6s or have adopted less costly framing techniques, such as the ones suggested by NAHB's OVE (Optimum Value Engineering), you'll have to extrapolate at times from the more traditional methods explained here.

Getting things ready—The carpenter I apprenticed to would begin wall framing on a Thursday so he could recover over the weekend from those first two grueling days of keeping up with the hot young framers. You don't need to plan things to this degree, but

you do need to make sure that the right tools and materials are at hand. Leave the detailed plates tacked in place on the deck for the moment (or up on foundation bolts in the case of a slab). Make sure the deck is clean; if it's not, sweep it. This surface is going to be the center of your universe for the next few days, so keep it spotless and plan how you will use each inch of it.

While you are setting up on the deck or slab, a helper can be cutting headers to length if this hasn't already been done. If the ground is flat, set up on sawhorses; if not, use one corner of the deck or slab. A cutting list can be made directly from the layout on the plates. I usually number each door or window opening sequentially around the deck with keel (lumber crayon) on the top plates, and then use these numbers to identify the headers as I cut them. This way I can easily find the piece I need for a particular wall and snake it out of the pile of corners, trimmers, channels and headers stacked on the deck. The information for cutting rough sills, sill jacks, and blocking is right there on the layout too, and your helper can make up a package for each opening.

If you're working by yourself or with only one other carpenter, it's just as easy to cut this 2x material in place when you're framing. The only exception is trimmers, which can be counted up and gang-cut if the headers are all to be at the same height. If you aren't using standard 92¼-in. precut studs, now is also the time to cut studs to length. Gang-cutting them is easiest, but precision at this stage is still important. If a few studs cut short of the line happen to get nailed next to each other in a wall, a dip will be left in the floor above.

You should also count up the total number of corners and channels you will need, and nail these together up on the deck or slab. Many framers don't bother with this step. They nail their corners and channels together when they're framing, but pre-assembly avoids having to sneak edge nails into a channel that faces down and is crowded by a regular layout stud.

Before you litter the deck or slab with any more material, you first must figure out how much of the frame you are going to nail together before you raise some walls, and where you are going to begin framing. On some second-floor or steep-site jobs where the plan is very cluttered, it pays to stack the frame. This

means nailing together all of the walls and then raising them at one time. But this requires a lot of planning since the walls literally have to be built on top of each other (often three walls deep), and a big enough crew to lift and carry the walls into place. Big production jobs use framing tables—essentially huge wall jigs—or the flat ground around the house to complete all of the walls on one level before raising them. But usually it's best just to frame as many walls as your deck or slab will accommodate (figure that you can frame an 8-ft. wall if you've got at least 100 in. of room on the deck for its height), and then raise them, repeating this procedure until there aren't any sets of plates left.

The layout will have a lot to say about which walls get framed first. Exterior walls get priority, so you can save the precious working room near the center of the deck as long as possible. Of the exterior walls, you will be framing the by-walls first and then the butt-walls so that you can build as many walls in place as possible. Pick one of the longest exterior by-walls to begin; the back wall of the house is the traditional place to start.

Studs and plates—The best place for a lumber drop is right next to the slab or deck. This way you can literally grab a stud when you need one. But on a steep site or second story you'll need to pack the lumber to the deck as

you need it. A laborer will help in this situation, but don't give in to the temptation to stockpile studs on the deck—you'll only end up having to move them again. You can keep plate stock handy, though, by leaning it up against first-story framing. Spot two or three bunches of 10- to 20-footers around the building for double top plates. It's pretty easy to take the bows and crooks out of a double top plate when you're nailing it, but if it's real bad, cull it out and start a pile that you can cut up into blocking and short jacks.

Nails—The only other items on the deck should be a skillsaw and a 50-lb. box of nails. The size and kind of nails a framer chooses seem largely regional. The allowable minimal size depends on whether you are *face-nailing* (through the face of one board into the face of another, such as nailing down the double top plate), *end-nailing* (through a board face into the end of board, such as nailing through the bottom plate into the studs) or *edge-nailing* (through the face of a board into the edge of another, such as a channel)—see the drawing, facing page, top. But rather than carry 8d, 10d, 12d, and 16d nails, it's easier just to carry a handful of eights in the small pocket of your nail bags for 1x let-in bracing and toe-nailing, and sixteens in the big pocket for everything else.

There are lots of choices when it comes to

nail coatings. Brights (regular steel nails without any coating) and even galvanized nails are okay, but I like *sinkers*. Sixteen-penny sinkers are a hybrid nail made in heaven (actually in Asia) for framers. Because of their coating, it takes half as many swings to drive one as an uncoated nail, and they don't crumple when you take a healthy swat. Their shank diameter is larger than a box nail, but not as thick as a common, which will split the ends of dry plate stock. They are also slightly shorter than 3½ in. (the length of the usual 16d nail), and have a thicker head. Sinkers can be either cement coated or vinyl coated.

In much of the far West, green vinyl (sold as g.v.) sinkers have become the predominant framing nail in the last five years. Although the vinyl does reduce the friction when they are being driven, these nails don't seem to offer much resistance on their way back out compared with other varieties. They're not nearly as bad, however, as nails that have had a treatment tract piece-workers affectionately call "gas 'n' wax." This is a coating made from kerosene and beeswax that is applied on site away from the eyes of building inspectors. You can whisper to these nails and they will drive themselves, but they unfortunately withdraw with the same ease.

My favorite nails are cement-coated (sold as c.c.) sinkers. The resin coating heats up from the friction of entering the wood and

Stocking the wall. **Some framers stock the wall with studs before they split the plates apart (facing page), so they know how far back to carry the top plate. But separating the top plate and stocking it with headers first will define the openings right away and save having to flop heavy headers down in a sea of 2x4s. Either way, crowning the studs pays off.**

makes the nail slippery. But unlike the green vinyl sinkers, once the nail is in place, the coating bonds with the surrounding wood, and holding power increases several times over brights. The disadvantage of c.c. sinkers is that the black resin accumulates on your fingers, nailbags and hammerhead. If you're nailing finish boards with these sinkers (such as 2x tongue-and-groove roof decking over exposed beams), coat your hands with talcum powder before starting to keep from leaving black fingerprints all over the ceiling.

Building the walls—Unlike Jud Peake, who details plates flat in his system (see pp. 13-21), I'm used to detailing plates up on edge. To frame a wall in place, I take the tacked-together plates and lay them just inside the snapped layout (wall) line with their inside edges down. Then I drive a 16d toenail through the bottom face of the bottom plate into the deck every 10 ft. or so. This way, when the top plate is separated, the bottom plate is already in position for framing. And once framed, the wall can be raised in place as if it were hinged to the floor.

Using this system, I separate the top plate with my hammer claw, walk it back on the deck and stock it with its headers first thing. This sets the location of wall openings early so you don't stud over one by mistake. It also means you don't have to flop a long, heavy header down in a sea of studs. But lots of other framers like to stock the wall with studs before separating the top plate so they know how far back to lay it (photo facing page).

Whichever order you use to stock the studs, cull the ones that are really crooked, and make sure that you lay all of the keepers crown up. Do this by sighting along each one before laying it down. Once you've got a stud on every layout mark, you can add corners and channels if you've made them up as units. Make doubly sure that the flat stud in each channel is facing as it should by looking at the layout on the floor as well as at the marks on the plate.

So far you haven't driven a nail, but soon that's all you'll be doing. If I'm using 4x12 headers and don't have to deal with head jacks, I like to nail the top plate to the top edge of the header first thing. This adds a lot of weight to the top plate and keeps it from moving around. Also, top-plate splices often come in the middle of a header, and you can begin making the wall a single unit by connecting the top plate to the header. Make sure the plates butt tightly so you don't lengthen the wall, and drive two nails into each plate end. At each end of the header, you'll also need two sixteens. In between, you should

Types of nailing

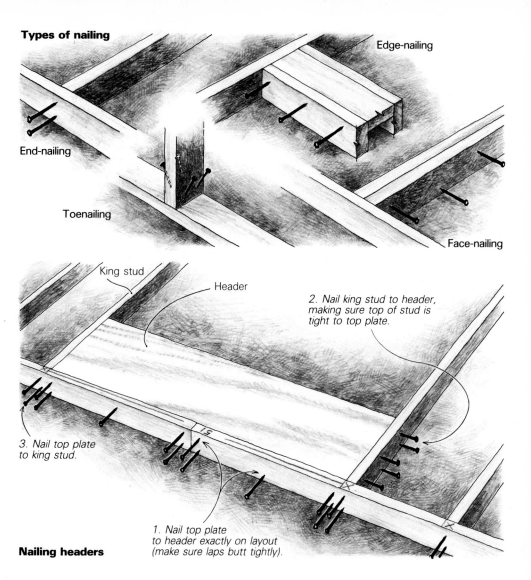

Edge-nailing

End-nailing

Toenailing

Face-nailing

King stud

Header

2. Nail king stud to header, making sure top of stud is tight to top plate.

3. Nail top plate to king stud.

1. Nail top plate to header exactly on layout (make sure laps butt tightly).

Nailing headers

stagger the nailing to each side at 16-in. centers. If the plate runs through without a splice, use two 16d nails at each end of the header, and then stagger nails every 16 in. in between, as shown in the drawing above.

Next, you should take care of the king stud to make sure that you have room to swing your hammer. Drive at least four nails through the face of each king stud into the end of the header. With a 4x12 header I use six 16ds. This is an important intersection. If the king stud doesn't stay tight to the header, it will pop sheetrock nails and leave a crack radiating away from the corner of door and window openings. Last, end-nail the top plate to the king stud with two nails. If you are using a header that doesn't reach the top plate and therefore requires head jacks, you can drive two nails into the top of each one through the top plate as you would a regular stud, and then toenail the bottoms to the top of the header with four 8d toenails each.

Now you can start nailing off everything that you've laid between the plates. Stay with the top plate and begin nailing at one end or the other. If you are right-handed, you'll find that working from left to right will probably be most comfortable and help you establish a rhythm. You'll be working bent over from the waist—one foot up on the edge of the plate, and the other foot nudging the stud onto the layout line and bracing it from twisting (photo

next page). Each 2x4, whether it's a layout stud or part of a corner or channel, gets two 16d nails driven into it through the plate.

Your first nail should be near the top edge of the plate, where the pencil layout line is marked. Set the nail with a tap of your hammer, then line the stud up on the mark and drive the nail through the plate and into the stud with your next couple of blows. Be careful to split the difference if the stud is narrower or wider than the plate to which you are nailing it. This may mean having to hold the stud up with your nail hand until you get the first nail in. Pay close attention to the layout; it's surprisingly easy to lose your concentration and begin nailing on the wrong side of the line despite the *X* on the edge of the plate that indicates the stud location. The only trick to the second nail is to make sure the stud is square to the plate. Judge this by eye.

Production framing requires a strong hammer arm and a dexterous nail hand. The only way to develop your arm is to drive a lot of nails, but there are some tricks to fingering the nails. With 16d nails, you need to orient the heads all in the same direction. I like to do this when I'm over at the nail box to refill my bags. This way, each time that you reach down you can pull out a large handful of nails ready to drive. Then, without dropping the nails cradled in your hand, use your index finger and thumb to reach into your palm to

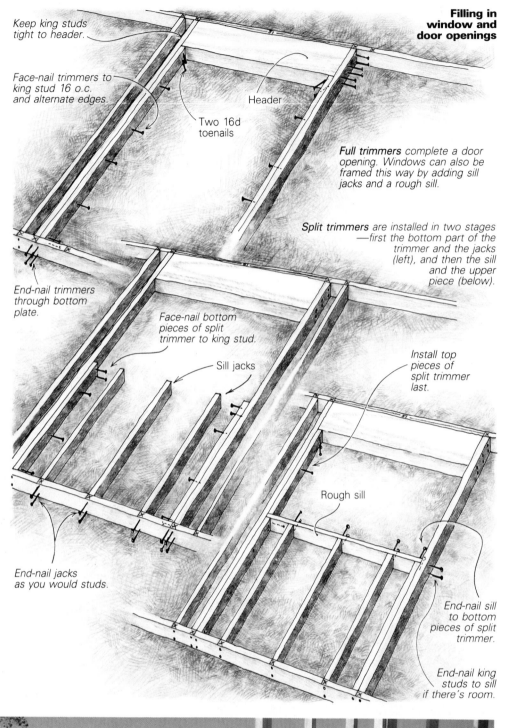

Keep king studs tight to header.

Face-nail trimmers to king stud 16 o.c. and alternate edges.

Header

Two 16d toenails

Full trimmers complete a door opening. Windows can also be framed this way by adding sill jacks and a rough sill.

Split trimmers are installed in two stages —first the bottom part of the trimmer and the jacks (left), and then the sill and the upper piece (below).

End-nail trimmers through bottom plate.

Face-nail bottom pieces of split trimmer to king stud.

Sill jacks

Install top pieces of split trimmer last.

Rough sill

End-nail jacks as you would studs.

End-nail sill to bottom pieces of split trimmer.

End-nail king studs to sill if there's room.

pinch a single nail. Extend the nail and rest it point down on the plate for the hammer while your other fingers regrip the nails in your palm. This should all happen while you are backswinging your hammer.

The first swing should be just a tap to start the nail. All carpenters hit their fingers occasionally, but you learn to keep your fingers out of the way when you are swinging hard enough to do any damage. This is particularly important with framing hammers, which can tear as well as bruise.

Once you've finished nailing off the top plate, move to the bottom plate and do the same thing all over again. You may want to reverse this process if you're framing exterior walls on a slab. The big problem here is that the anchor bolts invariably fall on the stud layout, and this requires chopping out some of the stud bottom. This is a lot easier if the top of the stud isn't already nailed.

When you are nailing the end studs on walls that butt by-walls on both ends, hold this last stud back from the end of both plates about ¼ in. This way if the stud bows out slightly, it won't prevent the wall from being raised. You can drive it back to its proper position when you nail the intersection together.

Another potential trouble spot is plate splices. If they come at a stud, you will need four 16d nails in the end of that stud—two from each plate end (photo below left). The other place that plates often join is in a doorway. Although the bottom plates will eventually be cut out, on long, heavy walls make sure that they stay butted by nailing a block on top of the plates at the joint.

Windows and doors—Once all of the headers, studs, corners, and channels are nailed in, you can complete the openings. Doors are easy. Fill in under each header with one trimmer on each side (two on each side if the opening is 8 ft. wide or more), as shown in the drawing at left. Trimmers need to fit snugly between the header and the bottom plate. You shouldn't have to pound the plate apart to get them in, nor should they be short. At its bottom, nail the trimmer like a stud. Then face-nail it to the king stud at 16 in. o. c., with the nails alternating from edge to edge. At its top, drive two 16s up at an angle to catch both the king stud and the corner of the header.

Taking some care with the trimmers will really pay off when it comes time to hang and case the doors. Make sure that the bottom of the trimmer is nailed right on its line—this will ensure a plumb and square opening. Also make sure that the edges of the trimmer and the king stud are even—this means that the

Framing. **The top plate being nailed at left is a splice over a stud, which requires four nails driven on a slight angle. When space gets tight on the deck, two framers often end up working on the same wall. One should take the entire top plate; the other, the bottom plate. If they are both right-handed, they will begin on opposite ends of the wall since they will be moving left to right for comfort and speed.**

mitered door casing will sit on a flat surface instead of a hump or dip. When you're finished nailing the trimmers, go back to the top of the king stud with your hammer and give it a bash or two so that it isn't separated at all from the end of the header.

Windows are a bit more complicated than doors because you've got the rough sill to contend with. Here you've got a choice. The sill can be cut to butt against both king studs, with *split trimmers* nailed up above and below it, or you can nail up full trimmers and toenail the sill into them. I was taught to use the split-trimmer method (see *FHB* #15, p. 43), and prefer it because you can end-nail the sill into the lower trimmers and even end-nail through the king stud for a very solid connection.

If you're using split trimmers, begin by cutting the rough sill to length. Instead of measuring between the king studs, hold your sill stock up to the bottom of the header and mark it. Here the framing will be tight, so you'll get an accurate measurement even if the studs bow out down where the sill will be installed. Next, cut the sill jacks if their length is given in the layout, nail them at the bottom plate and drop the rough sill on top. If the sill-jack measurement isn't given, the size of the rough opening will be, and you can use your tape measure to mark the top of the rough sill. Allow 1½ in. for the sill, strike a line, and use this lower mark to cut in the sill jacks.

Remember that the lower part of a split trimmer is just another sill jack except that it also gets face-nailed to the king stud using 16d nails on alternating edges at 16 in. o. c. (drawing, facing page). Once all the jacks and lower trimmers are installed, you can end-nail the rough sill to them with two 16d nails at each intersection. If there's room, drive a couple of end nails through the outside of the king studs into the sill. If you can't do this, drive a 16d nail at a slight angle down through each end of the sill to catch the king stud.

The last step is to cut the top half of the split trimmer and face-nail it in. The only other complications are double sills and trimmers for very wide windows. In this case, install the first trimmer on each side full, and split the inside one around the doubled sill.

If you're not using split trimmers, you can treat the opening as you would a door at first by installing the trimmers full length underneath the header. Then mark the sill for length, and cut it. Follow the same procedure that was outlined above to set the sill and its jacks. But when you set the sill, use at least four toenails from the sill into the trimmer, along with a 16d toenail through the thickness of the trimmer into the end grain of the sill.

Rake walls and specials—Rake-wall (gable-end wall) framing is a little different because the tops of the studs have to be cut at the roof-pitch angle. But if the wall has been laid out right, cutting shouldn't slow you down too much. Mark the studs in place and cut their tops with the shoe of the saw set at the roof-pitch angle.

The bottom plate nails to the stud normally.

Blocking. Horizontal blocking for tubs and showers on interior walls goes faster if you let it in rather than cutting and nailing short blocks in the stud bays. Make sure to set your saw depth accurately to ensure that the blocking ends up in the same plane as the rest of the wall.

At the top plate, I usually drive a toenail through the toe (long point) of the angle at the top of the stud into the plate to hold the stud on the line. Then I drive two 16s down through the top plate into the stud, as you would in a standard wall.

The last step before double top plating any wall is to take care of the specials, which usually means different kinds of blocking. Interior soffits (drop ceilings) and walls over 10 ft. high will need fire blocking or stops—horizontal 2x4s nailed between studs to delay the spread of fire up the wall. Chalk a line across the studs and stagger the blocks on either side of the line with two 16d nails driven through the studs into each end of the block.

Mid-height flat blocking may also be required on exterior walls that will be covered with stucco. Production framers often raise this blocking a few inches on 8-ft. walls to 51 in. or 52 in. off the floor. This makes it easier to duck through the stud bays when working, more convenient as a shoulder support for lifting the wall onto foundation bolts on a slab, and keeps the blocks just slightly higher than electrical switch boxes.

Blocking for tubs and showers is a bit easier since it is usually installed with the face of the blocking nailed flush with the edges of the studs. It can either be cut in short blocks (14⁷⁄₁₆ in. for 16-in. centers), or a length of 2x can be let into the studs (you can do this only on interior walls where the side that requires blocking is facing up). To let in blocking (photo above), position the 2x4 on the wall where the blocking will be needed and scribe

Bracing and raising. Let-in bracing can be cut into the top edges of the studs by setting the saw depth to 1½ in. and letting the shoe ride on the bracing (photo right). This procedure is not without risk of kickback. The safer alternative is scribing against the brace and cutting to the pencil mark. Facing page: The let-in brace in the foreground of the photo shows a forest of 8d nails that will be driven once the wall is raised and plumbed. Two keys to raising walls safely are even distribution of the weight and vocal coordination of effort.

Let-in bracing

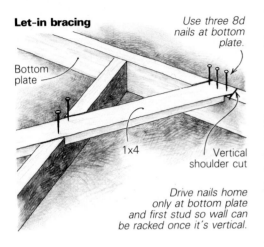

Use three 8d nails at bottom plate.

Bottom plate

1x4

Vertical shoulder cut

Drive nails home only at bottom plate and first stud so wall can be racked once it's vertical.

Double top plating

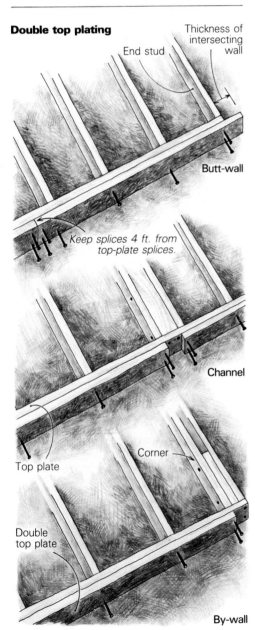

Thickness of intersecting wall

End stud

Butt-wall

Keep splices 4 ft. from top-plate splices.

Channel

Corner

Top plate

Double top plate

By-wall

---

a pencil line on either side. Cut to the inside of each of these lines with your saw set at 1½ in., remove the resulting scrap and nail the 2x4 in place.

Double top plate—The key to double top plating is the channels and corners. When two walls intersect, one of the double top plates acts as a tie between them. The double top plate on a butt wall will overhang its end stud 3½ in. for a 2x4 wall, and the double top plate of the corresponding by-wall will be held back 3½ in. to receive it. Double-plate splices should be held back at least 4 ft. from the end of a wall or from splices in the top plate.

Pull your plate stock up on the deck and lay it down where it will go. You may be tempted to use a tape measure to mark the overhang or hold-back on butt-walls. Instead, hold the stock to the far side of the channel and scribe the other end of the stock with your pencil held against the end of the wall. As long as the walls are all the same thickness, you will get a double top plate that's the right length by cutting at the pencil line.

Double top plating can go very fast. The ends of each piece require two 16d nails (drawing, left). In between you'll want to drive a 16d at every stud, alternating sides of the plate. Hitting the stud layout with these nails allows you to let in bracing in any of the bays without worrying about hitting a nail with your saw. You will only get to nail double top plates on butt-walls that are built out of place, since this plate has to project beyond its end stud by the thickness of the by-wall it will intersect. If it's easier to build the wall in

place, cut the plate anyway and nail it once the wall is raised.

Accuracy in cutting and nailing double top plates is essential. A double top plate that projects a little beyond a corner will drive you bananas when you're out on the scaffolding later nailing the sheathing. Be careful with the width of channels too—as soon as you leave a 3½-in. slot for the double top plate of an intersecting wall, it will invariably be cut from dripping wet stock and measure 3⅝ in. You can trim the double top plates on each side to make it work as it should, but you'll have to lay the wall back down to do it.

There's another precaution you should take just before raising a butt-wall that has been framed out of position. Give the overhanging double top plate a couple of blows from your hammer to drive it up off the top plate a half inch or so. This way when you are sliding the butt wall into place, the projecting double top plate won't hang up on the top plate of the already standing by-wall.

Bracing—All walls need some kind of corner bracing to prevent them from racking. There are lots of ways to get this triangulation. Sheathing and finished plywood siding provide excellent resistance if the nailing is sufficiently close. Cut-in bracing (flat blocks cut at an angle and nailed into each stud bay along a diagonal line) and metal X-bracing (long 16-ga. sheet-metal straps that are nailed to the studs under siding) are both quite effective. But for maximum strength when you are not using plywood, let-in bracing is usually specified. This 1x4 brace is mortised flat into

the exterior edges of the studs. It should extend from near the top corner of the wall down to the bottom plate at about 45°. In an 8-ft. wall with studs at 16 in., the brace will cut across six stud bays. You can get one into five bays by increasing the angle a bit.

Not every stud space needs to have a brace; the minimum standard for exterior walls and main cross stud partitions is one brace at each corner, and one at least every 25 lineal feet in between. But the more braces you use, the easier the wall will be to plumb and line, which in turn will create square rooms with plumb door jambs. And each wall that gets a let-in brace will act as a single unit, which in turn will increase the strength of your frame when it's subjected to wind or seismic loads. On a quality house, the small price you pay for 1x4 and the couple of minutes it takes to let it in allow you to brace even short walls. This goes for interior partitions as well. I even like to shear panel (⅜-in. plywood with close nailing) long cross walls.

A typical medium-length wall (say 30 ft. long) contains two let-in braces. These should form a V with an open bottom, since the tops of the braces start high on each corner. Of course it's not always this simple, because most walls contain window and door openings that have to be dodged.

These braces are best let in before the walls are raised, but you must make sure that the wall is close to being square. If you're framing the wall in place on the deck, it should be okay. But if it's been moved around some and is visibly racked, then you need to rough-square it by getting the diagonal measure-

ments approximately the same. Then lay a 1x4 across the top edges of the studs at an approximate 45° angle. Both ends should overlap the plates between studs. On an 8-ft. wall this will require a 12-footer. Now scribe each side of the 1x4 on every stud and at the top and bottom plates. Cut on the outside of each of these lines with your saw set at a fat ¾-in. depth. This will produce a slot about ¼ in. wider (two saw kerfs) than the brace—the extra width will accommodate the racking of the wall left and right during plumbing and lining. Remember, it's the number of nails and their shear strength that will keep the wall square eventually, not the fit of the brace in its slot.

Production framers eliminate scribing by cutting with the 1x4 in place (photo facing page). This procedure works best with a worm-drive saw, and requires a lot of experience with the saw because of the danger of kickback. Even then, safety takes a back seat to speed here. The brace is held down in place with one foot, and the saw is run along one side of it set to a depth of 1½ in. The shoe of the saw rides right on the brace. The framer will then change directions and saw back along the other side of the 1x4. The ends of the 1x4 are cut off in place at the bottom of the bottom plate and about ¼ in. shy of the top of the double top plate. The last step is to chop out the little blocks of wood between the saw kerfs.

Install the brace while the wall is still flat on the deck. Drop the brace into its slot, holding it flush with the bottom of the bottom plate, and drive three 8d nails there. You can also nail it to the first stud since this is still low

enough that it won't interfere with racking the top of the wall during plumbing and lining. You should also start two nails in the face of the 1x4 for each stud so that they can be easily driven home with one hand later. Start a total of five nails at the top of the brace—two into the double top plate and three just below this at the top plate. A lot of framers also start a nail in the top plate and bend it over the brace to keep it from flopping around when the wall is being racked.

Let-ins are most effective if they have a vertical shoulder at the bottom plate rather than coming to a point. This means making a plumb cut on the last inch or so of the brace measured along the angle, and a corresponding slot in the bottom plate (top drawing, facing page). Not all building inspectors will insist on this, but it's a good idea anyway.

If you're building on a slab, don't allow the let-in brace to sit on the concrete. End the brace in the middle of the 1½-in. thickness of the redwood or pressure-treated bottom plate.

Raising the walls—This is the best part, but also the point at which a lot of backs get wrecked (for the mechanical alternative, see the sidebar on the next page). Move anything you could possibly trip over out of the area. If the wall is a rake or is particularly tall or heavy, toenail the bottom plate to the wall line if it isn't already, or use lumber strapping (see *FHB* #8, p. 12). For a standard 8-ft. wall, nail stops—short lengths of 2x4—to the joist just below the deck every 6 ft. to 8 ft. so that they stick up where the wall is going to be raised, preventing it from skidding over the

Mechanical wall raising

Every carpenter has at least one story about a former partner or laborer who was 6 ft. 4 in. and immensely strong. However, being able to lift twice your body weight is of little consequence when you consider the weight of construction materials and the power of machines. While hiring a crane or forklift is usually not an economical option on small jobs, using simple mechanical advantage is, and it can change the way you work.

Wall jacks are a good example. With a pair of them, two people can easily raise a long wall full of solid headers weighing 2,000 lb. From there, the logical step is to sheathe exterior walls—and even add windows, siding, trim and paint—down on the deck before they are raised.

There are two kinds of wall jacks, but they work similarly. The first looks and operates much like a scaffolding pump jack. By pumping the handle of its ratchet winch (the same mechanism a car jack uses), the jack walks up a 2x4 (or 4x4), carrying the top of the wall to be raised with it on its horizontal bracket. In fact, these devices are often called walking jacks. The 2x4, which begins in a vertical position, is held from skidding by a block nailed to the deck. As the jack makes its way up, the 2x4 is allowed to get less and less vertical so that the wall will continue to bear on it. These jacks are relatively inexpensive, ranging from $60 to $125 apiece. Two brands I know of are Hoitsma walking jacks (Box 595, River Street Station, Paterson, N. J. 07524) and Olympic Hi-jacks (Olympic Foundry, Box 80187, Seattle, Wash. 98108).

The other kind of wall jack (the one that I've owned) is manufactured by Proctor (Proctor Products Co., 210 8th St. South, P. O. Box F, Kirkland, Wash. 98033). It consists of a metal boom that is fitted to a hinged plate at the bottom that nails down to the subfloor and joists below. What amounts to a ¼-ton come-along is mounted on the boom just below waist height. The ⁵⁄₃₂-in. galvanized aircraft cable is threaded through a sheave (pulley) at the end of the boom and is fitted with a nailing bracket that attaches to the double top plate of the wall to be raised. The boom begins in a vertical position, and begins to lean as soon as the ratchet winch is put to use and the wall begins to come off the deck. An

adjustable stop prevents the wall from going beyond vertical. Proctor wall jacks come in three lengths (16 ft., 20 ft. and 23 ft.) and all of them telescope for carrying or storage. Although this system is more expensive— the smallest pair of jacks retails for $445— it's very safe and can handle walls as long as 75 ft. with just two carpenters on the job.

The real advantage to wall jacks is that they give you the freedom to finish a wall completely while it's flat on the ground. This means money saved even if you're building a one-story house on a flat lot. On high work, you often can eliminate the cost of scaffolding and speed the usually slow progress of sheathing, windows, siding, trim and paint by doing most of the work right where the frame sits on the deck. On the houses where I've chosen to do this, I've also been able to use less skilled labor to complete the walls. For instance, any careful person can apply a full-bodied stain with a roller when the wall is flat and accessible. It doesn't have to be sprayed by a painting sub. Trimming out windows can be left to an apprentice, since having to recut a piece for fit isn't a big deal when you're not climbing down off scaffolding to do it.

Tilting up finished walls isn't for every house. It is best used on modest rectangular plans with long walls where the siding and the trim detailing are simple. In any case, corners have to be finished off from ladders or simple wood scaffolding suspended from nearby window sills.

The only real trick to completing a wall on the deck is to rack the framing so it's exactly square before you sheathe it. This is worth checking several times. The alternative is spending half an hour with a cat's paw removing all that shear nailing so that the ends of the wall will sit plumb once in place. To check for square, toenail the bottom of the bottom plate of the wall to the deck right on the snapped wall line and then use a sledgehammer to bump the top of the wall until the diagonals measure the same. The only other complication is with butt-walls once the by-walls are up. Butt-walls have to be finished slightly out of position (left and right) to buy clearance for the sheathing or siding that will overlap the corner framing of the by-wall, or the end panel will have to be left off until the wall is raised. —P. S.

edge. Also nail a long 2x4 with one 16d nail high up on each end of a by-wall. The single nail will act as a pivot so that the bottom of the brace will swing down and can be angled back to about 45° alongside the deck. It can then be nailed to the rim joists when the wall has been raised to approximately plumb. More brace material should be stacked nearby so that it can be grabbed quickly.

To get the wall in a position where you can get your hands under it, lean short lengths of 2x against the face of the double top plate every 12 ft. or so. Then, standing inside the wall itself, bury your hammer claw in the double top plate with a healthy swing. When you lift the top of the wall just a few inches, the blocks will fall beneath the top plate and you'll have enough room to get a grip.

Now it's time to gather your crew. Most carpenters can lift a good 12 lineal feet of 2x4 framing—more if there aren't a lot of 4x12 headers, less if the wall is 2x6 or framed with very wet lumber. Spread people out along the wall according to where the weight is. Headers are the worst because the weight is all at the top. The ends of the wall are almost always the lightest. The first maneuver in lifting the wall (photo previous page) is called a clean and jerk in weightlifting. If you don't bend your legs it's a sure road to a hernia or a bad back. The second stage—where you've got the wall to your waist and you are pushing with the palms of your hands—is basically a press and should be done with your legs braced behind you. Don't make it a contest. Raising a heavy wall requires staying in sync with everyone else.

If you're raising a by-wall, at least one of the crew should let go once it's up—with fair warning—and nail the outside braces. The wall should lean out slightly at the top to leave a little extra room for the butt wall that will intersect it.

Raising walls on a slab is slightly different. Here you've first got to raise the wall to a vertical position, and then lift and thread the bottom plate onto the foundation bolts. Long 2x4s on edge can be used effectively as levers under the bottom plate to lift the wall up above the bolts, as long as you have someone steadying the top of the wall. End braces can be nailed off to stakes driven in the ground right next to the slab. Once the wall is steady, beat on the sill at several spots to make sure that it's down, then put washers and nuts on the foundation bolts and screw them down finger tight.

If you're raising an exterior butt wall on either a slab or a deck, you'll be nailing the end stud to the corner of its corresponding by-wall, rather than using a brace. Make sure that the bottom plate is flat on the deck and that the two top plates match in height. Also align the outside face of the corner and the edge of the end stud so that they are in the

same plane all the way up. Alternate 16d nails on each side of the end stud every 16 in. If you are raising a partition or interior wall, nail its end stud to the channel in the same way, with the same kind of care.

Long walls may require an intermediate brace or two before everyone can let go. How much bracing you need to add depends in part on how soon you'll be going home for the day. Braces take up a lot of space on the deck when you're trying to frame the rest of the walls. But when you leave the site, it's a whole different story. Figure that a hurricane will strike that night and brace everything off accordingly, especially if you have already sheathed your walls.

If you're bracing off walls on a concrete slab for the night or weekend, you can use a "te-pee" on exterior walls. To make one, take a 14-ft. piece of plate stock and run it through a stud bay of the wall to be braced so that half

of the plate stock is cantilevered out beyond the building, and the other half is on the inside of the slab. Then nail a 2x4 brace from the top of the wall down to the end of the plate stock on both the inside and outside. This kind of brace allows a little give, but the wall won't go anywhere.

A precaution you want to take on a plywood deck is to nail down the bottom plate. First make sure that the ends of the wall are where you want them—on a layout mark, flush with the perimeter of the deck, or butting another wall. Then use a sledgehammer to persuade the bottom plate into a straight line that sits just at the edge of the wall line established in chalk during layout. After that, drive one 16d nail per bay to keep it there.

The last thing that you have to do to connect the walls is to nail down the double top plates that lap over intersecting walls. Remember that the walls don't have to be

plumb; you'll take care of that later. You are just making sure that the walls are nailed together exactly as they were laid out from the bottom plate all the way up.

To nail off the double top plates that lap, you can claw your way up a corner and walk the plates between channels. But you'll probably get it done more safely by moving a 6-ft. stepladder around. First, make sure that the walls are driven together tightly all the way up, and that they are aligned with each other vertically. Then you can finish off the double top-plate nailing using four 16d nails for each 2x4 plate lap.

When all the plate pairs are nailed down, your frame will be complete. Plumbing and lining it will make it ready for joists or roof rafters. The feeling at the end of a day of this kind of work is unparalleled. You are surrounded by the tangible evidence of your progress and the worth of your labor. □

Plumb and Line

Without this final step of straightening the walls, the care taken during framing will have little effect

by Don Dunkley

It's strange to think of a completed wall frame as being a kind of sculpture that needs final shaping. But that's just what it is. Until the walls have been braced straight and plumb, they can't be sheathed or fitted with joists or rafters without producing crooked hallways, bowed walls, ill-fitting doors and roller-coaster roofs.

The production name for getting the frame plumb, square and straight is *plumb and line.* The job doesn't take very long—three to six hours for most houses—but it's essential. After the frenzied pace of wall framing, plumb and line can be a welcome relief. It requires at least two carpenters (three's a luxury) working closely together. The work is exacting, but not hard, and there's a sense of casual celebration in having finished off the wall framing.

Stud-wall framing is based on things being parallel and repetitive (see pp. 13-21). If you plumb up the end of a wall, then all the vertical members in the wall will be plumb in that direction. And if the bottom plate of the wall is nailed in a straight line to the floor, then getting the top of the wall parallel to the bottom is easy: just plumb the face of the wall at both ends, and make the top plates conform to a line between these two points.

Plumb and line is a fluid process in which walls are braced individually, but the sequence of operations is important. Although you can start at any outside corner, walls should be plumbed and then lined in either a clockwise or a counterclockwise order, since each correction will affect the next wall. Once the bottom plates have been fully nailed to the floor or bolted and pinned to the slab, the exterior walls are plumbed up and the let-in braces are nailed off, or temporary diagonal braces are installed to prevent the wall from racking (the movement in a wall that changes it from a rectangle to a parallelogram, throwing the vertical members out of plumb). Then line braces are nailed to the walls to push or pull them into line at the top and to hold them there. Last, the interior walls are plumbed and lined with shorter 2x4 braces.

Plumbing and lining can begin once all the intersecting walls are well nailed to their channels or corners. On each of these walls,

Holding wall intersections tight while the lapping double top plate is nailed off can make the difference between a plumb frame and having to make a lot of compromises later on.

the end stud has to line up perfectly with the channel flat or corner studs. It's also wise to make sure that the heights of the walls match up. If they don't, it's usually because the end stud isn't sitting down on the bottom plate. Also, all double top plates must be nailed off where they lap at corners and channels. Be sure that there aren't any gaps here (a common defect in hastily framed houses), and that the end studs and plates are sucked up tight (photo facing page).

Fixing the bottom plates—The next job is to toss all the scraps of wood off the slab or deck and sweep up. This way you'll be able to read the chalklines on the floor. If the frame is sitting on a wood subfloor, the bottom plates have likely been nailed off. On a slab, though, you'll have to begin by walking the perimeter of the building putting washers and nuts on all the foundation bolts. Tighten the nuts a few turns before using a small hand sledge to knock the plates into alignment with their layout snaplines. The slight compression created by the nut will keep the plate from bouncing when you hit it.

Getting the plates right on the layout is critical, because the bottom of the frame determines the final position of the top of the wall. Once the plates are where you want them, run the nuts down tight. The fastest tool for doing this is an impact wrench, found mostly on tracts and commercial jobs. Next best is a ratchet and socket, with last place going to the ubiquitous adjustable wrench.

Wherever bottom plates butt end-to-end on a slab, both pieces must be fastened down with a shot pin (see pp. 86-90 for more on powder-actuated fasteners). The gun I like the best is the Hilti. Unlike many of the older guns, which have space-hogging spall shields, it is slim enough to get into tight spots between studs. The model that I've used (DX-350) is fitted with a multi-load magazine, so it can be reloaded in seconds. It also uses a pin and washer that are made as a single unit—a big advantage.

If there are two of you working, one person can set the pins into the top of the plate with a tap of a hammer. The other person can follow, slipping the gun barrel over the shaft of the pin and firing. The pins should be placed close to the inside edge of the plate so that you don't blow out the side of the slab. To prevent the bottom plate from splitting out at the ends, you can nail a ⅜-in. or ½-in. plywood scrap over it and then fire the pin through it. Since doorways take a lot of abuse, a pin should be shot into the sill at each side of the opening.

After all the outside walls are fastened down, you can line the bottom plates of the inside walls and shoot or nail them down. Be especially careful on long parallel runs like hallways—any deviation will really shout out when the drywall is hung and the baseboard installed. On interior partitions, one drive pin or nail every 32 in. should be sufficient. Make sure to hit the flanking bays of doorways, and both sides of bottom-plate breaks.

Setting up—Once all the bottom plates are secure, spread your bracing lumber neatly throughout the house. Interior walls will need 8-ft. and 10-ft. 2x4s. Exterior walls will need 14 and 16-footers for line braces, and shorter diagonal braces if the wall doesn't have let-ins. You'll have to guess how much bracing you'll need. More is always better than less, but figure that you'll need a line brace every 10 ft. and a diagonal brace on any wall over 8 ft. long without a let-in.

To plumb walls I use an 8-ft. level with two small aluminum spacer blocks that fit over the lip of the level near each end, and are tightened with handscrews. When the level is held vertically against a wall, the spacers butt the bottom plate and double top plate and prevent bowed studs from interfering. You don't need an 8-footer; any level that is dead-on accurate will do if you extend it a bit. Select the straightest 2x4 around, cut it to 8 ft., and tack a short piece of 1x to one edge of the 2x4 at each end for a spacer. Then attach the level to the other edge of the 2x4 using duct tape or bent-over 8d nails. (See *FHB #23*, p. 80, for a report on an extendable level.)

Plumbing the walls—Exterior walls are plumbed from their corners. Choose any one as a starting point. Position the level at the end of the wall so that you are measuring the degree to which it's been racked. The bubble will indicate which way you have to rack the wall back to plumb. To do this, you can make a push stick out of a 9-ft. to 10-ft. 2x4 that is largely free of knots. A more flexible push stick can be made from two 1x4s nailed together face-to-face. Cut a slight bevel on the bottom end to keep it from slipping. To use the push stick, place the top of it in a bay about midway down the length of the wall to be racked, up where the stud butts the top plate. The stick should angle back in the opposite direction from which the top of the wall should be moved. Plant the bottom end on the slab or deck right next to the bottom plate. Pushing down on the center of the stick will flex it, exerting pressure on the top of the wall. Keep pushing and the wall will creak into a plumb position.

Racking walls takes coordinated effort. If you're on the smart end of the things (using the level), you need to shout out which way the wall must move and how far. Although overcompensation is usually a problem at first, soon your partner (who's on the push stick) will get used to what you mean by "just a touch" and be able to make the bubble straighten up as if he were looking at it.

Once you've racked the wall to plumb, hold it there permanently by nailing off the let-in bracing (photo right). But before you do that, have your partner keep the same tension on the push stick while you plumb the other end of the wall as a check. If it's plumb also, nail off the braces on the wall as fast as you can. Here it's handy to have a third person to do the nailing while you keep your eye on the bubble. If for any reason (poor plating or a gap where the top plates butt) the wall isn't

Plumbing. The length of the level you use is a lot less important than being sure that it is dead-on accurate. Here a 4-ft. level is attached to a straight stud with bent-over nails. Using 1x spacers at the top and bottom of the stud (or manufactured aluminum spacer blocks on an 8-ft. level) will ensure an accurate reading despite bowed wall studs.

Keeping it plumb. While the let-in brace that will keep the wall plumb is being nailed off, this short wall is being held plumb by tension on a push stick made of face-nailed 1x4s.

plumb on the opposite end, you'll have to split the difference.

Walls that are shorter than 8 ft., walls with lots of doors and windows, and exterior walls that will get shear panel, sheathing or finished exterior plywood won't have let-in bracing. For these walls, a temporary 2x4 brace must be nailed up flat to the inside of the wall at about 45°. Be sure that none of these braces extends above the top plate. Use two nails at the top plate, one nail each into the edges of at least three studs and two nails into the bottom plate. As with all plumb-and-line bracing, drive the nails home (you won't be pulling these braces until joists, rafters and sometimes sheathing are in place). Continue this process until all exterior walls are plumbed.

Lining—The next step is to get the tops of the exterior walls straight. If you've got a practiced eye, you can get the best line of sight on a wall by standing on a 6-ft. stepladder with your eye right down on the outside edge of the double top plate (photo above). This is the way I usually line, but for absolute accuracy you can't beat a dryline.

Set this string up by tacking a 1x4 flat to the outside edges of the top plates at each end of the wall. Then stretch nylon string from one end of the wall to the other across the faces of the 1x4s, and pull it tight (see *FHB* #11, p. 27 for a twist knot that will hold a taut line). The 1xs hold the line away from the wall so that if the wall bows, the string won't be affected. To determine whether the wall is in or out at any given point, use a third scrap of 1x as a gauge between the string and the top plates. If an inside wall intersects with the outside wall, I plumb the end of the interior wall, and then check the rest of the wall with the 1x gauge.

To make corrections in the wall and then hold it there, use line braces. These are 2x4s on edge that are nailed high on the wall and angle down to the floor or the base of an interior wall. To be most effective, they should be perpendicular to the plane of the wall. If a wall is at least 8 ft. long, it gets a brace—even if it's already straight. Braces not only correct the walls, but also make them secure enough to walk on while laying out and nailing joists or rafters, and keep these ceiling members from pushing the walls outward.

Usually, I begin lining a wall by sighting it quickly for bows. My partner nails braces where the deviations are worst. Then we brace off the rest of the wall where needed. If the wall is pretty straight, we scatter braces every 10 ft. or so, and do them in order.

The top of the line brace is attached to the wall first. Face-nail it to a stud just under the top plate with two 16d nails, making sure that it doesn't run beyond the exterior plane of the wall. If a header is in the way, face-nail a vertical 2x4 cleat to the header with three or four 16d nails, and then edge-nail the brace to the cleat. Line-brace nailing is shown in the center and bottom drawing panels on the facing page. Trimming an eyeballed 45° angle on the top of the brace will allow it to hug the wall and give you better nailing.

Setting the braces also requires good communication. While you pay attention to the string, your partner either pulls or pushes the brace according to your directions, and then anchors the bottom end when you say that it's looking good. You can usually find an inside wall to tie the bottom of the brace to. This won't affect plumbing the inside wall (which will be done later) as long as you nail the line brace to the bottom of a stud down where it butts the bottom plate. Use at least two 16d

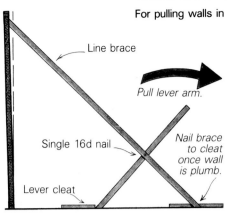

For pulling walls in

Line brace

Pull lever arm.

Single 16d nail

Nail brace to cleat once wall is plumb.

Lever cleat

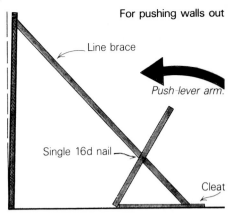

For pushing walls out

Line brace

Push lever arm.

Single 16d nail

Cleat

Persuading stubborn walls. Using a 2x scrap (a 3-footer is ideal) for a kicker under a line brace will bring a wall in at the top. Straightening walls quickly requires the carpenter setting the line braces to make adjustments in or out a little bit at a time with the constant direction of his partner who is eyeballing the wall.

Lining. For absolute accuracy, there is no substitute for a dryline. But if you've got a good eye and an experienced partner you can get straight walls in a hurry (facing page). Either way, use the top exterior corner of the double top plate to gauge the straightness of the wall.

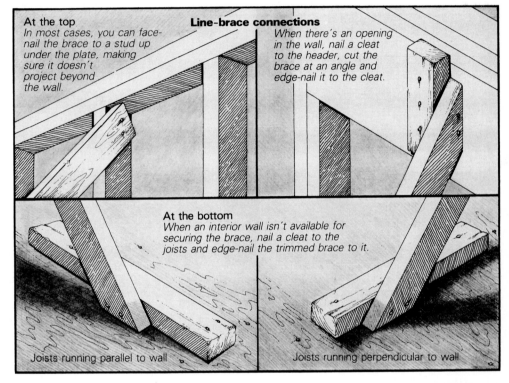

Line-brace connections

At the top
In most cases, you can face-nail the brace to a stud up under the plate, making sure it doesn't project beyond the wall.

When there's an opening in the wall, nail a cleat to the header, cut the brace at an angle and edge-nail it to the cleat.

At the bottom
When an interior wall isn't available for securing the brace, nail a cleat to the joists and edge-nail the trimmed brace to it.

Joists running parallel to wall

Joists running perpendicular to wall

nails (or duplex nails) and drive them home. If an inside wall is not available, nail or pin a block to the floor, no closer to the outside wall than the wall's height. This will leave the brace at an angle of about 45°. Be sure to find a joist under the floor, since a floor cleat that is nailed only to the plywood will pull out. Trim the bottom of the brace in the same way you did the top, so that you can drive two good 16d nails through it into the cleat.

Most walls are relatively easy to move in or out, but every house has a notable exception or two. Walls full of headers can be a real pain to line because they are so rigid. Bringing the top of a wall back in is usually more difficult than pushing it out because you have to work exclusively on the inside of the wall. One effective technique is to use a *kicker* (photo above). Toenail the line brace flat (3½-in. dimension up) to the header and nail a block above it so it won't pull out under tension. Then toenail the bottom of the brace into a joist, and cut a 3-ft. 2x scrap. Toenail one end of it to the floor and wedge the top under the line brace so you have to beat on the kicker to get it perpendicular to the line brace. This bows the brace out, bringing the wall back in

(although occasionally it pulls the toenails loose on the line brace). When the wall lines up, end-nail the brace to the kicker.

You can also use a parallel interior wall to help pull in an exterior wall (see *FHB* #8, p. 10) or a scissor lever that nails temporarily to the line brace. Scissor levers can be used to push or pull, depending on where the blocks are nailed (top drawing panel, above).

Line the exterior walls one at a time until you come full circle. Try not to block off entries and doorways with braces, but don't skimp either. I've learned to double the normal bracing on exterior walls that will carry rafters for a vaulted or cathedral ceiling. Also, if this ceiling includes a 6x ridge beam or a purlin 18 ft. or longer, I like to place the beam on the walls just after I've plumbed them. This way you don't have to pick your way through a forest of braces with that kind of load.

Interior walls—The last step is to plumb up the inside walls. Start at one end of the house and work your way through. You'll develop a sense of order as you go, and soon find that the remaining walls are already nearly plumb, and that the frame is beginning to become rig-

id and to act as a unit. Quite a few of the interior walls will have let-ins; if they don't, use the 8-ft. and 10-ft. 2x4s as diagonal braces on the inside of rooms. Where there are long hallways, I usually cut 2x4 spreaders the exact width of the hall measured at the bottom, and nail them between the top plates every 6 ft. to 8 ft. The spreaders will keep the entire hall a uniform width, and when the wall on one side is plumbed, the wall opposite will be too.

Once all the interior walls have been plumbed, go back to any long walls to check for straightness, and then throw in one more line brace for good measure. This is also the time to make a last check on exterior walls to make sure the interior-wall plumbing you did hasn't thrown a hump in the works. Then go around the entire job, shaking walls to make sure there's no movement. This kind of precaution means that when you're rolling joists or cutting rafters you won't have to measure the same span or run every few feet along a wall for fear of a bow, or worry that the walls have acquired a lean over the weekend. □

Don Dunkley is a framing contractor and carpenter in Cool, Calif.

Framing a Conical Roof

by Geoff Alexander

The frequent appearance of the turret in Victorian houses is symbolic of the playful freedom with which Victorian architects and carpenters approached their work. But underlying that playfulness was a great skill for manipulating wood into unorthodox forms—forms whose creation would baffle many a modern carpenter. A typical turret consists of a cylindrical tower capped by a cone-shaped roof. This article will detail one way to build a conical roof, and suggest some techniques that are useful for any conical construction, or reconstruction.

If you take a circle of paper, cut out a wedge and bring the cut edges together, the result is a true cone. This is an invaluable tool if you are framing or finishing a conical roof, because you can work out all of the layout details in two dimensions with the paper flat, or you can cut it out and bend it into shape to see how the roof will look in three dimensions. To make a model of a conical roof, you need to know the length of the rafters and the radius of the cylinder (the base of the cone). A vertical section through the center of a conical roof is a triangle, as shown in the drawing below left. Thus, if you know the radius of the turret and the height of the cone, you can solve for the rafter length. Then draw a circle whose radius (in scale) is the rafter length, and draw

the wedge. Here are two methods you can use to lay out the wedge.

1. Calculate the circumference of the base of the turret (C = 2πr). Measure this distance around the circumference of the circle you drew and connect the two endpoints of the arc to the center of the circle.

2. The ratio of the turret radius to the rafter length is the fraction of the whole circle required for the cone model. For example, let's consider a roof with a 6-ft. radius and a 9-ft. rafter length. Since 6 ÷ 9 = ⅔, we need two-thirds of a circle. Since there are 360° in a circle, we need two-thirds of 360°, or 240°. Therefore, the central angle of the wedge equals 360° − 240°, or 120°. The middle drawing shows the flat layout of this roof.

The roof in our example sits atop a cylindrical tower 6 ft. in radius, with bandsawn 2x4 top plates. We'll use 12 main rafters with conventional bird's mouths and short tails. The rafters meet at the top with a plumb cut fitted to a 12-sided wooden plug. The length of the rafters without the tails but adding on for the thickness of the wooden plug at the top is 9 ft. All of the rafters can be 2x4s since the longest horizontal span is 6 ft. (the cylinder radius). Between each pair of full rafters is a half-length rafter capped by a bandsawn block.

To cut these blocks, set your bandsaw to the angle of the pitch

of the roof, and draw arcs whose radii are the inside and outside radii of the cone at that height. You can read these radii right off the flat layout. If the blocks are halfway up the rafter length, the radii are one-half the inside and outside radii of the turret. The spacing between the main rafters and half-rafters at the base is about 19 in. o. c. Just above the blocks that cap the half-rafters, the main rafters are also 19 in. o. c.

The dimensions of our sample roof provide an easy layout. The 12 main rafters split the cylinder into 30° arcs, and the flat layout of the cone into 20° arcs. Thus each full and half-length rafter marks off 10° in the flat layout.

To find the radius of the wooden plug where the rafters meet at the top of the cone, you'll have to do some figuring. The plug has the same number of sides as the cone has rafters, in this case 12. Each side will be as wide as the rafters, or 1½ in. Each interior angle will be 360° divided by the number of sides, or 30°. With these givens, we can construct a right triangle whose angles and one side are known, and solve for its hypotenuse (the radius of the plug). The calculations are shown below right.

Sheathing and shingles—Victorian conical roofs were usually sheathed in vertical boards tapered to fit the roof. Today, I would use

wedge-shaped pieces of ⅜-in. plywood bent to fit. The shapes of the cuts can be taken directly off the flat layout. You could use trammel points and a rod to draw the bottom curve, and if possible, use plywood long enough to reach from the rafter tail to the peak (the turret in our example would require 10-ft. long sheets of plywood).

Shingling can be handled in various ways. You need to experiment with ideas on the two-dimensional paper circle. Cut it out and bend it into shape to help visualize how it will look in three dimensions. I like the shingle lines to maintain horizontal rows in the finished roof (concentric arcs in the flat layout). For this, each shingle has to have its sides tapered, and the amount of taper increases with each course. These details, too, can be read off the flat layout.

Three-tab composition shingles would need to be cut into individual tabs to make the tapers work. Installing three-tab shingles vertically like overlapping radial spokes is an interesting alternative. The roof peak is usually crowned with a copper dunce cap, or a weathervane. A sheet-metal shop could easily fabricate a simple cap. □

Geoff Alexander is a carpenter and woodworker in Berkeley, Calif. For more on shingling conical roofs, see New England Builder *(Box 97, East Haven Vt. 05837), May 1984.*

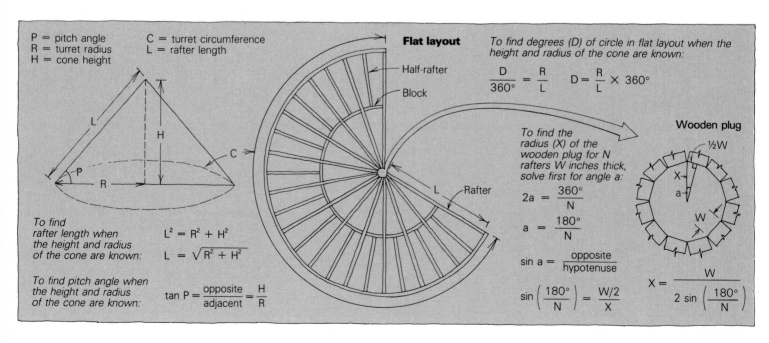

P = pitch angle
R = turret radius
H = cone height

C = turret circumference
L = rafter length

Flat layout

Half-rafter

Block

L — Rafter

To find degrees (D) of circle in flat layout when the height and radius of the cone are known:

$$\frac{D}{360°} = \frac{R}{L} \qquad D = \frac{R}{L} \times 360°$$

To find the radius (X) of the wooden plug for N rafters W inches thick, solve first for angle a:

$$2a = \frac{360°}{N}$$

$$a = \frac{180°}{N}$$

$$\sin a = \frac{\text{opposite}}{\text{hypotenuse}}$$

$$\sin\left(\frac{180°}{N}\right) = \frac{W/2}{X}$$

Wooden plug

½W

X

a

W

$$X = \frac{W}{2 \sin\left(\frac{180°}{N}\right)}$$

To find rafter length when the height and radius of the cone are known:

$$L^2 = R^2 + H^2$$

$$L = \sqrt{R^2 + H^2}$$

To find pitch angle when the height and radius of the cone are known:

$$\tan P = \frac{\text{opposite}}{\text{adjacent}} = \frac{H}{R}$$

Tools

Framing Hammers

Longer handles, heavier heads and milled faces can make a surprising difference

by Paul Spring

Generations of beautiful homes have been built with the classic 16-oz. curved-claw hammer, and most carpenters have a beat-up favorite that they use for finish work. But with the advent of production framing in the early 1950s, a new kind of hammer appeared. Heavier than finish hammers and with handles that are 2 in. to 4½ in. longer, framing hammers have straight rip claws and faces that look like deeply etched checkerboards.

These long-handled nail drivers (they can be up to 17½ in. in overall length) are well designed for their task. Most framing these days is done flat on the deck or slab, and unlike the toenailed studs of a stick-nailed wall, almost everything is end-nailed or face-nailed. That calls for longer nails. This means many blows with a light hammer, but just a few whacks with a heavier hammer.

When you are framing a wall flat on the deck, you work bent over at the waist with your legs slightly flexed. Your hammer makes nearly a 180° arc before hitting the nail (photo left, p. 54). As a result, gravity does a lot of the work, so you can handle a hammer with more ounces of metal in the head and a longer handle for balance, leverage and reach.

Head weights and handles—Framing hammers are made with various head weights and handle materials. The usual weights are 20 oz., 22 oz., 24 oz., 28 oz. and 32 oz. The most common is 22 oz., although Vaughan makes a wood-handled 24-oz. model that has become very popular in the last few years. A heavy head has its advantages when it comes to coaxing the bottom plate of a wall onto its chalkline. Extra head weight also helps if you're trying to toenail the edge of a trimmer back into line with its neighboring king stud. But you can overdo it too. If your hammer is too heavy, it's doing you a disservice.

Save your heaviest hammer for framing walls and nailing off subfloors and roofs. Use something a bit lighter for toenailing. I used to use a Plumb 22-oz. hammer with a wood handle for rafters and joists, and then I dropped down to a 20-oz. smooth-face hammer for siding. Exterior trim and siding don't really call for a framing hammer, but I liked the grip on the long wood handle better and it gave me some extra extension when I was nailing out in thin air from a ladder or scaffolding.

What's important is comfort. If you don't drive nails all day long, stay with an all-

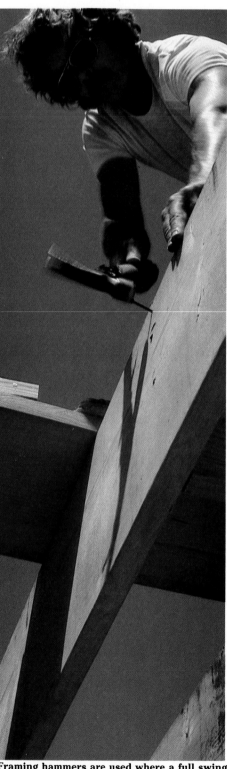

Framing hammers are used where a full swing can be taken, letting gravity do some of the work. A mill-face head delivers this force to the nail even when the blow is misdirected.

purpose 20-oz. or 22-oz. hammer. How you swing and how well you keep that form after driving a pound or two of nails is what counts. Few carpenters I know consistently swat in 16d nails with one blow. Two or three strokes is more typical, depending on how wet the wood is and how the nail is coated.

Handles are made of fiberglass, wood, tubular steel or solid steel. Fiberglass was championed by Plumb for years and has now been put to use by all of the other major manufacturers. Its advantages are that the handle and head are permanently joined, and that it is nearly impossible to break the handle. Solid-steel hammers have these same advantages, but opinions about them run strong. I've asked carpenter friends from all parts of the country about them and get one of two responses each time—they either love them or hate them. I hate them. Swinging a framing hammer all day long puts a considerable load on your wrist and elbow. This isn't the muscular fatigue that you feel in your forearm after returning from a few months' layoff from heavy nail driving. It is the shock that runs up your arm and is absorbed by your tendons and joints. A wood handle absorbs some of this shock and vibration, but a solid-steel shank, even with a rubber grip, feels like it sends every bit of that shock right up my arm. My elbow gets tender after just a morning with one. But the slender shank profile of these hammers (Estwing makes the most popular model) gives them wonderful balance. They are also almost indestructible.

I don't have the same vibration problem with tubular-steel handles (True Temper's Rocket and Stanley's Steelmaster are two brands in this category), but I've never found one that feels right. My choice is wood. Wood handles need a steel shim occasionally to cure a loose head, and you have to be careful about putting your body weight behind the handle when you're pulling a nail. But for me, the resilience of wood outweighs its disadvantages. You can feel the nails enter the wood every time you swing because you are holding the hammer itself and not a rubber sheath. Yet the sound and shock of each blow is softened and dispersed.

All of the major tool manufacturers make wooden-handled framing hammers, with Plumb and Vaughan being some of the best known. The first 20-oz. Vaughan hammer I owned vibrated at an audible pitch every time it struck a nail. Although it's a little disconcerting to swing a tool that peals like a bell, it was an excellent hammer. I've compared notes with other guys who have had the same experience in the past, but tool-store managers just look blankly at me and mumble about improper tempering of the hammerhead.

Another long-handled hammer worth mentioning is the rigging ax or rig hatchet. This is also a tool that carpenters tend to feel strongly about one way or the other. To a large degree, it's a matter of image and fashion. In the 1960s, the rigging ax was one of a handful of well-balanced heavy hammers on the market. Once they became popular with West Coast piece-workers, they rapidly got an undeserved reputation as quick-and-dirty tools. It got to the point that if you showed up on a custom job with an ax, you were branded as someone whose skills were limited to framing and who wasn't interested in anything but speed. But I was taught to frame with one by a carpenter who appreciated production tools and techniques, and yet was uncompromising on quality. I continued to use a rigging ax throughout my career when I framed walls, and I did only custom work.

A rigging ax combines a typical hammerhead with a hatchet face instead of a claw. This distributes a lot of weight well back of the head for excellent balance. This is really the main advantage of the blade, and for safety's sake, the hatchet edge should be kept less than razor sharp. A notch in the bottom edge of the hatchet is designed for pulling nails, although it's not very good at this job. I've seen tract framers weld claws on their axes to try to get the best of both worlds.

The rigging ax I've seen used most often is a Plumb. It has a slender handle that is slightly contoured a little more than halfway down and is almost oval in section. You don't hold the ax right at the end, but up around the contouring. This sounds clumsy and amateurish, but held this way it has marvelous balance. Of all my framing hammers, my rigging ax delivers the most force for the least effort, although the Vaughan 24-oz. hammer seems to have a similar combination of power and balance.

Be selective about where you use a rigging ax. For instance, because of its broad profile, I found it clumsy for joisting and blocking. But it's perfect for unconfined framing, and for nailing off roofs and subfloors. It's also especially useful for applying 2x6 T&G decking on a cathedral ceiling. Here you use a 16d toenail down into the rafter on the leading (tongue)

A long–handled sampler
The popularity of different styles of framing hammers varies from region to region. Of the three hammers above, the 22-oz. Estwing, shown with a smooth face at the top of the page, seems to be the Eastern favorite. The Plumb rigging ax in the middle is still common among production framers on the West Coast, while the 24-oz. Vaughan, above, has gained wide acceptance throughout the country for its power and balance.

Framing walls on a deck (left) lets you use a 24-oz. or even a 28-oz. hammer because much of your swing is assisted by gravity. A rigging ax (above) has a similar weight, although it is distributed differently. If you work overhead a good deal (top), a 22-oz. hammer is often more comfortable to swing for a long period of time. In any case, the long handles (up to 17½ in.) of framing hammers are designed to be gripped higher than a typical finish hammer.

edge of each board to drive it tight. Doing this often mashes about 1 in. of the tongue. A couple of swings on either side of the damaged part of the tongue with the ax edge and it will fly off, with no damage showing below.

Mill face or smooth face—Most manufacturers of long-handled hammers offer a choice of a traditional smooth face or a checkered (or waffle) face. This milled grid forms squares of hardened steel that look like little truncated pyramids. These will dig into the softer steel of the nailhead and grip it, transferring the force of the blow even when the hammer doesn't hit the nail squarely. Even if only a few of these teeth make contact with a nail head, the nail will be driven straight. The mill face will sink the head of the nail slightly below the surface of the framing lumber, which is the way it should be. Don't worry about the hammer dings—you're not building a piano bench. In fact, the rough surface created by the waffle head is a great way to knock down the edge of a stud when you want to start a toenail on the corner of the angle.

A mill face is almost a necessity if you're driving cement-coated nails (sinkers). On a smooth face, the build-up of resin from these nails will lubricate the striking surface and can cause the hammer to glance off the nail heads. You'll feel like a dog with fleas as you try to scrub the black stuff off by dragging the hammer face back and forth across the plywood deck or slab.

Hammerheads are forged from tool-steel blanks; thus they are harder than the milder steel used for nails, prybars and other tools meant to be struck by a hammer. But with this hardness comes a degree of brittleness. The face of a good hammer is slightly convex, and the rim is beveled. These features help to prevent square edges from fracturing off upon impact and whizzing through the air at a dangerous velocity. But still, bits of steel can fly off.

Under no circumstances should you strike two hammerheads together. This isn't an idle safety precaution. A former partner of mine once had to take a fellow worker to the hospital with a piece of steel lodged in his chest. Instead of using a prybar, he had been driving the claw of a hammer between two boards by pounding on its head with a second hammer. I have also chipped hammer faces by using them to enlarge the hole in a concrete slab for a bathtub drain.

Try not to use your mill-face hammer when you strike harder steel tools like cold chisels, nail sets or cat's paws. This flattens the little surfaces and eventually turns the hammer into a smooth face. It also chews up the surface of the tools you strike, making them look like the victims of tiny meteor showers. If you find yourself carrying only a mill-face hammer, use the side of it (the *cheek)* to deliver an occasional blow. You can recover the mill pattern once it becomes worn by running a thin file in the grid or turning it over to a saw sharpener for a facelift. The result doesn't work as well as a new hammer, but it'll save your aging favorite from an early retirement.

Using a rip claw—The claw on a framing hammer has several uses; the most obvious is pulling nails. Although you'll seldom bend more than a few nails in a day if you're an experienced framer, when you do need to pull a nail it will likely be a 16d, and could very well be hot-dipped galvanized or a coated sinker. Both of these are tough to pull even with the leverage a long-handled hammer gives you. With smaller nails you can slip the claw under the nailhead and rock the hammer forward. But with big coated nails, that technique will produce a tug of war that is likely to have you staggering backward with the headless nail shank still firmly embedded in the wood. And if you are using a wooden-handled hammer, you also risk leaving the hammerhead back with the nail shank. I learned this lesson the hard way as an apprentice when I broke my framing hammer, bought with my first paycheck, on the day I got it. I also nearly went flying off the second-story subfloor.

The trick, I learned, is to ignore the head of the nail, and use the sharpness of the claw to grip the shank of the nail tight (you can keep the claw sharp by pulling nothing but nails with it, and even then you should avoid hardened concrete nails). Hold the hammer down as close to the surface of the wood as you can, and slide it along toward the nail in a quick, hard motion—this will literally trap the nail in the claw edges. Then lever the nail by pulling the hammer handle down on its side in a single motion. This will pull less than an inch of the nail, but can be repeated another time or two. Using the side edge of the hammer as a fulcrum, you've got plenty of leverage and you're less likely to break the handle.

In addition to pulling nails, the claw on a framing hammer ends up being used often as a combination peavey and crowbar. Although it sounds a little extreme at first, the safest, most efficient way of moving a heavy header or wall a few feet on a deck or slab is to bury your rip claw in the wood and pull the piece toward you. This will prevent the strained back and crushed fingers that can result if you try to pick the object up and move it.

The swing—Swinging a framing hammer is not much different from swinging a finish hammer, except that the stroke is designed for more power and is therefore exaggerated—a bigger backswing is coupled with a more forceful follow-through. Assuming that you are framing in a bent-over position, you begin by raising your arm with your elbow fully bent so that the hammer hovers around your ear. Then you should bring the hammer almost straight down—first by lowering your arm, and then by extending your elbow—in a slow, smooth motion. Your hammer should make contact with the nail about the same time that you begin to extend your wrist. If you are really swinging hard, you'll end up following through with your wrist and the body English that comes from straightening your knees somewhat.

The grip you use on the hammer handle is also important. First, it should be fairly loose; a white-knuckle grip will only wear your forearm out faster. Your thumb should be wrapped around the handle, not sitting up on top of it pointing to the head. You may have to break yourself of this habit at first, because there is a natural tendency to want to direct the hammer with your thumb. But your control should come from your forearm and wrist, which will also make your swing more fluid. It's also safer, since your thumb is the opposing lock that will make your grip effective, particularly when your hand is sweaty.

Where you grip is also important, and will surprise a lot of people. Most hammers are held near the bottom of the handle to get the best leverage when swinging. Choking up on the handle is one of the classic errors of novice carpenters. But heavy framing hammers are not always held out on the end. Because of their long handles, you can achieve the best compromise between power and balance when you frame bent over by gripping the handle just down from the midway point. Most wooden handles are contoured, and swell at this point. Grip the narrower part lower down if you need to generate a lot of power, and are not too concerned with accuracy—when you're driving a beam, for example, or making sure a king stud is tight to a header.

What your nail hand (the one that isn't clutching the hammer) is doing while all of this is going on is most important. How to feed the nails to the hammer is a matter of practice (see pp. 40-41), but coordinating when that hand should be in the line of fire and when it shouldn't is something you can't afford to get wrong. Big hammers with milled faces can literally tear away part of a finger, so be careful to keep your nail hand far away when you are swinging hard. If you find it necessary to get in the habit, put that hand behind your back except when you're actually starting a nail.

Buying a framing hammer—A new framing hammer or ax will cost you between $15 and $25. It pays to buy one of the major brands. Still, check to make sure that the hammerhead is mounted straight—it's surprising how often you will find one that's slightly cocked.

If you're looking for a used one at a flea market or secondhand shop, check that the head is secure, the inside edges of the claw are still sharp, and that the mill face isn't badly worn. In the case of a wood handle, it pays to buy new. Handles eventually break, and even if you spend a lot of time getting the replacement handle just right, it never seems to feel the same. Lightly sand the grip of a new wood handle from the halfway point down to the butt where you hold the hammer most. This will take off the finish and give you a better grip when you're sweaty and swinging hard. Some carpenters carry a paraffin block for rubbing down the handle to increase their grip. Wiping your hand across a pitch pocket on a stud will also give you that George Brett advantage. Even then, lots of framers wrap tape around the butt end of the handle as a stop for the heel of their hand.   □

The Worm-Drive Saw

How this powerful tool compares with a sidewinder saw

by Jim Picton

When I saw my first worm-drive saw on a job in Brattleboro, Vt., the foreman said it was a left-handed saw, and I carried that belief with me for many years. The sidewinder, or contractor's saw, is the preferred tool here in the East, so it wasn't until I moved to Anchorage, Alaska, that I really got to use a worm-drive saw and to see the advantages it offers.

I saw a lot of production framing in the West, where the worm-drive saw was used almost exclusively. I also built my share of expensive custom homes in the Chugach foothills near Anchorage, and the worm-drive saw was king there too. Before I moved back East, I bought one, and I still rely on this kind of saw for most of my cutting.

It's impossible to mistake a worm-drive saw for a sidewinder. The motor for the worm-drive saw is parallel to the blade, instead of perpendicular to it, and the blade is mounted on the left side of the body. The grips are also different. Worm-drive saws have a trigger switch mounted on the hand grip, which sits just behind the saw. There's another full-sized grip—angled slightly toward the user—right on top of the saw. In contrast, sidewinders have their handle and trigger switch at the two-o'clock position (Porter-Cable features a top-mounted handle that gives you a grip straight up, at twelve o'clock), and a sort of "saddle horn" at ten o'clock. But what really distinguishes the worm-drive saw from all of the others is the worm gear inside (sidebar, facing page).

Although my construction crew here in Connecticut claims that my worm-drive saws are ungainly and outpaced by their own light, tight Makita and Porter-Cable sidewinders, my spare worm-drive saw seems to get quite a lot of use when there's heavy work to do. Worm-drive saws do take some getting used to. General cutting should be approached in a surprisingly different manner, so you almost have to re-educate yourself about circular saws in order to use the worm-drive saw's full range of capabilities.

One of the fastest and safest ways to cut framing lumber with a worm-drive saw is to hold the work at an angle off the ground with your left foot. Your right foot should be well back. This way the saw is held out to the right of your body, and the weight of the saw literally makes the cut. In this position, the offcut falls only a few inches to the ground, and if the saw should kick back, it won't be into your body.

Sawing stance—What bothers people most about worm-drives is the weight: 15 lb. or more versus 11½ lb. or less for garden-variety sidewinders. There are some jobs for which a worm drive is just too heavy, but in most cases the extra weight is useful if you know how to take advantage of it.

The trick is to cut downhill. The position of the rear handle lets you stand above the work, the saw practically dangling in your hand with little or no pressure on your wrist. The left-mounted blade allows right-handers to see the line of cut clearly from as much as an arm's length away, so you can start the cut close and just gradually extend your arm. The saw's own weight moves it through the cut.

When right-handers operate a sidewinder they are positioned to the left of the saw, whose blade is mounted on the right. To watch the path of the blade directly in front of the cut, a right-hander must look over the top of the saw or lean forward to see in front of the motor. This means staying right on top of the saw with your elbow bent and your wrist crooked. It's a position you get used to, but if you try to assume this stance with a worm-drive saw, you wind up holding it in your left hand. This makes the slight angle of the top grip feel strange.

Instead, worm-drive saws are intended to be held with the right hand on the trigger grip. With this stance, you're on the same side of the saw as the blade, and your view of the cut is wide and clear.

You get used to using your body to support the work, as Jud Peake's article on worm-drive techniques (*FHB #3*, p. 33-35) points out. Setting up material so it can be cut in a downhill direction would be a pain in the neck if it weren't for the well-documented fact that carpenters stand perpendicular to the horizon. This happy circumstance, combined with the rear grip and arm's length view of the blade, means you can eliminate the need for sawhorses or even flat areas to make your cuts—a real advantage in site work.

Carpenters who aren't used to worm-drive saws may consider this practice dangerous, but it's not, provided you keep a firm footing and don't walk into obvious traps. In general, you lean the work in front of you, and cut forward. Make sure the area behind you is clear, so if you need to step back you won't trip. You cut with the saw a comfortable distance away and to your right (photo facing page). This way, your eyes aren't so vulnerable to wood chips thrown by the saw, and if kickback occurs, you're not in the path of the blade.

Cutting techniques—For crosscutting anything from a 1x4 to a 2x12, I support the board on edge with my left foot, step back with my right, and drop the saw through the line. The offcut falls only a few inches, so it doesn't bounce or get damaged. Ripping boards is even easier. With one end on the ground and the other end supported on a window sill or bucket, I stand next to the saw, pushing ever so slightly as it rides downhill.

For single thicknesses of plywood, whether

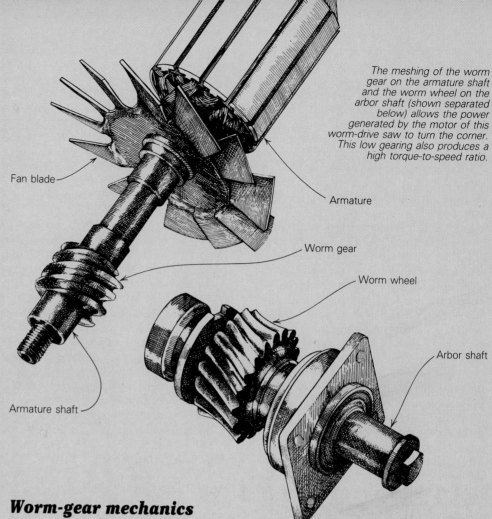

The meshing of the worm gear on the armature shaft and the worm wheel on the arbor shaft (shown separated below) allows the power generated by the motor of this worm-drive saw to turn the corner. This low gearing also produces a high torque-to-speed ratio.

Fan blade

Armature

Worm gear

Worm wheel

Arbor shaft

Armature shaft

Worm-gear mechanics

All saws use gearing to lower the high rpms of the motor before they are delivered to the blade. A worm-drive saw accomplishes this with two gears at right angles to each other, as shown above. The *worm gear* is about 1½ in. long and ¾ in. in diameter, and is attached directly to the splined armature shaft on the long axis of the saw. The armature shaft runs through a ball bearing to turn the worm gear, and the other end of the worm gear is supported by a ball bearing at the front of the saw. The threads of the worm gear engage the *worm wheel;* it is mounted on the arbor assembly, which runs across the saw, at right angles to the armature shaft. The arbor is supported by ball bearings on each side of the worm wheel, and runs out to the blade on the left side. This setup results in the saw's elongated shape. Since the mechanical energy "turns a corner" at the junction of the arbor and the worm gear, the motor sits behind the blade, not next to it.

But the worm gear's primary advantage isn't that it changes the shape of the saw. Instead, the worm gear is the key to the saw's power and durability. Gears can be cut to a number of different configurations that reduce relatively high motor speeds to acceptable rpm levels at the arbor. Some consumer tools just use straight-cut gears. The shoulders of these gears have very little support and can wear quickly and even shear off when subjected to the kind of shock load that comes with hitting a nail or making a pocket cut into a large knot. A helical gear—particularly the worm gear—is a great deal more durable because the teeth on the driving end are threads and so supported on all sides. But there is a trade-off here. Non-worm gears can be made from high-carbon alloy steel, which can be heat-treated for tremendous wear resistance. Worm gears are

made from a less durable material, like bronze, that has a low coefficient of sliding friction. This is necessary to minimize the heat and wear caused by the peculiar meshing of these gears.

Even so, the worm gear takes a high motor rpm and delivers it to the arbor at a much slower speed and with greater torque. Each revolution of the worm on the armature shaft turns the worm wheel on the arbor by only a few notches. The unrestricted arbor speeds on worm-drive saws are 4,300 rpm for the Black and Decker Model #3051, and 4,400 rpm for Skil's Model #77 and Milwaukee's Model #6377, as opposed to an average of 5,400 to 5,800 rpm for most sidewinder saws. Given the size of the motors in worm-drive saws, this slower speed means an amazing amount of torque behind a given rpm.

In framing, you run into heavy work requirements like ripping through several thicknesses of plywood, cutting a stack of rafters, or ripping 2xs over 16-ft. lengths. Worm-drive saws practically walk right through this kind of work. The high torque-to-speed ratio of worm-drive saws means that as you load the tool down, the speed of the motor changes little, so you can push the saw just about as fast as the blade will cut.

Unlike worm-gear saws, the initially high rpms of most sidewinders come way down in heavy work. And while the worm-drive saw will continue to cut smoothly even if it's pushed hard enough to bring the motor down to half its usual pitch or less, the sidewinder at this point is ready to bind up in a full stop. This is a dangerous situation, because the torque of the blade against the wood as you ease off on the feeding pressure can result in kickback. At best, you get a bumpy ride when you bear down on a sidewinder, and the only alternative is to cut more slowly. — *J. P.*

Illustration: Elizabeth Eaton

it's a rip or a crosscut, I like to hold one end up with my left hand while the other end rests on the ground. I let the saw ride down the line as far as I can reach, then drop the work and finish the cut from the other side.

Despite the advantage of making the weight of a worm-drive saw work for you by cutting downhill, it's often better to work flat on finish work where you need more control. If you are finish-cutting plywood, you don't want to stop the cut and pick it up from the other side. Instead, lay the work on a level, well-supported surface, and climb right up onto it if you are cutting across a full 4 ft. The setup technique is the same as with a sidewinder saw. Although in this case the weight of the worm-drive ceases to be an advantage, you can use the rear-grip handle to advantage by locking your wrist. This way, your wrist can be used for fine course corrections, and the full force of your upper arm and shoulder is used to power the saw through the work. Locking your wrist like this can relieve a lot of the muscle fatigue in long, laborious cuts like sawing a stack of curved plates out of ¾-in. plywood. I sometimes lock my arm straight and just lean into the work, using my entire upper body to drive the saw.

For fine cutting, the full-sized upper handle on worm-drives gives you a secure and comfortable grip. It helps control direction when you're freehanding a scarf joint in fascia, mitering a soffit or making an in-place cut with the saw held sideways out in front of you (photo facing page, right). I don't have any objection to the horn-type grips on sidewinder saws, but a full hand-width handle suits me better. My brother's old Stanley worm-drive has a rear grip that wraps right around to form the top grip—like some chainsaws—and a horn grip at the front left corner to boot.

One of the toughest ways to cut with any kind of circular saw is straight up. Of course it's best not to get into spots like this, but more than once I've let a piece of fascia run wild on the end of a building, only to realize that I must now get below it to make the bevel cut required for the return. Actually, this is a good argument for having both a worm-drive saw and a cheap sidewinder knocking around among your tools. Because the blades of sidewinder saws and worm-drive saws are on opposite sides, their shoes tilt in opposite directions. A sidewinder saw would make the bevel cut on the fascia I was talking about from above—a real improvement over having to undercut a suspended board.

But for times when you have no choice, you will have to support the offcut end of the board and make the cut as fast as possible before the weight of the offcut collapses the kerf, binds the blade, and causes kickback. Although a lightweight saw seems like the best choice for overhead work, I worry about relying on the muscles in a bent wrist to power the saw through with no hangups. My choice is to get a straight arm locked behind a worm-drive, and press it.

If you can't cut downhill or support the weight of the saw on the work and push from behind, you might as well set the worm-drive aside. I once stood on a steep roof steadying myself with my left hand to cut siding for an upper wall, with my right hand holding my worm-drive saw in front of me unsupported. If you want to sample the strain this puts on muscles and tendons, try picking up a suitcase and holding it out in front of you with your wrist bent at 90° to your arm for a few minutes. Handling my saw like this for a couple of hours put my wrist out of commission for a week. Next time, I'll use a sidewinder.

Safety—Everyone I know who works with tools for a living is vitally concerned with safety. Like professional athletes, tradespeople can't do their jobs without healthy bodies. Using any saw in a cavalier manner is foolish. If you are a novice builder or just unfamiliar with the saw you are using, you should start out slowly and as always, concentrate completely on what you are doing. You need to know the basic rules of saw safety listed in the owner's manual of any saw you use, and follow them. One oft ignored imperative, for instance, is unplugging your saw before changing the blade. Another common mistake is to burn through a cut with a dull blade instead of taking the time to change it.

But for me, safety sometimes goes beyond the usual admonitions, and this brings up the subject of blade guards. Worm-drive saws have a spring-return guard identical to the ones you find on sidewinders. The shape and operation of the guard are based on a set of design compromises that keep the blade covered when it's not buried in the work, and still allow it access to the work—most of the time. It is the exceptions that create truly unsafe situations. When you use any portable circular saw to cut an acute angle on the face of a board, the opening action of the guard is a lot less positive than when the shoe of the saw is perpendicular to the work. This often causes the saw to slide down the edge of the board before engaging. Also, if you are cutting with the shoe of the saw set at a bevel angle approaching 45°, the problem gets worse. At this point it's necessary to hold the guard up somehow in order to make the cut. This is also true if you are ripping or crosscutting less than ¼ in. from a board (the photo on p. 56 shows a board being trimmed by a sliver).

The procedure taught by experts in the field of safety is to hold the guard back with one hand while operating the trigger switch of the saw with the other. This requires two hands on the saw, which in turn means clamping the work to a horizontal surface before making a cut. This is something I do if I'm making a critical cut on a piece of finished woodwork, but I can't afford the time and frustration of getting out a C-clamp for every 2x4 I have to shorten. As a result, I block up the guard for the cuts I mentioned above. I know that doing this is a bone of fierce contention among safety experts, tradespeople and manufacturers. But I think that as an alert and well-informed operator, I am freer to give my full attention to the saw I am handling and the work beneath it when I am not distracted by having to struggle with the guard. I have had too many near-accidents with saws that have hung up on the ends of boards because the guard didn't retract, or kicked back on me because the guard snagged on the work and made the saw veer off course. However, I want to stress that I have come to these conclusions only for myself after using professional-duty saws daily for nearly 15 years. This is not a procedure a novice should use.

There are lots of ways to keep the guard retracted, but I think that the safest is to use a small wedge or shim shingle between the lift lever for the retractable guard and the body of the fixed upper guard. This way, you can remove it easily after you've completed the cut. Then make sure that the saw is to the side of your body when you are cutting, and that your elbow remains locked until you have finished the cut and taken your finger off the trigger.

Some portable power tools now have automatic electric brakes, and in this situation, I think a brake would contribute to safety. I've used Black & Decker's Sawcat, which is equipped with a brake, but I don't know of any worm-drive saws with this feature.

Maintenance—Any good power tool has replaceable parts, and worm drives are no exception. There are over a hundred pieces to my Skil Model 77, all accessible with a screwdriver and pliers. Although I have taken saws apart myself, I find it more economical to send them into service through a dealer. They come back quickly and in perfect shape. In most cases, the repairs that I have ordered were for damage inflicted on the job site, and not for mechanical failures in the saws themselves. One day, I tried to cut overhead costs by using a motorcycle as a work vehicle. My worm-drive saw flew off the back at the first good bump, and I had to gather it up in a shopping bag over a quarter-mile stretch. I sent it into repair with dim hopes, but two weeks later it returned completely rebuilt, like new, for just over half the cost of a new saw.

To keep the worm gears in good condition, you are supposed to change the gear lubricant about four times a year. Most manufacturers sell a gooey kind of oil especially for this. It comes in something like a toothpaste tube, and you just unbolt the oil-fill plug, drain the old stuff, and squirt in the new. I confess I'm pretty bad at following this schedule. In fact, I haven't changed lubricant in more than four

Cutting techniques. A worm-drive saw has an advantage in ripping sheet stock because you don't have to lean over the blade. The rear-mounted trigger handle allows you to extend your arm—a safer stance that doesn't require constant repositioning as you rip long pieces (photo, facing page, top left). By using two hands on the rear of the saw and locking your wrists and elbows, you can actually push the saw with your upper body (bottom left). This lets you power through gang cuts or wet lumber without sacrificing control or fatiguing your arms. The top handle on a worm-drive saw can be useful for more than carrying when you're edge-cutting out in front of you (right).

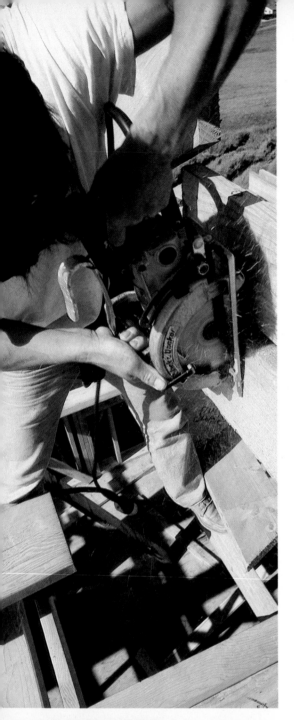

years. Nonetheless, our saws work fine winter and summer under daily use. A properly maintained worm-drive should last for decades.

Blades—The only other maintenance required on worm-drive saws is occasional replacement of the brushes—standard procedure for all electric motors—and, of course, changing blades. Some worm-drive saws have an integral positive stop to lock the blade while you loosen the arbor nut. It is a spring-loaded button located at the front of the saw, and you push it in and turn the blade until the stop clicks into place. An offset blade wrench supplied with the saw fits the arbor nut, which is reverse-threaded (photo bottom left).

To prevent blade slippage, the arbors on all the larger worm-drive saws are diamond-shaped in section rather than round. Blades that fit this arbor are readily available—they have a diamond-shaped knockout that has to be removed before the blade can be mounted. This diamond knockout feature is not universal, so check before making the purchase. It also makes sense to save one or two of those knockouts in one corner of your tool box—you'll need them to return to the round arbor fitting if you are going to use one of these blades on saws with round arbors.

Almost any kind of circular-saw blade will work in a worm-drive saw. I keep a carborundum blade for masonry and a planer blade for fine work. For general purposes, carbide-tipped blades work well, but they cut a pretty wide swath. Also, the cheapest ones still cost over $5, and they're ruined after two or three encounters with a nail. I buy "throw-away" combination blades in bulk for their finer cut, and they cost about $2.60 each. They can be resharpened, but the service costs as much as the blades themselves.

Models, brands and prices—My saws use a 7¼-in. blade, and this is the most common size. With the base retracted to its fullest, a new 7¼-in. blade will yield a depth of cut of a little less than 2½ in., depending on the brand of the saw. This depth is reduced by any bevel you may set on the saw. At 45°, most 7¼-in. saws can take a little more than 1⅞ in. of bite.

The other common sizes of worm-drive saws are 6½ in. and 8¼ in. The 6½-in. saw (both Skil and Black & Decker make one) has enough blade to cut through a 2x at 45°. This sounds like a good choice until you realize that it weighs only ½ lb. less than a 7¼-in. saw and takes up about the same amount of space. You can also go upscale with worm-drive saws to

an 8¼-in. blade size, but here you'd have to lug around four extra pounds that will buy you only another ½ in. in depth of cut. On the small end of things, Porter-Cable makes a trim and paneling (or siding) worm-drive saw that uses a 4½-in. blade. At 7 lb., this saw is a tough little alternative to the usual worm-drive saws for lightweight materials.

All the worm-drive saws I've used are good, and all of them are expensive. In my opinion, you can't buy a bad—or cheap—worm-drive (sidebar, facing page). With the exception of the little Porter-Cable, worm-drive saws are nearly identical in design. All of them have a wraparound shoe that measures 1½ in. to the blade from the edge on one side, and considerably wider on the other. The depth-of-cut adjustment is on the left rear of the saw, and is similar to most sidewinders. The rear of the shoe drops down to limit the blade exposure. The setting can be made quickly because it isn't geared or restrained in dovetail ways. Instead, a thumb lever operates an eccentric, which holds the adjustment by compression. The bevel adjustment is similar, only the shoe pivots from the front of the top blade guard. The angles stamped into the pivot guide are seldom reliable for accurate cuts.

My favorite worm-drive saw is the Skil Model 77, because its size (17 in. long) and weight (15¾ lb.) are about as conservative as you can get in a 7¼-in. worm-drive saw. I also give it high marks for good behavior and durability. This saw lists for $247, but it's not hard to get one for $180 to $200. I like to buy my gear from a dealer who specializes in contractors' supplies. By waiting for his promotional sale, I've been able to get a Model 77 for $150. I've seen lower prices, but I like the fact that my dealer does his own repair, and gives me the attention a regular customer deserves.

The 7¼-in. Milwaukee Model 6377 worm-drive lists for $230. At 17 lb., it is a bit heavier than the Skil, and it is also 1¼ in. longer. The differences are hardly noticeable in operation though, and the saw has an extra 2 amps of draw and the solid, well-built feel that's typical of Milwaukee tools. The closest thing to a Skil 77 in size, weight, and appearance is the Black & Decker model #3051. This 7½-in. worm-drive saw weighs 16½ lb. and is 17½ in. long. We have one that handled well through some rough treatment, but it was in for repairs twice, and it is the only worm-drive saw that ever completely stripped its gears on one of our jobs. The Black & Decker lists for $232.

We used to have a Rockwell worm-drive saw, but currently the only worm-drive saw made by Porter-Cable (Rockwell's successor) is the 4½-in. trim saw. None of the major Japanese tool manufacturers I know of makes a worm-drive either.

Whichever worm-drive saw you choose, you can be secure in the knowledge that it is one of the most reliable, well-constructed and repairable saws available today. Once you get used to it, you'll never want to switch. □

Using the weight. By cutting down on the work, both hands can be used for control in a deep pocket cut like the one on this exposed ridge beam, above left. The aluminum hook wired to the saw solves the problem of where to set it down when everything is at a pitch.

Changing blades. The blade lock makes this procedure easy. With the saw unplugged, left, unscrew the reverse-threaded arbor bolt to the right, and insert the blade with its teeth up.

Jim Picton is a contractor who lives in Washington Depot, Conn.

Power saws: rating the ratings

Label reading has become a necessary sport for cost and quality-conscious buyers, and shopping for power tools is no exception. The serial-number plate on every portable circular saw lists electrical and speed ratings that should provide a basis for comparison between types of saws, and between the different brands within each category. Some manufacturers' catalogs list even more specs on their tools—things like horsepower and torque. But even after you read these figures, it's hard to know what you're getting.

The biggest problem is the proliferation of rating categories. Long ago, the marketing divisions of the major tool manufacturers discovered that the buying public was more impressed by hard numbers generated by engineers than they were with the war of superlatives waged by the advertising agencies. But since the different brands within each category often came in with similar numbers, new ratings (or the same general rating taken farther back on the power train or under a higher load) were used. As a result, ratings that described only a minor part of a tool's performance were often touted because they generated bigger comparison numbers than the rating the competition was featuring that year.

Recently, there has been a trend toward offering professional-duty power saws with less hype and fewer ratings. Manufacturers of worm-drive saws are beginning to list just the amperage and the arbor speed. To get a better idea of how these and other terms could be used reliably to determine saw power and quality, I talked to several engineers, service mechanics and a customer-service representative at the companies that make full-size, worm-drive saws. I began by asking what power or quality rating they would look for first in choosing a saw. Most replied that no single rating or any combination of ratings really speaks to the total quality of a saw, but that amperage, arbor speed, and to a lesser degree, horsepower, hold the greatest promise for describing what the saw will actually do.

Amperage—Since most of the portable power saws in the United States operate on 110-120 volts, one of the differences between saw motors will be evident in how many amps each draws. Although an inefficient electric motor or gearing system can take a large amperage draw and use it to generate more heat than power, generally the greater the amp rating, the more powerful the motor.

Taken by itself, a saw's amperage rating won't really tell you about how it cuts or about the quality of the components like gears and bearings. But the amperage rating can help you determine the general quality of a saw. Worm-drive saws in the 7¼-in. category are pretty closely grouped. Milwaukee's worm-geared workhorse draws 15 amps, while Skil's and Black & Decker's 7¼-in. worm-drive saws use 13 amps. These figures are higher than those for sidewinders generally. There are two reasons for this. The first is evident: worm-drive saws are designed in every way as professional-duty saws, and they have to be capable of heavy cutting. There is a much wider range of sidewinders, from low-amperage consumer saws to heavy-duty saws with about the same amperage draw as worm-drives. The second reason involves the action of the worm-drive itself. Because a worm gear slides rather than rolls like a spur or helical gear, its efficiency is reduced. In fact, the reason a worm-drive saw uses a bath-type lubrication is to minimize the effect of the heat produced in the gears.

Knowing the amperage rating of a electric tool is important in using it safely since it will determine the size drop cord you will need and the circuit you're going to tap into. Because the amperage figure is important in these ways, the methods for determining amperage are standardized and have to meet both government and private specifications. For instance, manufacturers applying for a U.L. listing are required to stamp the amperage rating right on the tool, and it must be a figure that is generated during continuous operation. The continuous-operating range of electrical tools in this kind of rating is defined by a maximum rise in temperature in the windings of 65°C during operation.

Arbor speed—The arbor speed of a saw is how fast the blade will turn measured in revolutions per minute (rpm). Generally, smoother cuts are produced by higher rpms. Most inexpensive power saws run at high speed, but they slow way down when their blades meet the work (remember that advertised arbor speed is a no-load measurement). At this point the advantage is lost. This is where amps need to be considered again. A high-amperage saw (13 to 15 amps) that is geared down to a low rpm (4,300 to 4,400 rpm) has the torque to maintain much of this speed even during heavy cutting. These qualities are particularly valuable in a saw that has to make long rips in green lumber or will be used in unsupported, freehand crosscutting where speed is of the essence and binding is much more likely—typical uses for worm-drive saws.

There are several professional-duty sidewinder saws, such as Black & Decker's Sawcat and Milwaukee's 7¼-in. contractor's saw, that draw similar amperage to the worm-drive saws, but turn at a high (5,800) rpm. An exception that is new to the market in the last year is Porter-Cable's 7¼-in. Speedtronic, which draws 14½ amps and operates at a constant 4,500 rpm, regulated by a microprocessor-based control.

Torque—One rating that in part measures the cutting ability of a power saw is *torque*. Although this term is often used interchangeably with power, they are not the same. Torque measures the turning force of the tool (this can be felt in the jerk a high-torque saw makes when you first pull the trigger), while *horsepower* (defined below) takes both torque and arbor speed into account.

Although torque sounds like a useful category for comparison because it's measured in concrete units—foot-pounds—it can be deceptive because of the different ways it can be measured. The first thing to distinguish when describing a saw using torque is whether work was actually being done when the rating was taken. *Stall torque* and *locked rotor torque* both measure the maximum torque that a motor can deliver, but the saw can't perform at these limits. Another much lower figure is *rated torque,* which unlike the other two, is measured at a more normal level of output, at which the tool can operate indefinitely. Unfortunately, many ratings aren't labeled as to which kind of torque they measure.

Horsepower—Few ratings have the reputation that this one does for certifying the power of a machine, but the difference between the way a gasoline engine and a universal-wound electric motor do their work is considerable. Electrical horsepower is based on a formula: output watts divided by 746. But there is more than one way to calculate horsepower, and many engineers use the following formula:

$$hp = torque \text{ (ft.-lb.)} \times rpm \times K$$

(K, a constant, is .1904 $\times$ 10^{-6}). This one sounds a little less abstract because it deals in terms (torque, rpm) that can be experienced.

Horsepower is actually measured with a dynamometer, which charts the complete output curve of the saw from zero to the maximum. Where the horsepower is measured and how high the saw is revving have everything to do with the number that is generated.

Motor horsepower is measured at the motor itself rather than at the arbor. It isn't very significant to the saw purchaser since it isn't an end-of-the-line number, but is measured back behind all the power transfer at the gears. A more useful figure is horsepower measured at the arbor at *operating speed*. For worm-drive saws, you'll get a rating of just over 1.0. This is *rated horsepower*, which is also confusingly called *full-load horsepower*. This is the actual horsepower you will be using since it's measured at a level where the tool can operate continuously. Rated horsepower is somewhere between 40% and 70% of *maximum horsepower*, which means that the saw is pushed to its very limits. When a saw is said to "develop" a certain horsepower, it is this maximum rating that is being used. Although you can use the maximum-horsepower figure for comparison, remember that your saw wouldn't be able to deal with this loading for more than a few seconds before its windings began to burn. Worm-drive saws (7¼ in.) develop between 2¼ and 2¾ hp at this level of operation.

Reputation and price—The best determiners of quality in a portable circular saw may be the tool's list price and its reputation among people who own one. Pick the most expensive saw you can afford, and then try to get it as cheaply as possible. A well-wound motor, solid castings and ball bearings are expensive, and the price of a good saw will reflect this. Price is a less important consideration when you are choosing within a category, such as full-size worm-drive saws, where all of the brands are similarly priced, and of correspondingly high quality.

A saw's reputation in the field also helps fill in information that isn't available elsewhere. Electrical and work ratings don't measure the durability of a tool; that's a matter of the quality of the parts and the craftsmanship used in putting them together. The reputation that a tool gets with rental yards and repair shops will give you a good insight. Convenience is something else that is best rated subjectively. And in this area there is no substitute for talking to people who already own the saw and use it for a living.

—*Paul Spring*

On the Edge

An overview of grinding and honing for the carpenter, wherein the author argues for more skill and fewer gadgets

by John Lively

I revered my Arkansas stones, until I met Apolinar. He worked as a carpenter in a Texas shop that made custom doors, windows, interior trim and furniture for rich clients. He came across the river from Mexico, where the guild system that followed the conquistadores into the New World still spawns craftsmen who serve out formal apprenticeships. Apolinar at the age of 50 was a master of his trade, with a measure of self-esteem to match his skills. He could not be coddled or coerced into sweeping the floor, he thought it entirely beneath his station to maintain or adjust any of the machines in the shop and he steadfastly refused to use sandpaper. He got paid by the piece, worked his own hours, and made about six times the money he would have earned in his native Mexico City. Not bad for an illegal alien.

Apolinar didn't have any whetstones. Yet his chisels and plane irons were always impeccably sharp. Nobody ever saw him sharpen a single tool, yet every morning there they were, freshly honed and ready for another day's service at the bench. One night when I was working late, I noticed Apolinar fussing around his bench in an unaccustomed way. He was getting ready to go home, and kept acting like he'd forgotten something. Finally he gave a demure grin and a resigned shrug, and said he'd touch up the edges of his tools before he left. So at last I got to witness the Apolinarian mysteries of sharpening.

He disassembled his three planes and lined up his six or seven chisels. From under his bench he produced a dirty 8/4 board, about 16 in. long and 6 in. wide, with a cleat nailed along the bottom, like a keel. He reached under his bench again and came up with a small cardboard box, which was fairly filled with neatly cut strips of, God forbid, sandpaper—240-grit, 400-grit and 600-grit aluminum-carbide wet/dry paper, to be precise. He clamped the board up in his vise, squirted some viscous fluid (which I later found out was a mixture of motor oil and kerosene) on the surface and spread it around with his fingers. Then he took a strip of 240-grit paper, patted it flat onto the oozing surface of the board and secured it there with two thumbtacks, using the same timeworn holes. He squirted some more fluid on the paper and began to hone one of his plane irons.

He was fast, taking no more than 15 seconds or so to have done with the bevel. He

followed with his other plane irons. Next came his chisels. When the paper stopped cutting, he'd put on a fresh piece. Stage two—400-grit paper. He went through the same routine as before, making sure the paper was kept wet with the oil/kerosene mixture. Then came a final round with 600-grit paper, which went the same as the first two stages. This time, though, he finished each tool by flipping it over and backing it off—holding the back of the tool flat against the abrasive surface and rubbing back and forth for several strokes to remove the wire edge that results from honing the bevel.

Ten minutes from the time he started, he was rubbing the oil off his fingers with a handful of jointer shavings. I stood there aghast. It seemed almost immoral that with all the honing jigs and grinding rigs on the market and with all the hush-toned lore surrounding this sacred subject, Apolinar should get such serviceable edges from strips of sandpaper and a grungy board.

The lesson is clear and direct. You don't need expensive sharpening equipment to get sharp edges. What you do need is to know what you're doing, to understand what happens when steel is rubbed against an abrasive surface and the effect this has on the edge. Once you have this understanding and have practiced the mechanics of grinding and honing, sharpening tools stops being an onerous, contraption-cluttered task that's consigned to weekends and rainy Monday afternoons.

The fundamentals of sharpening are learned early in a society that still values apprenticeships, but in our own country many builders and craftsmen are self-taught, and so seek instruction from tool catalogs and other self-taught tradesmen. There's no shortcut to acquiring these skills, but once you've got them you won't have to put up with dull chisels and planes that jump and chatter down a board. If your kind of carpentry involves considerable joinery, finish woodwork and cabinetmaking, you need efficient sharpening skills because your work requires using chisels, gouges and planes, and because these tools just don't work unless they are sharp. And even if most of your work is confined to rough framing, nothing takes the place of a chisel or plane when you need one.

To maintain edge tools you have to know how to grind and how to hone. Grinding shapes the bevel on the tool, and honing fin-

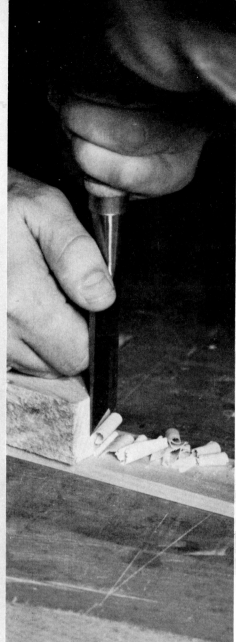

How sharp is sharp?

Look at the forearms of some timber framers and cabinetmakers, and you might see patches of scabby-looking bare skin where they've pared the hair away, testing their chisels and plane irons for razor sharpness. As a dramatic gesture (and testament to one's sharpening acumen), nothing beats showing the amazed onlooker that a fat hunk of tool steel can shave one's arm. While testing your edges on your arm can give you an indication of relative sharpness, it's not a method a dermatologist would recommend. A better way to assess the sharpness of an edge is to see how easily the blade will slice tiny slivers of wood from the edge of a board. The smaller the slivers, the sharper the edge. Another way is to pare away wispy curls of end grain, as shown above. If the wood powders and flakes, your edge is dull, but if you can slice off translucent, tissue-thin pieces that stay in one piece, you've got a sharp, serviceable edge. A third way to test for sharpness is to rest the edge, bevel up, on your thumbnail and slowly raise the back end of the tool. The lower the angle at which the edge will bite into your nail, the sharper it is. —*J. L.*

ishes the edge by removing the small ridges and ruts left by grinding. The edge can be further refind by buffing or stropping.

Grinding—Shaping the bevel of a tool by grinding (backs of blades are never ground) is something that should happen infrequently. There are only three reasons to take your tool to a grinder. The first and most common is that the edge has been nicked or dinged, and you need to regrind the bevel back into undamaged steel. You can minimize dinging up your good chisels by keeping them sheathed and using a junk chisel for nasty work that doesn't call for a sharp edge. A second reason to regrind is that the width of the micro-bevel (more about this later) has become too wide, and you have to restore the original 25° bevel. Lastly, you might want to regrind a new tool to alter the angle the factory put on it.

Most grinders can be taken onto a building site if necessary, but it's usually more convenient to do your grinding at home, where the machine and its accessories can be permanently set up. To avoid investing in a commercial grinding machine (and to make do in emergency situations), you can improvise. I know a cabinetmaker who uses a plywood disc faced with 100-grit sandpaper and fitted with a mandrel so it can be chucked in a drill press. I've seen others use a belt sander turned upside down and clamped to a table.

The two most common types of grinders have abrasive wheels (bench grinders) or abrasive belts. There are rules of use that apply to them both. The abrasive surface—whether an emery or vitrified aluminum-oxide wheel or an aluminum-oxide belt—has to be kept clean. Clogged wheels and glazed belts don't cut effectively, and they cause you to apply excessive pressure to the tool. This overheats the steel and ruins its ability to hold an edge. To get around this, dress the wheel often or replace worn belts. They're a lot cheaper than new chisels and plane irons.

While grinding, quench the tool often in a can of cold water. Watch the little beads of water at the edge of the tool as you grind. When they begin to fizzle or evaporate quickly, dunk the tool again. All the grinding and honing you do will be wasted if you overheat the tool and ruin its temper, and the edge—no matter how sharp you get it—won't last. A brittle, distempered edge will fray, splinter off in tiny pieces, and soon become too dull for anything but spooning yogurt.

If the steel turns blue, the edge has been heated beyond its original tempering temperature. The only way you can salvage the tool is to grind past the blue area into good steel. So take your time when you grind, holding the tool against the wheel or belt with a firm but

---

All your grinding and honing will be wasted if you overheat the tool and ruin its temper, and the edge—no matter how sharp you get it—won't last.

---

light touch. Keep the steel cool, and inspect the bevel often to make sure you're removing metal in the right place, in the right amount.

Bench grinders—Most motorized bench grinders accept 7-in. or 8-in. dia. wheels, one coarse-grit, the other fine. Many people think that you shouldn't grind a good tool on a coarse wheel, and so make the mistake of using the coarse wheel only for the rough shaping of metal parts. The truth is that a coarse wheel (say, 36-grit), freshly dressed, will cut faster, cleaner and cooler than a fine wheel. After all, the point of grinding is to shape the bevel, not to finish it. A fine wheel will glaze over quickly, and even if it's clean, you'll have to apply more pressure to remove the same amount of steel that requires less pressure on the coarse wheel. This means you have to spend a lot more time grinding, and that you

risk burning the tool. A coarse wheel will yield an edge that's quite fit for honing.

Because of the relatively small diameter of grinding wheels, the radius of the hollow grind they produce is correspondingly small. This deeply hollowed bevel, as opposed to a bevel with a shallower concavity or a flat bevel, is fragile at the edge, and requires considerable honing to get a flat surface wide enough to give support to the cutting edge.

The purpose of grinding is to produce a uniformly flat or hollow-ground bevel on the tool. But holding a chisel or plane iron at the proper angle, using nothing more than the little tool rest that comes with most bench grinders, and being able to return the tool to the same position repeatedly after numerous quenchings isn't easy. And those unpracticed at grinding often end up with multifaceted bevels and cutting edges that aren't square to the sides of the blade. That's why there are several sliding tool rests on the market that will attach to a grinder, hold the tool at the angle you choose and guide it past the wheel. But these devices have drawbacks. Using a sliding tool-rest attachment can double your time at the grinding wheel. To set one up, you have to adjust the angle of the tool rest, clamp the tool on the slide and then unclamp it when you're done. The ones that have a rack-and-pinion feed mechanism keep you from feeling what's going on between the wheel and the edge of the blade because you're turning a knob instead of holding the tool.

Freehand grinding—For grinding on a wheel, the best method I know uses nothing more than the stock tool rest on your grinder, yet gives you a high degree of control over the grinding angle and lets you regulate the pressure against the wheel with considerable sensitivity. Quite simply, it involves using the tool rest as a fence rather than as a surface to rest the tool on. Set the angle of the tool rest so that it's slightly lower than the bevel angle on the tool. With the motor turned off, place the edge of the tool against the wheel so that grinding will happen across the full width of its bevel and so the blade contacts the outer edge of the tool rest.

Now grip the tool firmly, thumb on top and forefinger underneath, perpendicular to the blade, and push your forefinger smartly against the edge of the tool rest (photo left). This grip turns your hand into a jig and lets you use your forefinger to gauge the distance from tool rest to wheel. You just slide your entire hand back and forth, keeping your forefin-

Freehand grinding uses the standard tool rest as a fence rather than as a flat surface for supporting the tool. The blade is gripped so that the right forefinger passes underneath and perpendicular to the side of the tool. With the forefinger pressed firmly against the edge of the tool rest, you can move the blade from side to side to grind the full width of the bevel. The distance from the tip of the tool to the forefinger determines the grinding angle. Fingers on the left hand deliver pressure to the cut, and sense when the steel heats up.

Belt grinders, like Woodcraft Supply's Mark II sharpener shown at right, can cut cooler and faster than abrasive wheels. The 11½-in. urethane contact wheel grinds a shallow, sturdy hollow bevel, and the 2½-in. wide belt means plane irons and big chisels can be ground without having to move the tool back and forth across the abrasive surface.

ger pressed against the tool rest and using your thumb to deliver pressure to the cut. If your grip remains firm, you can remove the blade from the grinder as often as you wish to inspect the edge or quench it.

Dressing the wheel—The point of dressing a wheel is to remove a very small layer of clogged and dulled abrasive particles from its edge, and thus expose a new surface of clean, sharp particles. Some diamond-tipped wheel dressers lock onto a jig that slides on the grinder's tool rest. The jig ensures that you dress the edge of the wheel at 90° to the sides. But you can use a diamond dresser freehand by making several deft, deliberate passes across the edge of the wheel with the tip. Excessive pressure can cause you to remove too much material, gouge the wheel and end up with an uneven grinding surface.

Other kinds of dressers—star dressers and carbide dressers—are also available, but a diamond-tipped dresser lasts longer and cuts cleaner. Dressers are sold by industrial-supply houses and mail-order tool companies.

Belt grinders—There are quite a few belt grinders on the market. Rockwell makes two different types—one is a 7-in. grinder/finisher equipped with a grinding wheel on the right of the motor unit and a platen-backed, 2-in. wide belt on the other side. It's more expensive (about $600) than their sander/grinder ($120 to $235, depending on the model), which comes with a single 1-in. wide belt, a tilting table and a platen to back up the belt.

The Mark II Sharpening System sells for about $500 at Woodcraft Supply Corp. (41 Atlantic Ave., Box 4000, Woburn, Mass. 01888). Like the expensive Rockwell belt grinder, the Mark II would be a good investment for professional shop carpenters and timber framers. But unless you do a lot of grinding, a cheaper machine would be a wiser choice. The Mark II (photo above) has an 11½-in. dia. cast-urethane contact wheel that drives a 2½-in. wide, 60-in. long belt around an idler that contains the belt-tracking device. Because of the wheel's large diameter, the hollow grind it produces is minimal, and the edge is therefore more substantial than one ground by a small-diameter wheel. The grinding side of the machine is fitted with a sliding outrigger arm that has a block at the end for holding the butt of a chisel or plane iron. By sliding the arm in or out, you adjust the grinding angle the tool makes with the contact wheel. On the left side of the arbor there's a muslin buffing wheel, something I'll comment on a little later in the article.

Rockwell's belt grinders, because the belts are backed up by a platen, produce a flat bev-

el on the tool. And both come with standard tool rests, rather than with the arm-and-block rest that's part of Woodcraft's machine. The craftsmen and tradespeople I've talked to who own belt grinders prefer them to bench grinders. They say that the belts cut faster and cooler, and that they have more control over the tool. The wider the belt, the better, as you can grind butt chisels, framing chisels and plane irons without having to slide them back and forth across the cutting surface.

As with wheels, coarse-grit abrasive belts will cut faster and clog less quickly than fine-grit belts. Don't use anything finer than 80-grit or 100-grit belts, or you risk overheating the steel. You can clean grinding belts with the crepe sole of an old sneaker, or you can go buy a bar of the stuff (sold as a "dressing stick") for about $9. But after several cleanings the abrasive particles will have been dulled, and many of them worn away; so it's best to replace the belt.

One last thing about grinding. Wear goggles.

Honing—With a magnifying glass, you can see that grinding gouges the steel. Even to the naked eye, a freshly ground edge looks a little ragged out on the tip. Such a sawtooth edge might even be sharp, but after a couple of cuts into wood the fragile slivers on the end will bend and fracture, and you'll have a dull edge. The purpose of honing is to polish the bevel, to smooth out all the ridges and trenches left by grinding and to produce an edge that ideal-

ly is straight across. A polished edge is sharper and sturdier because there are no unsupported slivers of steel to break off.

Honing is done on stones that are usually about 2 in. wide and 8 in. to 12 in. long. Arkansas stones are natural; other stones—Carborundum, Crystolon, India, and Japanese waterstones—are manmade. You begin with a fairly coarse stone, progress to a medium-grit stone, and end on a fine-grit stone.

The more finely honed an edge is, the sharper it will be and the longer it will last. And, paradoxically, the more frequently you hone a tool, the less time you spend at the stones. Because honing is tedious, messy and repetitive, and because it requires discipline and practice to hold a tool at a constant angle while rubbing it on a stone in a back-and-forth or figure-eight motion, many carpenters never develop an effective, reliable technique. This is why the tool catalogs peddle honing guides. These roller devices hold the tool at the proper honing angle while you move the edge over the stone. As with grinding jigs, these things just get in the way, slow you down and discourage you from developing a valuable skill.

Posture and grip—Stance and grip are critical to proper honing. The grip I'll describe enables you to hold the tool at a constant angle while honing, and it lets you work without getting tired. Grab the blade in your right hand with your forefinger extended down the right side of the tool so that your fingertip is about

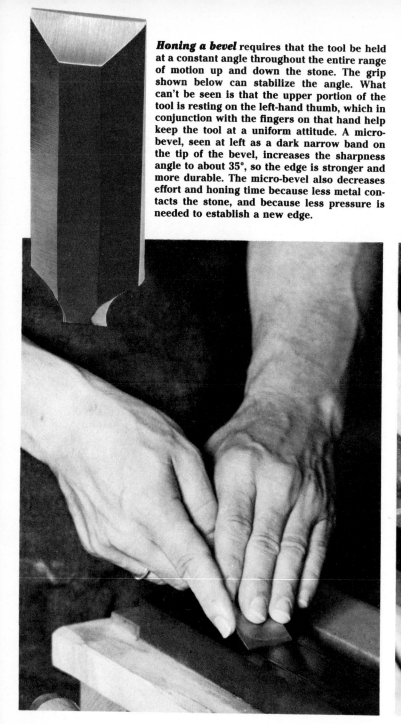

Honing a bevel requires that the tool be held at a constant angle throughout the entire range of motion up and down the stone. The grip shown below can stabilize the angle. What can't be seen is that the upper portion of the tool is resting on the left-hand thumb, which in conjunction with the fingers on that hand help keep the tool at a uniform attitude. A micro-bevel, seen at left as a dark narrow band on the tip of the bevel, increases the sharpness angle to about 35°, so the edge is stronger and more durable. The micro-bevel also decreases effort and honing time because less metal contacts the stone, and because less pressure is needed to establish a new edge.

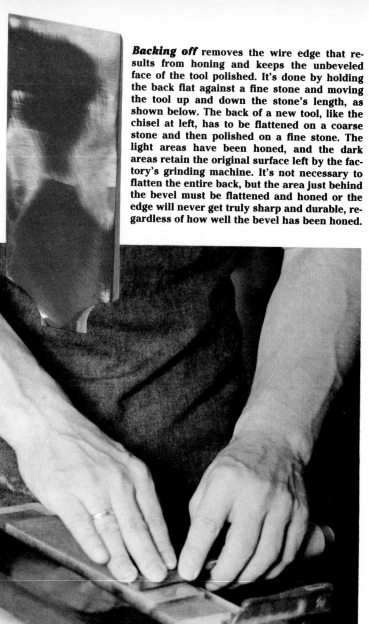

Backing off removes the wire edge that results from honing and keeps the unbeveled face of the tool polished. It's done by holding the back flat against a fine stone and moving the tool up and down the stone's length, as shown below. The back of a new tool, like the chisel at left, has to be flattened on a coarse stone and then polished on a fine stone. The light areas have been honed, and the dark areas retain the original surface left by the factory's grinding machine. It's not necessary to flatten the entire back, but the area just behind the bevel must be flattened and honed or the edge will never get truly sharp and durable, regardless of how well the bevel has been honed.

an inch from the edge (photo above left). Your right thumb sits on the blade. Place your left thumb under your right thumb and across the underside of the blade; then put your left-hand fingers on the face of the blade about ½ in. up from the edge. Wide tools like plane irons can accommodate all four fingers. Narrow blades take fewer fingers. The thumb on your left hand acts as a fulcrum, while the fingers on that hand serve to deliver even downward pressure across the width of the blade. Your right hand stabilizes things laterally and delivers energy to the stroke of the cut.

Position your feet a comfortable 18 in. or so apart, bend at the waist and touch the bevel to the front end of the stone, which should be about 10 in. in front of your belly. Work the tool up and down the length of the stone. Let your arms (not your body) do the work. Flex only at your elbows, keeping your back and shoulders still. If you shift your upper body, you'll alter the angle of tool on stone, and end

up with a convex bevel. To hold the tool at the proper angle while you're learning this technique, cut a scrap block at a 35° angle and set it by your stone as a visual reference. With practice, you'll be able to find the correct angle automatically.

Try to distribute your strokes over most of the stone's length and width. If you don't, you'll gouge a rut down the center of the stone. Keep the stone amply lubricated. If you're using oil, wipe the stone clean when the oil gets thick and black from metal filings and apply fresh oil.

Micro-bevels—Most edge tools are ground to a sharpness angle of about 25°, just fine for a nice easy cut, but not so good for a durable edge. A second angle of about 35° honed at the tip of the bevel, called a micro-bevel, can lengthen the life of an edge without adversely affecting its cutting efficiency.

Whether the bevel is flat ground or hollow

ground, working with a micro-bevel increases the speed of honing because substantially less steel contacts the stone. And honing is less tiresome because it takes less pressure than honing the entire face of the bevel. With each honing the micro-bevel widens, and once it gets so wide that you can't get a new edge in a minute or so at the stone, you have to regrind a fresh 25° bevel on the tool.

You can tell when your honing should stop by running your finger along the back side of the edge and feeling for the wire edge—that ever-so-small flap of steel that's produced when the surface of the bevel collapses or wears through to the back side of the tool. If the wire edge is so small (or your fingertips so calloused) that you can't feel it, eye the face of the bevel in raking rays of light. If it's time to stop honing, the micro-bevel will be flat and even from its heel out to the edge (inset, above left). But if you see a line of light reflected off the edge, you're not done yet.

Backing off—The back side of a new tool must be honed absolutely flat and kept that way. This is necessary because the backs of new tools are ground more or less flat (inset, facing page, right), and the trenches left by grinding will form little sawteeth at the edge. It doesn't matter how finely polished the bevel is; if the back of the blade isn't flat, and if the grinding lines haven't been leveled, you'll never get the edge really sharp or durable.

For a new tool, begin backing off by holding the blade flat on your coarsest stone (photo facing page, right). Keep working until you've honed a flat surface at least ⅛ in. wide behind the edge. Don't worry if you can't get the entire back flat; it's only the area immediately behind the edge that counts. Next, proceed to your medium-grit stone, and finally polish the back on your fine-grit stone. Now that the tool has been properly flattened and polished on its back side, never back it off on anything but the fine stone. The deeper the scratches, the duller and less durable the edge will be.

You should back off a tool every time you hone its bevel; eight or ten strokes up and down the length of the stone should remove the wire edge. It's not unusual, though, to go from backing off to stroking the bevel a couple of times, back to backing off again, possibly four or five times, to remove a stubborn wire edge. The lighter your touch when honing the bevel, the thinner the wire edge, and the easier it will be to remove.

Oilstones vs. waterstones—Oilstones (Arkansas stones) are novaculite. Mined from pits and caves near Little Rock, they come in four grades (grits). The coarsest is a reddish-looking Washita stone (about 800 to 1,000 grit). Next comes the mottled slate-colored soft Arkansas (about 2,000 grit), followed by the hard Arkansas (about 3,000 grit), which looks like white marble. The finest stone is a black hard Arkansas. Because these are natural stones, the grit in each grade can vary from stone to stone; indeed, from spot to spot in the same stone.

Arkansas stones are fast cutting and durable, but have to be kept clean, or their pores clog with a paste of oil and metal filings. These stones should be cleaned after each use by flushing the surface with fresh oil, rubbing it into the stone and wiping it clean with a lint-free cotton rag. Badly clogged stones can be cleaned by soaking them in mineral spirits or some other solvent that will soften and remove the gunk.

It's pretty easy to gouge a Washita stone

Buffing the edge further refines and strengthens the bevel by removing grinding ruts and honing scratches altogether. The machine in the photo is equipped with a hard-felt buffing wheel that's charged with a grey buff compound. As with honing a micro-bevel, it's best to buff just the tip of the bevel to increase the sharpness angle and reduce working time. But any buffing wheel will round over the edge slightly, and it will have to be rehoned after a number of buffings have made it too convex and too steep to be effective.

(especiallly with the roller on a honing guide), and all the natural stones will get uneven after years of use. You can flatten them anew with a diamond whetstone (from Diamond Machining Technology, Inc., 34 Tower St., Hudson, Mass. 01749).

Japanese waterstones are made from abrasive particles that are bonded into bricks. The binder, which softens with water, lets worn particles on the surface float away so that the steel always contacts new sharp particles. For most honing you need only two waterstones—a coarse (say 1,000 grit) and a fine (about 4,000 grit). Before honing can begin, you have to immerse the stones for several minutes in water to let the pores fill up. Otherwise, the water you try to pour on the surface to lubricate the cut will get soaked up, and you'll have a dry stone. But it's not good to keep the stone submerged for a long time because the binder will soften to considerable depth and the stone will wear out too quickly.

Waterstones cut fast, don't clog and make less of a goupy mess than oilstones, especially if you do things right and fix the stones in a rack over a trough of water. This way you can scoop handfuls of water over the stones as you hone without getting spills all over the table and floor. But waterstones will wear faster than oilstones, and these days they are not much cheaper (a black hard Arkansas sells for about $48, and a fine-grit waterstone for about $47). Because waterstones are manmade and therefore consistent all the way through, you get a more uniform cut with them than you do with natural oilstones, which vary in density, porosity and cutting ability. People who use waterstones like them and say that they wouldn't go back to oilstones.

Buffing—To go that extra step toward the perfect edge, you might want to polish the bevel of a tool by buffing, a refinement that almosts gets rid of honing scratches altogether. A buffed edge is sharper and stronger than a honed one, and you can keep it sharp by frequent buffings, though eventually you'll have to re-establish the bevel by grinding and honing. Buffing is done by loading a cloth or felt wheel with a polishing compound (usually grey buff that comes in ingot-size bars) and applying the edge to the wheel just as if you're grinding it, though it's a good idea to adjust the angle to 35° to produce a micro-bevel.

A muslin buffing wheel (a stock item with the Woodcraft Mark II sharpening system) consists of 80 plies of cloth sewn together. The edge of the wheel yields quite a bit under pressure, and so can round over the edge of the tool at an angle much steeper than 35°. I think that a hard felt buffing wheel (photo below) is a better choice because its edge is firm and it won't round over the bevel appreciably as it polishes. Also the hard-felt wheel smooths out honing scratches better than a muslin wheel.

I've made the mistake of trying to buff bevels straight from the grinder, something you can do if you're in a hurry, but not if you want a good edge. You should hone the edge first, then buff it, and buff it thereafter as often as you need to. A buffing wheel shouldn't be used for backing off the blade. Several strokes on your finest stone will remove any wire edge that comes from buffing the bevel.

Compared to grinding, buffing produces very little heat, but you can lean on a tool hard enough to burn it, so take it easy, and load the wheel often with fresh compound. Because buffing is so fast and gives such good results, it's something that site carpenters, as well as shop woodworkers, ought to try. A good setup, one that could easily be made portable, is to take a common bench grinder and equip one arbor with a coarse grinding wheel and the other arbor with a hard felt wheel. Used in conjunction with bench stones, buffing can be part of a format for quick, effective sharpening.

But if you're traveling light, far from home, you can always use sandpaper, kerosene and motor oil.   □

Handsaws
Care and use of a tool
that gets little attention these days

by Tom Law

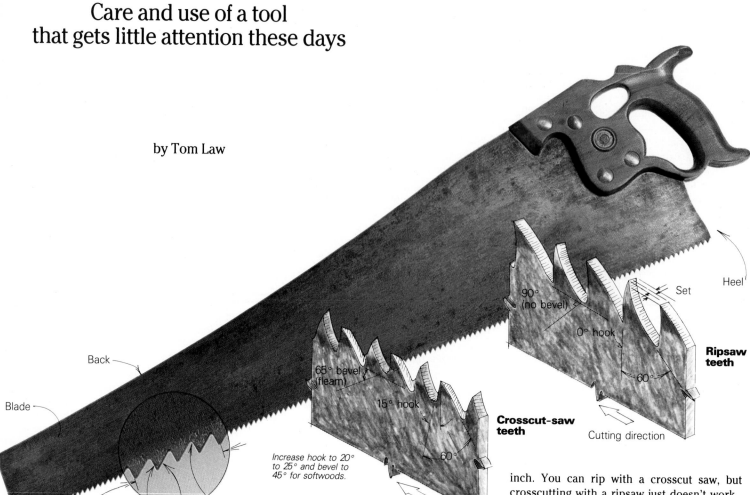

Back

Blade

Toe

1 in.
5 points,
4 teeth

Gullet

Back of tooth

Face of tooth

65° bevel
(fleam)

15° hook

60°

Increase hook to 20°
to 25° and bevel to
45° for softwoods.

Cutting direction

90°
(no bevel)

0° hook

Set

Heel

Ripsaw
teeth

60°

Crosscut-saw
teeth

Cutting direction

I have a special affection for handsaws. I was taught by carpenters who used handsaws almost exclusively. My first framing job as an apprentice was a highway bridge in a remote location with no electricity; all the cutting was done with handsaws. I learned to keep mine sharp, and ended up filing saws for the entire crew. In later years, I sharpened the saws for crews of more than 25 carpenters on large commercial projects.

Even though most of the cutting I do these days on the houses I build and remodel is with power saws, I still use my handsaws. It's surprising how often their slender profile, depth of cut, and lack of power cord make them handy. And the finest scribe-fitting I do almost always calls for a handsaw.

But because of the dominance of power saws today, few in the generation after me have learned handsaw skills. With incorrect technique and an inferior saw that is dull or badly sharpened, handsawing can be pure drudgery. But it doesn't need to be. The difference is in knowing how to pick out a good saw, how to joint, shape, set and sharpen the teeth, and how to use it once it's sharp.

Basics—Most handsaws are about 26 in. long. Shorter ones (24 in., 22 in. or less) are called bench saws or panel saws. The top edge of the saw, called the back, can be straight or skewed. Skew-backs taper from handle to toe in a gentle S-curve; they were favored in the first part of the century. Skew-backs are better at cutting a curved line, but I still prefer square-back saws. They make good medium-length straightedges, and you can even scratch a square line across the blade and use it as a framing or combination square. The front end of the saw is the toe; the rear, down below the handle, is the heel.

One of the first things to learn is the difference between a crosscut saw and a ripsaw. Crosscut saws are made to cut across the grain, and their teeth act like a row of knife points, severing the fibers as they cut. Crosscut saws come with 7, 8, 9, 10, 11 or 12 points per inch. The more points, the finer the cut. A 7-point, for example, is used for wet rough framing; an 11 or 12-point for fine trim work.

Ripsaws, like the Disston D-8 skew-back shown above, are made to cut along the grain. Their teeth act like a row of chisels and remove small chunks of wood as they go through the cut. Ripsaws have larger teeth and deeper gullets than crosscut saws, and 4½, 5, 5½ (the most common) or 6 points per

inch. You can rip with a crosscut saw, but crosscutting with a ripsaw just doesn't work.

Points per inch and number of teeth per inch are not the same. As shown in the small drawing above left, a 5-point saw has 4 teeth per inch, a 10-point has 9 teeth, and so on.

Viewed from the side, each tooth on a handsaw forms a 60° angle. But how these cutters are tilted forward and aft (called pitch, or *hook*) is different for the two saw types. Still viewing the saw in profile, imagine a line that connects all the gullets. A line drawn down the face of the sawblade perpendicular to this represents zero hook. It is this lack of pitch that gives the teeth of a ripsaw their chisel-like quality. The tops of crosscut teeth, however, are pitched back from this perpendicular line (about 15° for cutting hardwoods).

Another angle to consider is the one across the tooth face. This *bevel,* or *fleam,* is determined by whether the file used in sharpening is held perpendicular to the sawblade or askew. Viewed from above, ripsaw teeth are filed at a right angle to the sawblade. Crosscut teeth are alternately beveled 65° to the face of the blade, producing a knife-like point on the leading edge of the tooth.

For crosscutting softwoods, the hook should be increased to 20° to 25°, and the bevel should be closer to 45°. This way the saw will cut the wood rather than tear it, which would cause the blade to bind in the kerf against the torn fibers.

One thing ripsaws and crosscut saws have in common is *set*—the alternate bending of

You can learn a lot about a saw by making it sing. The tone and its duration are good indicators of the quality of the steel and its thickness. To produce a note, the author has thumped the blade near the handle with his thumb to set the metal vibrating, and is varying the sound by increasing or decreasing the curve of the blade with finger pressure on the toe of the saw.

the top half of the teeth. Each tooth is bent either to the right or left of the body of the sawblade. Setting the teeth makes the cut, or kerf, wider than the blade thickness and reduces the friction of metal against wood. Good blades are also taper-ground; that is, the blade is thinner in section at the back than at the teeth. Taper-grinding improves the balance of the saw by lowering the blade's center of gravity, and works in the same way as set to reduce friction in the kerf to all but the teeth themselves. The thickest part of the blade goes into the new wood along the kerf, while the trailing metal is thinner.

A little history—Today I carry two saws—an 8-point crosscut for general work, and a 10-point crosscut for trim. The old-timers carried five or six saws in their own box. This "nest of saws" might include a 5½-point rip, a 7-point crosscut or an 8-point with wide set for wet lumber, an 8-point for general work, a 9-point for outside trim and a 10-point or 11-point for fine trim, and maybe an over-the-hill favorite for tight places or when there was danger of dulling against nails or masonry.

Turn-of-the century saws were wider than saws made today; the extra metal added weight to help make the cut. Moderate-width blades called *lightweights* were also made. Saws of this period were generally good, but the quality of the steel was sometimes inconsistent. A single blade could have some very soft teeth, while others were so hard they would ruin a file. Too, the metal could be so brittle that the teeth would snap off when being set. The best use for one of these is to hang it proudly on the wall as old grandpa's.

One of the best blades I've seen was a Disston made in about 1941 with a V for victory and some patriotic words about the war effort printed on the side. Disston also made excellent saws after the war and into the 1950s. One of the best of these was the D-95, with its incongruous plastic handle. Atkins also produced excellent saws during this period.

While domestic manufacturers were going downhill in the 1960s, Sandvik of Sweden was producing strange-looking but marvelous saws. They came with plywood handles embossed with sea serpents. The best grade had a plastic handle—this at a time when quality was symbolized by walnut. The line of teeth was also peculiar; it was convex. The natural arc that is produced by the motion of sawing is the reason for the convex curve. When the saw is progressing toward the middle of the cut, the blade curves down to meet the increasing pressure most effectively.

The quality of American-made handsaws deteriorated rapidly during the 1960s and 1970s, as manufacturers responded to the market with lower quality and higher prices. Today's saws, wrapped in plastic and covered with promotional claims, are sad remembrances of what saws were 30 years ago.

Finding a good saw—My advice is not to get a new saw—buy an old one and fix it. Literally hundreds of saws have passed through my hands, and I can get a good idea of their quality just by filing a few teeth. The expression "They don't make them like they used to" is certainly true, but just because a saw is old doesn't necessarily mean it's good. If you have an old saw that is deeply pitted by rust or has sharp kinks in it, get rid of it. Although I've heard that slapping a saw on the surface of a body of water will straighten out a bend, you can't prove it by me. A kinked blade is damaged goods, and serious pitting means you're getting less steel than the original sawmaker thought you should have. But if a saw is only rusty, give it a few tests.

Take it by the handle and shake it back and forth; the front half should whip. If it moves very little, the metal is too thick and heavy, and the saw will be clumsy to use. If it moves so much that it's flimsy, it is cheap and too thin, and will be hard to control in the cut. If it moves just right, try this—pass your fingers through the hole in the handle and hold it by the cheeks; hold the toe end with your other hand, and bend the blade into an S-curve toe up, heel down (photo above). While starting the bend, thump the blade near the center of the handle with your thumb and it will emit a musical tone. The pitch and duration of the sound are indicators of the quality of the metal and balance of the blade. A dull sound of short duration indicates an unworthy blade; a high pitch of long duration indicates good quality and balance. There are some exceptions to this test. Some excellent old blades that are quite wide sound rather gutteral.

I've heard old-timers say that a good spring-steel blade should snap back straight after the toe has been passed through the hole in the handle. I think this goes a little far in inviting a brittle blade to snap, or even good steel to retain a bend, but you do want to flex the saw to check the quality of its steel.

If you've found an old blade, don't worry about the teeth—they can be recut. Clean off the rust by any method from sandblasting to hand sanding, and polish the blade bright with fine abrasive paper. Then make a new handle. Today's handles are just chunks of wood with scratches on the side, screwed onto the blade. Those scratches are barely recognizable as heads of wheat, an ancient symbol that proclaims the virtues of labor. Old handles were made to fit the hand and were scooped and curved in styles from graceful to grotesque. It may take a day or longer to make a nice handle, but that won't be much time if your grandchildren inherit the saw.

If the line of the teeth on your old blade is crooked or concave from toe to heel, it's because it has been improperly filed. Some old carpenters touched up their saws before they left the job each day, and filed only the dulled teeth. That practice has drawbacks. A sawblade doesn't get equal wear along its length as it moves through the cut. It's the center portion that gets the most work and wears the fastest, but it's a mistake to file only the teeth that are dull. This condition will disrupt the rhythm of each tooth contacting the wood when sawing and will give you a ragged cut.

Getting the teeth in shape—A saw with broken or very uneven teeth may have to be retoothed. This is best done by a saw shop,

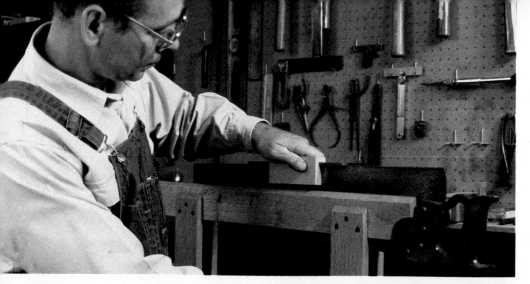

Sharpening a handsaw

Jointing. A flat mill bastard file is used to joint (level and align) the tops of the teeth, left. This creates a line of saw points of equal height (or slightly convex on some saws) so that each one will be brought to bear on the wood when sawing. Here a simple L sawn from a pine block is used to hold the file perpendicular to the face of the sawblade. Manufactured metal jointers that clamp over the file and act as a guide on the blade give the same results. Even when the height of the saw teeth doesn't need correcting, the flattened tops of the teeth that result from a light jointing are a useful guide when shaping and sharpening the teeth.

Setting. A saw-set is used to bend the top half of the teeth alternately to the left and right. As shown above, the plunger pushes on the saw tooth, forcing it against the anvil each time the pistol grip of the handle is squeezed. All saw-sets are adjustable for width of set. A handsaw used mostly on wet lumber should get a wider set than one used mostly on dry wood to reduce friction on the blade and to evacuate the moist sawdust more easily.

Shaping and sharpening. Using a triangular taper file held at a 65° angle to the face of the blade, the author sharpens a crosscut saw (right). On a ripsaw, the angle is 90°—perpendicular to the sawblade. In both cases, the file cuts only on the push stroke. Unless the saw has been abused or sharpened badly, shaping the teeth can usually be a part of the sharpening process. Every other tooth is sharpened on the first run down the blade, and then the saw is turned around in the vise to get the remaining teeth. You'll have done the job well if the gullets are equally deep, and if none of the saw points reflects light. If any of them do, they either haven't been filed yet (below), or they need another stroke or two.

whose machines will punch or grind out a new line of teeth in minutes—something that would take hours by hand. But when it comes to sharpening the teeth of a saw that hasn't been badly abused, I like to do it myself.

Sharpening a saw involves four operations—*jointing,* which makes the teeth the same height; *shaping,* which evens out the size of the teeth and the depth of the gullets; *setting,* which bends the teeth alternately to either side of the blade; and *sharpening,* which gives each tooth its precise point. When saws are maintained in good condition, not all of these operations have to be performed each time. That's one of the problems with sending your saws out to be sharpened. Often, all you'll need is a light touchup with a file, but most saw shops have just one price—$3 to $7 for the full gamut of operations. And they'll end up keeping your saw for a week.

Also, sharpening machines that use a file will produce the same big-teeth little-teeth pattern as hand filing (sidebar, above right). Corrective filing for this condition is a matter of judgment, not what a machine is good at. Even a saw fresh out of its wrapper needs some hand work. It will have the right angles on the teeth, but they probably will not be very sharp. I don't like machine setting either, because it tends to push the whole tooth out rather than bending just the top half.

Learning to hand-file a saw takes some practice because the result has to be near perfect. But don't hesitate to try it. You can always correct your errors. After all, each time you file a saw, you are making new teeth.

Hand-filing a saw that is just slightly dull takes me about 15 minutes. No jointing, shaping or setting is required, just two or three lightly controlled strokes on each tooth. Lightly jointing, resetting and sharpening a saw take me about half an hour; reconditioning a misshapen saw takes me about two hours. But I've had a chance to practice some. When I was filing for a crew, I would average three saws an hour, or 24 a day. A 26-in., 8-point saw has 184 teeth; 24 saws a day is 4,992 teeth.

Before you begin filing, the first thing you need is good light; your eyes will be in close concentration for a long time. The best is natural sunlight on a bright but cloudy day. When I'm inside at my bench, I work under four 40-watt fluorescent lamps.

The next thing you need is a saw vise. You could make yourself a wooden one, but manufactured metal ones are more common. Most of these are about 12 in. long and hold the sawblade, teeth up, with an eccentric roller-bar or cam lock. Saw vises are short enough to be carried in the toolbox, but you have to keep sliding the saw along in the vise as your filing progresses. The vise I like best is a Lodi, made in California, but unfortunately, it's not being manufactured any longer. The jaws hold the saw along its entire length, with clamps at each end. In any case, the vise should be set up so that it's about 4 in. above your elbow.

With the exception of a flat mill-bastard file used for jointing, the tool that will be doing all

the cutting is an ordinary triangular taper file. These come in lengths from 5 in. to 8 in., and cross-sectional thicknesses called regular, slim, extra slim and double extra slim. Which one you use isn't critical, since all triangular files have equal 60° sides, but for most saws I use a 7-in. double-extra slim. The reason I use this thickness is that the apex of the triangle is sharper than on thicker files, and this serves to cut the gullet deeper. The narrow cross section also makes it easier to judge the angles on longer teeth. I like the longer files because I get more cutting per stroke.

Jointing is done to correct the line of teeth and make them all of equal height. This line should be either straight or convex. A saw in good shape won't need much jointing, but the resulting flat tops of the teeth will be useful later on in shaping and sharpening.

To joint, put the saw in the vise with the teeth up and about 2 in. of the blade showing. Using long strokes with a flat mill file, keep working the tops of the teeth, flattening the points until the top of each tooth has been struck (photo facing page, top). It's important that the file be perfectly perpendicular to the sawblade. You can make a simple wood block to ensure alignment, or buy a metal handsaw jointer that does the same thing.

Shaping corrects any teeth that have been deformed by bad sharpening or, more likely, contact with a nail. It ensures that the teeth are all of uniform size and spacing. If the saw is in good condition, some shaping can be done while sharpening. Hold the file at a 90° angle to the face of the sawblade, whether you are shaping a ripsaw, which will also be sharpened this way, or a crosscut saw that will get sharpened with alternating bevels. Sharpening will take a bit longer this way, but you'll be able to shape more accurately.

Place the saw in the vise with the handle to your right, letting the teeth show about ¼ in. If too much blade is out of the vise, the metal will bend when the file is in motion, causing the file to chatter and wear down quickly. Select the first tooth from the toe that is set toward you, and place the file in the gullet to the left of this tooth. Hold the file horizontally, an end in each hand, and push it straight across. As the file passes through the gullet, the left side is cutting the back side of one tooth and the right side is cutting the front side of another tooth. File until one half of the flat top made by jointing is worn away, skip the next gullet and go to the next tooth set toward you. File it as before, and then move down the length, skipping every other gullet. When you've finished that side, turn the saw around and again select the first tooth from the toe that is set toward you. This will be the one you skipped the first time through. File this side the same as the other, using the flat tops as a guide—when the tooth comes to a point properly, the top will seem to disappear, as it will no longer reflect light. Check the depth of the gullets as you progress to make sure that they are the same.

Setting tends to distort the leading edge of the tooth and should always be done before

sharpening. It is done with a tool called a saw-set, which works on the principle of hammer and anvil, although the hammer is called the plunger and is activated by squeezing the pistol or plier-grip handle. The anvil is beveled and adjustable for different-size teeth, usually with a dial that adjusts for the saw's number of teeth per inch.

As with jointing and shaping, setting is not always required. Under average conditions, a saw can be sharpened about three times before it needs resetting. At this point, the teeth have been filed so much that the set near the middle of the tooth is eliminated. Wet lumber requires wide set, while dry wood requires little. Set also affects the smoothness of the cut—a 10-point saw with almost no set will make as fine a cut as an 11-point saw.

To use a saw-set, index the plunger with the top half of a tooth that is bent away from you and squeeze the handle (photo facing page, center left). The plunger will bend the tooth over against the anvil. Take it easy, though; too much pressure will cause the plunger to slip over the top of the tooth. Move down the blade, skipping every other tooth just as in filing, then reverse the saw and set the other half of the teeth.

Sharpening. I was taught to sharpen by placing the saw in the vise with its handle to my right and to work from the toe to the heel, so I'll explain it this way, although you can start wherever you like. Hold the saw just ¼ in. above the top of the vise. Start with the first tooth that's bent toward you, and place the file in the gullet to its left. For a crosscut saw (photo facing page, bottom right), hold the file horizontal, point the front of the file toward the handle of the saw at a 65° bevel to the face of the sawblade, and tilt the file for

71

the hook to 15° away from vertical. For a sharper point, hold the handle of the file slightly lower than the tip. If you're cutting softwoods, change the bevel to 45°, the hook to 20° or 25° and keep the file handle lower.

File on the push stroke only, then lift slightly and return to start and stroke again. File until half of the flat top is removed if you are shaping and sharpening in one operation, or until half of a sharp point is produced if you are just sharpening. Once you're finished on this side, reclamp the saw so that the handle is on the left, begin again at the toe, and file the gullets that you deliberately skipped on your first run. When you have completed every other tooth with the saw in this position, examine the points for reflected light—sharp points will be invisible. Refile any that gleam.

Sharpening a ripsaw is considerably easier. In most cases, the file should be held horizontal, the bevel is 90° to the sawblade, and the there is no hook, which means holding the file face straight up and down. Some carpenters give the teeth an 8° hook so they can crosscut with their ripsaw if they have to.

Human eyes are not calibrating devices that display angles of bevel and hook on a scale, but when the eye and the mind form a partnership, you can make very fine judgments about consistency. The key here is practice, but as an aid in the beginning, crosshatch the top of your vise at the desired angle and hold the file parallel to these marks. There are commercial devices that hold the file at the correct angle, but I think these are a hindrance to learning the stroke freehand.

The final step in sharpening is side dressing. Old-timers did it to remove metal burrs from the sides of the teeth. Lay the saw flat on the bench or up high in the vise, and lightly run a file or oilstone down the side of the teeth, as shown in the photo below. Then repeat on the other side. I do this to reduce the set slightly.

When you have finished filing, hold the saw with the teeth up. They should feel sticky when touched with your fingertips. Sight down the line of teeth—it should be straight (or slightly convex), the points should look identical in shape and size, the gullets should form a straight line, and the set should be equal on each side. If the saw is filed and set correctly, a needle will slide down the valley formed by the set in the teeth. But the real test is how the saw cuts.

Take a flat piece of 1x6 and clamp it edgewise on a workbench. Mark a 45° line across the top, and a plumb line on the side. Start a cut across the board using long rhythmic strokes. Stop the saw in various positions and examine the kerf; you should be able to see the tiny V-cuts on each side, and they should be equally deep. If the points of the teeth are even, the saw will go right down the line with only a pushing-ahead motion. If the saw wanders and must be pushed to one side, the teeth are not even and it's back to the vise.

Using a handsaw—Hand-sawing is straightforward as long as you relax. Muscling the saw won't do you or the saw any good. With the stock laid on sawhorses, bring the saw to the cutline. Place the teeth on the waste side of the line, and brace the blade with the thumb of your free hand. Slowly draw the saw back, letting the teeth drag on the wood. This will start the cut. Push the saw lightly forward to cut a little deeper. Now move your thumb away; too much pressure on the first few strokes may cause the saw to jump out of the kerf, leaving you with some lifetime scars.

Now angle the saw about 45° above horizontal for crosscut saws and 60° above horizontal for ripsaws. Three or four moderate strokes should start a reasonable kerf so that you can begin using longer strokes. Most amateurs use short jabbing strokes. This wastes energy and sacrifices control. Instead, stand so that your shoulder, elbow and hand are in line, and the saw is an extension of the line. Use the full length of the blade with long rhythmic strokes. If the saw chatters on the return stroke, you are not pulling it straight back, or the blade is bent.

As you saw, the center portion of the blade does the most work. The toe end is just for starting and requires less pressure to keep it from jumping out of the kerf. Pressure is applied progressively toward the middle and then decreasingly until the heel is reached; then you lift and return for the next stroke. To me, this kind of motion feels more like slicing than sawing. The stroke motion is not a straight line, as it would seem. Because the elbow is lower at points than the hand and shoulder, the motion is actually a segment of an arc. Straight saws will have a barely perceptible rocking from toe to heel as they cut, while a convex line of teeth won't.

Generally, ⅛ in. is the shortest piece to cut off with a handsaw. When there is no resistance on one side of the blade, it will wander to that side. Backcutting is one alternative if you are fitting only one face of a board. The wood on the underside of the cut will provide the needed resistance. Another way is to block the end of the piece to be shortened. For example, if you have a piece of molding that is just one blade thickness too long, take a piece of the same molding and clamp or nail it in the miter box at the saw guide, butt the molding against the piece and saw through. The block provides the required resistance.

Some expert sawyers can saw square cuts and miters freehand without lines. The trick is to use the polished sides of the saw as a mirror. The reflected image in the sawblade tells you when the angle is right.

Base molding is more easily cut with a handsaw than by machine because you don't have to move the wood in and out of the box. Use a sawhorse with a 2x6 nailed onto the side of the top. Make a kerf in this 2x6 for a one-sided miter box. Then move the molding to the end of the sawhorse for any handsawn straight cuts, and finish the molded part with a coping saw. For flat moldings, perfect joints can be made by tacking a miter joint together and then sawing through the joint itself. □

Tom Law lives in Davidsonville, Md.

Side dressing. Several quick strokes with an oilstone or file along each side of the blade remove any burrs produced by sharpening. It is also useful in reducing the set of a saw slightly.

Site-Sharpening Saw Blades

It may not be a skill you use often, but it can sure come in handy

by Tom Law

It's bad enough when passersby suspect that there's a fire on your job site because of all the smoke in the air from your last skillsaw cut. But when you do finally dip into your sawbox for a sharp blade and discover that the blackened one you are using is the best you've got, you might want to overreact.

Jumping in your truck and driving all the way back to town for a case of throwaway blades isn't necessarily the answer. If you know the rudiments of sharpening a circular-saw blade, you can have your saw back in action in 20 minutes (less if you are just touching up the blade) and save yourself some money and a time-wasting trip into town.

I sharpen all of my own steel blades. Besides saving a few bucks and knowing that the blades will be consistently sharp, I get a kick out of filing as long as I can do them a few at a time during the evening. Hand-filing steel blades is not difficult, and it's not necessary to have memorized the textbook on tooth geometry to produce a better cutting blade.

Sharpening circular-saw blades is a lot like sharpening handsaws (see pp. 68-72), except that in most cases the teeth are larger and more widely spaced, so they are easier to see and file. And since the blade is driven by electricity and not your arm, getting the exact pitch and bevel angles right isn't critical if your goal is just to get the blade to cut.

Blade types—Carbide-toothed blades have become much more popular for table saws, radial-arm saws, power miter boxes and even for portable circular saws as the technology has improved and the price has come down over the last 20 years. The flats on carbide-tooth inserts can be touched up with diamond-dust stones, but once they reach the stage where they are burning through the work, they should be sent to a sharpening service for regrinding on a diamond wheel. Despite their popularity, I use carbide-tipped blades only for composition materials like particleboard, where the dulling action of the glue is best resisted by the harder teeth. I use high-speed steel blades for everything else because I don't have to rely on anyone else for sharpening. I make sure that the steel blades I buy are hard enough to hold an edge but not so hard that they destroy the file; this knowledge comes only with experience.

The types of steel blades most commonly used on portable power saws are rip, crosscut, combination and fine-tooth combination blades. There are a number of specialty blades including *planer* (a smooth-cutting, hollow-ground finish blade), *plywood* (a blade with lots of evenly spaced teeth that leave a smooth edge with very little chipping of the surface veneer), and *metalcutting* (small teeth with negative hook for cutting thin sheet metals and extrusions). Many specialty blades are just as easy to touch up with a file as common steel blades.

Basically, woodcutting blades do only two things, rip and crosscut, and the teeth are designed to do one or the other. Ripping is done parallel to the fibers, and the teeth act as small chisels plowing out the kerf. Crosscutting is done perpendicular to the fibers, and the teeth act as knives slicing through them. A combination blade has teeth that are a compromise between rip and crosscut teeth, allowing it to do both tasks reasonably well. It is used primarily for framing. Fine-cutting combination blades, such as the master combination, are used primarily for trim work. They have two types of teeth—spurs and rakers. The spurs do the cutting while the rakers remove the sawdust. These teeth aren't all at the same height—since the rakers do no cutting, they are held slightly below the spur

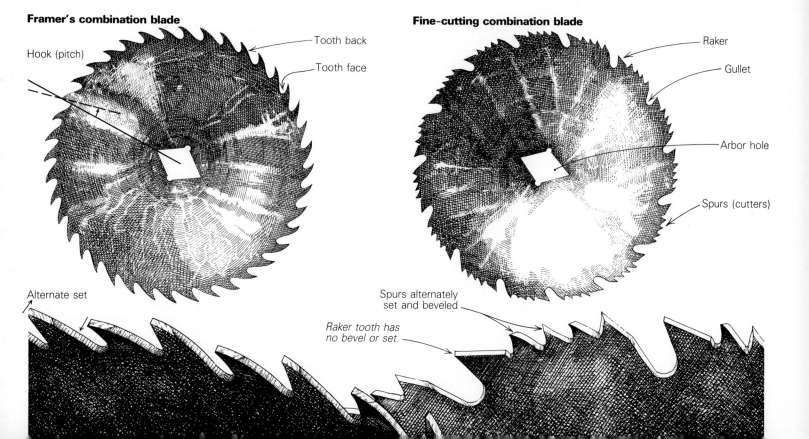

Framer's combination blade

Hook (pitch)

Tooth back

Tooth face

Alternate set

Fine-cutting combination blade

Raker

Gullet

Arbor hole

Spurs (cutters)

Spurs alternately set and beveled

Raker tooth has no bevel or set.

points. Spurs and rakers are usually arranged in segments with four spurs (set alternately) preceded by a raker. These segments are separated from each other by deep gullets.

The face of each tooth will be consistent with the others on the blade in pitch or hook angle, whether it be negative or positive (drawing, previous page). If the faces of the teeth are beveled, then these will alternate from side to side. With the exception of rakers, most circular-saw teeth also have some degree of set to reduce friction. On a planer blade, the teeth have no set, but the body of the blade is tapered toward the center (hollow ground).

Whether or not the backs of circular-saw teeth are beveled depends on the type of blade. In any case, the back of the tooth usually falls away rapidly in what's known as the clearance angle, which helps direct sawdust down into the gullet and out.

There are several reasons why circular-saw blades require attention. When teeth wear from wood cutting, the points become worn toward the center, leaving a sharper ridge on each side. Under magnification, the dulled area is blunt; the metal is rounded over rather than sharp. As teeth wear, their cutting action is slowed and more of a load is put on the saw motor. The accumulation of pitch on the side of blade can also cause a lot of friction; an all-purpose spray cleaner such as 409 or Fantastic is the easy solution for this. And then there's the enemy of any cutting edge—the hidden nail. When a blade strikes metal the points of the teeth can become severely distorted. Then as the blade turns in the wood, the jagged edges cause it to wobble and vibrate violently.

Several steps are required to sharpen such a blade. A full regimen would include jointing, setting, gumming (deepening the gullets), and shaping and repointing the teeth by filing their backs and faces. On a job site, you probably will only want to take the time to joint and repoint. But remember that repeating just these two procedures will eventually change the shape of the teeth.

Jointing—The first step in sharpening is to make sure that the teeth form a true circle that is concentric to the arbor hole. This way each tooth will be doing equal work. If you file only the teeth that have struck metal, for example, you may be making them shorter than the others, and they end up taking the day off.

I joint a circular-saw blade by removing it from the saw arbor and remounting it in the opposite direction of rotation. Then I place the saw's baseplate, in its lowest position, over an oilstone or sandpaper and slowly low-

Jointing. **With the blade mounted backwards in the saw and the shoe lowered all the way, the author just touches the edge of a sharpening stone with the blade at full speed. This guarantees a truly circular blade that is concentric with the arbor, and produces a flat spot on each tooth that acts as a sharpening guide.**

er the spinning blade (photo facing page). A light touch is critical. Too much pressure will gouge deep ruts in your stone, overheat the steel (possibly ruining its temper) and send chunks or grains of stone flying. Slowly lower the saw until the blade barely touches; when the sparks stop, lower it again. After doing this a few times, examine the teeth to make sure each one has a small flat spot on the top.

Jointing as an operation by itself can improve the cutting action of a blade that has just come into contact with a nail. When I'm cutting through old strip flooring (something I do often in remodeling), I expect to hit nails. Old floors were put together with cut nails—lots of them. I move the saw slowly along the cutline, and as soon as I hear the familiar screech of metal abusing metal, I lift the saw. Moving ahead slightly, I lower it again without trying to cut through the nail. When the blade will no longer cut I take it off, turn it around and joint it. This removes the flared burrs on the part of the tooth where the point used to be and restores the blade to round. Although I haven't really sharpened it, the blade will cut and I can finish the job, jointing whenever it's necessary. This trick allows me to use just one blade, and even if it is destroyed in the process, which is not likely, it's cheaper than sending a carbide blade out to be sharpened.

Repointing—Once you've jointed the blade and removed it from the saw, you're ready to repoint it. The blade has to be secured for this task. The handiest workbench on a framing job is a rough window sill. A scrap of 2x4 or 2x6 can be used for a saw vise. Drive a nail through the block and then through the arbor hole into the sill, sandwiching the blade. If you want to get a little fancier you can purchase a metal saw vise. The one I use is a small handsaw vise that clamps by means of an eccentric roller bar. The ideal height for filing the sawblade is about 4 in. below your armpit. Indirect sunlight is the best illumination for this.

When I file circular saws I do the whole tooth, trying to maintain the same shape each time. The idea is to reproduce the original configuration of the teeth as best you can. To get the best results, of course, you need specialized sharpening equipment and a knowledge of tooth geometry. But if you're just trying to turn a useless blade into a serviceable one, this method will work.

On rip or combination blades select a tooth that is set away from you, and file the face first. Although some larger-toothed blades will accommodate a mill file, the file I like best for this job is a crosscut. In section it is the shape of a teardrop with teeth all around, including the large round edge. Place a flat side of the file on the face (front side) of the tooth and cut only on the push stroke, two or three times (photo top right). If the file matches up in size with the gullet, you will be able to kill two birds with one stone with these strokes by gumming at the same time you're filing the face of the tooth. File straight across on the push stroke until half of the flat spot made by

Repointing. **Using a crosscut file (right), which looks like a teardrop in section, you can often file the face of the tooth and deepen the gullets (gumming) at the same time. Two or three long, even forward strokes should be enough. Make sure to follow the already established hook and bevel. Repointing (middle right) is completed by removing the flat spot created by jointing on the back of the tooth. When the gullets are small, a chainsaw file can be used to deepen them (bottom right). Although you may not find yourself doing this on site, if the gullets get too shallow sawdust can begin to accumulate in the blade, causing increased friction.**

jointing is worn away. On blades with a 90° tooth face (no bevel), such as rip blades and some framing combination blades, this stroke should be perpendicular to the body of the blade, but you'll find that holding the file at a very slight angle will reduce the drag on it considerably. If the tooth face is already beveled, duplicate the existing angle by eye.

Next, you're going to file the top of the tooth to bring it to a point (photo middle right). Whether or not this area is beveled depends on what kind of blade you are filing. Most cutting teeth (but not rakers) are beveled on top in the direction of their set. To reproduce this bevel, start the file in a horizontal position with its handle perpendicular to the blade itself. Then adjust the face of the file to the bevel by dropping the handle down and moving away from perpendicular. Continue with every tooth that is set away from you, and then reverse the blade and file the remaining half of the teeth.

For larger crosscut teeth or combination spurs and rakers, I use a cant file, which is triangular in section with angles of 30°, 120° and 30°. Adjust each segment in the vise vertically and file those spurs set away from you, keeping the hook and bevel correct. The rakers are filed straight across just $\frac{1}{64}$ in. below the spur points. I always file the rakers on the first half no matter which side comes up first. Fine-tooth crosscut blades (including planer blades) should be done with a taper file, which is an equilateral triangle in section.

A few refinements—Circular-sawblade setting is normally done with a sawset just like handsaws. Although you are not likely to be doing this on the job site, you can revert to the traditional hammer-and-anvil method if one or two teeth are way out of line. Use the edge of a concrete slab, or even better, a heavy steel hardware flange or an I-beam for an anvil. It isn't necessary to gum the blade every time you sharpen its teeth, but making sure that the gullets are all the same depth is essential. To ensure good results, draw a pencil line around the face of the blade with a compass to indicate where to stop filing. Make sure that this circle is concentric with the outer one that is formed by the tooth points. Round bottom gullets can be filed with the round edge of a crosscut file, a round end-mill file or a chainsaw file (photo bottom right). □

Tom Law is a contractor in Columbia, Md.

Power Miter Saws
Crosscutting trim with accuracy and speed

by Geoff Alexander

The best of the traditional miter boxes are beautiful and elegant tools. For generations, craftsmen have used a backsaw and miter box to trim out houses; and in skilled hands, these tools do nice work. But power miter saws have all but eliminated the miter box as a finish carpenter's tool. Sometimes called chop saws, power miter saws are faster, more accurate and better able to withstand job-site abuse than miter boxes. With a stiff-bodied, carbide-tipped blade, a power miter saw can shave a minute amount from a cut you just made to give you a precise fit. But with a traditional miter box, you've got to get your cut exactly right the first time, because a backsaw won't cut true if it's not supported on both sides by the wood.

There are several makes and models of power miter saws, but they all have the same basic setup—a table about 5 in. wide by 16 in. long and a vertical fence at the rear to position and hold the work. A sawblade and motor are mounted to a spring-loaded arm that terminates in a handle and trigger-switch. To make a cut, the arm assembly is pulled down so that the blade passes through the work. When downward pressure on the handle is released, the arm assembly returns to its rest position above the work. The entire arm assembly is pivot-mounted so that the blade can be set to cut any angle from 45° right to 45° left, or sometimes a little more.

Power miter saws find many applications on the construction site. Plumbers working with plastic pipe often use them to create fast, clean joints. Since non-ferrous metals can be cut safely on a power miter box fitted with a carbide blade, I frequently use mine to cut aluminum extrusions such as sliding-door tracks, shower-door parts and trim for a tub surround. For cutting exclusively non-ferrous metals, use a negative-hook carbide blade, wear safety glasses and cut slowly.

The power miter saw's intended function, and the job at which it is truly extraordinary, is cutting wood trim. Its portability and fast setup time, the ease of setting the angle of cut, and its ability to make tight-fitting joints quickly make it the best choice for cutting inside and outside miters, butt joints on baseboard, door and window casing, chair rails, crown molding and other pieces of wood trim.

Its chief virtue, however, is its ability to shave ⅟₃₂ in. off the end of a board, or to modify the angle of cut by a fraction of a degree,

Alexander keeps his Rockwell miter saw mounted on a short 2x12, which makes it easier to carry and gives it a stable base. He made a pair of work-support tables whose tops are level with the saw table. Cutting long unsupported pieces is difficult and dangerous.

and to do these things without sacrificing any of the crispness or accuracy of the cut. When I began working as a carpenter in 1970, one of the first things I learned was that professionals make almost as many mistakes as amateurs. But professionals are very fast at fixing their mistakes. In this context, the power miter saw is truly a professional's tool. Its capability of correcting errors and still achieving a perfect joint with little wasted time or material justifies the saw's cost to anybody doing professional-calibre finish carpentry.

What's on the market—When I bought my first power miter saw in the early 1970s, I got a Rockwell 34-010 because it was the only one on the market. It has a 9-in. blade, a particleboard table and a manual blade brake. By the late 1970s, Black & Decker and Makita were each making a 10-in. saw. Makita made several noteworthy improvements, first with its 2400B (10 in.) and then its LS1400 (14 in.). Both Makitas have a slotted steel table that pivots along with the arm assembly when setting the angle. Both have an automatic electric brake, and a dust bag. Accessories include table extensions, an adjustable stop, and a vise for clamping the work in place.

The Black & Decker (#3090) also has an automatic brake, but the saw itself is of lightweight construction and has a particleboard table. Black & Decker is discontinuing model #3090 and replacing it with an upgraded 10-in. saw (model #3091) that has a pivoting table. Hitachi's current models (the C10FA 10-in. saw and the C15FA 15-in. saw) have an automatic brake; a dust bag, table extensions and a vise are standard equipment. These saws have something unique—a phenolic-resin fiberboard insert for the slot in the table, and a carrying handle.

In my estimation, Sears power tools are designed for the home owner rather than the tradesman (although that wasn't always true), and judging by look and feel alone, their power miter saw is meant for the consumer market. Of possible interest, however, is Sears' radial miter saw—a 7½-in. compromise between a power miter saw and a small radial-arm saw. I think its design has definite potential. Rockwell's Sawbuck (see *FHB* #12, p. 14) is a much more fully realized and versatile version of the radial/miter saw concept. If the Sawbuck can withstand daily job-site use, it may well be as important an innovation as the original power miter saw.

The blade brake—At this point, all of the power miter saws except Rockwell's have an automatic brake on the blade. As soon as you release the trigger, the brake brings the blade to a stop within a few seconds. Rockwell's manual brake is operated by pushing a thumb button on the handle. The button actuates brake shoes that grip the spinning shaft. Having worn out three sets of brake shoes ($8 a whack if you replace them yourself, about $35 if the repair shop does it), I now try to minimize my use of the brake. But since I often have to work near the blade right after cutting to remove scrap or to align the next cut, and since the saw is a real screamer (5,000 rpm under load), I use the brake a lot anyway. And that's why I prefer automatic brakes.

Pivoting vs. stationary tables—Because a pivoting table (actually it's a round baseplate insert in the center of the table) stays in the same relation to the blade whatever the setting, it can have a permanent slot cut into it to receive the blade. With a wooden table, the blade actually cuts into the table surface. The advantage of the wooden table is that the slot is exactly the width of the saw's kerf (when the table is new), so the workpiece is firmly supported on both sides of the cut. This minimizes splintering on the bottom of the work where the blade passes through. The disadvantage of the wooden or composition table is that each time you set the saw to a new angle and make a cut, a little bit more of the table gets eaten away. Also, any minor widening of the existing slots in the table lessens the support behind the cut and increases the likelihood of splintering. If the only angles you cut are 90° and 45°, a table can last a long time, but I usually end up making mincemeat out of mine on each major job. Frequent replacement or resurfacing of the wooden table helps keep the splintering under control.

With pivoting tables, you don't have to worry about the blade chewing up the surface. But because the slot in the rotating metal baseplate is wider than the blade, the workpiece doesn't get support where the blade exits, and so your wood can splinter out on the back side of the cut. Hitachi addresses this problem directly by filling the slot with a replaceable, phenolic-resin fiberboard insert. Since this is a new feature I've never used, I can't say how long each insert will last. You will certainly need to change inserts if you change to a thinner blade.

Setting the angle—Each saw manufacturer handles the angle setting a little bit differently. Rockwell put positive stops (plunger in slot) at 45° and 90°, and at 22½° on the new machines. To change from one detent to another, you can move the handle to about the right place and let the plunger fall into place. After many years of heavy use, the settings are still reliable on my saw. To set any other angle on the Rockwell, you pivot the arm to the right spot and tighten a thumbscrew down on the protractor scale. Unfortunately, the screw has a tendency to pull the whole arm assembly slightly to one side as it tightens. This throws off the setting. I often tighten the thumbscrew lightly to avoid changing the

angle, and then move very carefully to avoid bumping the saw and throwing off the setting. This strategy usually works if I'm making only one cut, but isn't always successful for making multiple cuts at the same setting.

The Makita saws (except the model 2401B) have no positive angle stops. Each angle setting must be locked in place by tightening the same handle with which you set the angle. It clamps horizontally against the side of the protractor scale, and does so with no shifting. When I first used the Makita, I thought that the lack of positive angle stops would be a nuisance, but I have found that not to be so. The handle lock works quickly and reliably, and the protractor scale is easy to read. Best of all, it's easy to set the saw for 44½°—a real problem on the Rockwell saw because the detent-and-thumbscrew arrangement almost always makes the setting drift. Most other power miter saws do have positive stops, but I have not used any of them enough on the job to know their idiosyncracies.

Size of the saw—All of the manufacturers except Rockwell make a 10-in. saw. For the purpose of comparison, Rockwell's 9-in. saw performs like the 10-in. saws. Hitachi's 15-in. saw (C15FA) and Makita's 14-in. saw (LS1400) are in a class by themselves, however. Substantially larger and heavier than the other saws, they can also cut much larger stock in a single pass. But for ordinary finish carpentry, I don't see much use for the big saws. And as we shall see, even the 9-in. Rockwell can be used to make square cuts on a 2x6, and can cut any type of miter on a 2x4.

Without removing the blade guard on the 14-in. Makita, the widest board you can cut is

8 in. The portability of the smaller saws is an advantage while doing finish work, both for moving from room to room and from job to job. Because of their size, beefier construction and slow blade speed, the big saws are best for gang-cutting green 2x4s or 2x6s, or for cutting 4x4s and 4x6s. But Makita and Hitachi use the same motor on their big saws as they do on their 10-in. saws, so don't expect to get extra cutting power.

Several saws have means to lock the pivoting arm in a safe down position, which makes carrying the tool a little easier. Hitachi is the first to have a separate carrying handle. I often carry my Rockwell by the main handle if the blade is locked down, but the Makita is hard to carry one-handed. The arm on the new Hitachi can be held down by a flimsy chain—an arrangement that seems inadequate to me. The table extensions on the Japanese saws are sometimes useful, but I almost always leave them off because they are in the way when moving the saw, and they are too short to be helpful when cutting long boards. The Makita and the Hitachi have a stop gauge that fastens onto the extension arms for repeat cuts. But most often the stock you're cutting is too long or too short for the stop gauge to be of any use.

Setting up the saw—On the job site, everything that is done is a compromise between the right way and the fast way; setting up the miter saw is no exception. In the "right-way" universe, the saw is at waist height, there's 6 ft. of solid auxiliary support on each side, the blade has just been sharpened, the floor's clean, and there's a cold beer waiting at the end of the day. Back in the real world, though,

If Alexander is going to be on site for an extended period, he brings his two tool-drawer units, which sit on either side of the saw and support the work. Note that the plywood saw table is shaped to prevent the kerf from breaking through the edge. This keeps the wood in one piece and makes a better surface to cut on.

we don't always have the time to set it up that way, so here are some techniques that should help ease the compromise.

One thing you should never do is cut a long board without supporting it at the far end. In the hustling pace of the job, it's tempting to try holding a long board to the saw table with nothing more than hand strength. About eight years ago, I tried to hold a 10-ft. piece of trim with no outrigger support. My fingers were curled around the fence to clamp the board down, which worked fine until I made the cut. Then, having lost the counterweight of the off-cut end, the piece started to fall, and in an attempt to save it, I grabbed the spinning sawblade instead, and left a quarter inch of my middle finger behind in the sawdust.

When I'm moving from job to job, I always carry with me one or two little auxiliary tables with 6-in. by 10-in. plywood tops that are mounted on two lengths of 2x stock, as shown in the photo on p. 76. These tables are the same height as the miter-saw table, and let me use it anywhere I can find a continuous surface to work on (including the floor). My Rockwell saw has been mounted on a 2x12 for several years now. The board is easier to clamp down than the saw alone, and it lets me carry the saw on my right shoulder while I'm carrying the tables in my left hand. This way I can carry a basic setup on, off or around the job in one trip.

With the auxiliary tables, I can adjust to almost any situation. If there is a long, flat 2x12 and a pair of sawhorses available, it's fast to set the board across the horses and flop the saw down. Then, by positioning the auxiliary tables on either side of the saw, most lengths of stock can be handled easily. But if nothing springs to hand, and there aren't too many cuts to make, put the saw on the floor, get down on your knees, and get to work. By the time you make some elaborate setup, you probably could have finished all the cuts.

If I'm going to be on a job for a while, however, I want a comfortable setup. There are several ways to make one. When I'm really

moving into a job, I have several sets of tool drawers with casters mounted on one side that I take along. Each one is about 2 ft. square and about 3 ft. tall. I also have a table built to the right height so that when the miter saw sits on the table, its cutting surface is the same height as the tops of the drawer units. This setup (photo previous page) supports the workpiece on both sides, and offers small work surfaces and drawers for tools, hardware and sandpaper.

Sears and Rockwell both sell steel stands as accessories for their saws, but it's easy to build your own out of wood. With the power miter saw sitting on a stand, the saw is at a comfortable working height, but unless you're building doll houses, you'll need a way to hold up the ends of long boards.

One contractor I know built a stand for his Makita 2400B, and then made two detachable wings. Each wing has an oak 1x6 for a table and an oak 1x2 for the fence. To fasten the wings to the saw, he bolted short lengths of angle iron to the saw itself, and then tapped the angle iron to receive flathead machine screws through the table of the wing. Each wing then had one leg of appropriate length hinged to its table so the leg would fold flat for storing and carrying.

If you screw the miter saw to the center of a long board (say a 10-ft. 2x12), and then fasten tables to the board on each side of the saw to get support at the correct height, you can create a stable work surface to which you can affix stop blocks for repeat cuts. This system (see *FHB* #5, pp. 5-6) requires sawhorses to set the board on and takes up a fair amount of space, but it offers a solid work surface. Biese-meyer, Inc. (216 South Alma School Rd., Suite 3, Mesa, Ariz. 85202) makes an outrigger support with an integral stop gauge called the T-Square Miter Saw Stop System.

Choosing a blade—All of the power miter saws come with an ordinary steel blade. The first thing you should do is take it off and put on a carbide-tipped blade. The teeth on car-

bide blades are accurately aligned so they make cleaner, more precise cuts, and the tooth form stays practically the same after each sharpening. Carbide-tipped blades can cut aluminum, and some other non-ferrous metals and plastics, as well as other abrasive woods and wood products like particleboard. They stay sharp about 40 times longer than steel-tooth blades. If you plan to cut only wood, the alternate top bevel (ATB) is a good choice of tooth design, but if you expect to use the saw periodically for aluminum or plastic pipe, the triple-chip tooth design is stronger and more durable. With either style, a 9-in. or 10-in. sawblade with 40 to 60 teeth will give good results. Expect to spend $1 per tooth or more for a good carbide blade. Some cheap blades have improperly tensioned plates, or teeth that are poorly fastened to the blade. There should be at least ⅛ in. of carbide on each tooth of a 40-tooth saw. The price, the warranty and the manufacturer's reputation are the best indications of quality.

Making cuts—Keep the table clean. Watch out for sawdust accumulating where the table and the fence meet. Even a little dust here can cause your workpiece to misalign and give you a bad cut. Take special care to keep small offcuts out of the blade's path. If the blade catches a loose scrap, it can hurl it like shrapnel in almost any direction.

Mark your cut on the workpiece where the saw will enter the work. If you're cutting a 1x4 laid flat, the saw will enter the middle of the board; if you're cutting a board positioned on edge, you'll have to mark the cut on the top edge. As long as your pencil mark is near where the blade first enters the stock, there is no need to square it across the entire board. Make sure that the work is in good contact with the table and the fence, and that it is solidly supported at the end. To align the cut, I usually start the saw and then sight along the spinning blade as I pull it toward the work. As the blade enters the work, I move my head to the left so that the flying sawdust will miss my

Casing a window

In an ideal world, two 45° angles add up to 90°, but in the real world of construction, it doesn't always work out so neatly. Walls, ceilings and floors may all be out of square or out of plumb. One job that often reveals framing inaccuracies is casing a door or window opening, especially if the casing is a wide or complex molding. I correct the inaccuracies as I go, without wasting time or material. I always work from the bottom up. If the casing goes around all four sides of a window, picture-frame style (instead of sitting on top of a window stool), I cut and nail the bottom first, leaving a constant reveal between the casing and window frame. Then, for each side, I make a trial cut at 45° and test-fit it, holding it with the same reveal and snug to the wall and jamb. If it fits perfectly, I mark the length and cut another true 45° at the top.

If the angle isn't perfect, I recut the miter, leaving the saw at the 45° setting, but shimming the material at the fence. To help me gauge the thickness of the shim, I compare the length of the miter cut to the length of the fence on the miter saw, and adjust accordingly.

For example, let's assume I'm using my Rockwell to cut 1x4 trim. The length of the miter cut is about 4⅞ in. The length of the fence to the left of the sawblade is 8¼ in. Assume that the original cut left a gap at the

heel of the miter that I eyeballed to be ⅟₃₂ in.. Then, since the left-hand fence is roughly twice as long as the miter, I'd want to use a shim that was about ⅟₁₆ in. thick placed against the fence right next to the sawblade. This would change the angle of the cut slightly, just enough to shave the toe of the miter to close the gap at the heel and get a good fit. I very seldom measure any of these things. Eyeballing becomes much more accurate with practice.

If the angle looks correct, but the miter gaps open on the face side of the trim, I correct it by elevating the board with a shim placed on the table near the blade. Then I recut the miter—in effect, back-beveling the cut, taking care not to change the edge where the original cut meets the face of the trim. If the casing has a deep molded shape, I back-bevel with great care to avoid changing the shape of the face edge of the cut.

Once I've got both lower miters fitting properly, I cut the side casings to length using a true 45° cut, squirt a little glue into the joints, and nail them in place. I usually leave the top third of the trim unnailed until I have cut and fit the head casing, and then always put at least one nail across the miter itself. To fit the head casing, I cut one miter to a perfect fit first, then make the length cut and any necessary adjustments to the second miter. Then I glue the miters and nail the trim in place. —*G. A.*

face. Then I check the accuracy of the cut, and make an adjustment if needed. In any case, the blade guard is almost always in my line of sight, an issue I will discuss below.

If you're doing rough work, you can feed the blade through the wood as fast as you can go without taxing the motor. But if the appearance of the cut is important, it's better to use a slow, steady stroke. The blade can pull splinters out of the top of the cut on the upstroke, so if that's a concern, hold the blade at the bottom of the cut for a moment and pull the stock away from the blade before you let the blade come back up.

Sometimes, no matter how smoothly you cut, the bottom of the cut will splinter. The best way to minimize the splintering is to put a smooth piece of wood on top of the saw's table as an auxiliary table and place your work on that, show side up. For my Rockwell, I make up several auxiliary tables at a time, in a shape that allows the wood of the table to extend beyond the saw cuts.

Pushing the limits—The blade does not achieve maximum cutting width at table height. But adding an auxiliary table that's about 2 in. thick greatly increases the cutting width (photo top right). Rockwell claims a crosscutting capacity of 2½ in. deep by 4 in. wide, but by raising the table 2 in., you can cut a board 2½ in. by 5 in. wide; or with a thicker table, 1½ in. by 6 in. wide. But if you have only a few wide cuts to make, you can bury the saw in the work and then lift the front edge of the board up into the saw until the cut is complete. Or you can cut to full depth from one side of the board and then flip the board over, eyeball the kerf straight across, and finish the cut from the other edge.

If you're working on edge, there are two advantages to be gained by putting a spacer between your work and the fence. If you come out about 3 in. from the fence, you can fit a wider board under the saw when the saw is in its uppermost position. The very deepest cuts can be made with the stock positioned so that the arbor washer just clears the inside of the work (middle photo at right). The spacer must be a uniform thickness so it will be precisely parallel to the fence, and high enough to support the work on its back side.

Using the miter saw to work on the edge of a board, you can cut fairly intricate shapes. Say you've got to fit a piece of 1x4 to the base of a round lighting fixture, and your client, eager to help with the job, has just broken your last saber-saw blade. To make the cut on the miter saw, space the work away from the fence so the blade cuts horizontally at the deepest part of the cut. Then make repeat cuts just shy of the line on the face of the work (bottom photo at right). After removing most of the stock, you can sculpt the wood by using one hand to raise and lower the saw while the other hand slides the work back and forth along the auxiliary fence. Working this way, it's best to remove only about ¹⁄₁₆ in. of wood at a time.

One year, I did so many latticework jobs

The saw's width of cut can be increased by shimming the work up from the top surface of the table. This means wide trim can be cut in one motion, without having to be flipped over to finish the cut from the other side.

The saw's depth-of-cut capacity for working with thick pieces and framing members can be increased by spacing the work out from the saw's fence. The spacer should be sized to position the work so that the arbor washer just clears its back edge. The spacer should be a uniform thickness for an accurate cut.

Notching the edge of a board on a power miter saw can be done by making a number of downward cuts to a scribed line. The cut is cleaned up by raking off small amounts of wood with the blade.

that I wrote charts of lath-lengths on the top of one of my miter-saw work tables. It's easy to stack six or seven lengths of lath on top of each other and cut them all at once, and so I could just sit at the saw and make "kits" while somebody else nailed them off. It's a good idea to cut the very short pieces from longer pieces so you don't have to hold the wood too close to the saw.

Safety—For me, the thorniest safety issue concerning power miter saws is whether or not to use the blade guard. All the saws come from the factory with a blade guard that in some manner shields the blade. But in order to make accurate cuts, you must have a clear view of the blade at the moment it enters the wood. Unfortunately, most of the blade guards get in the way. On my Rockwell, the original guard was made of yellow translucent plastic. When it was new, it was very hard to see through, and after a few months of use, it was so dinged up that I took it off, and eventually threw it away. Rockwell's new guards are transparent, but even so, the visibility diminishes rapidly as you use the saw because fine sawdust clings to the plastic.

Even after shortening my finger on a saw without a guard, I still prefer working without one. Had I been using the blade guard at the time, it would not have interceded on my finger's behalf.

The blade guard is still on my Makita, but I keep a piece of wire handy to tie it back. One has to make one's own peace with guards. They probably have prevented some accidents, and perhaps caused as many by obscuring the blade. I don't like them, and I suspect that the things I do in order to see around them are at least as dangerous as working without them. But I wouldn't recommend removing a guard to a novice.

When the saw is cutting at 45°, it has a tendency to pull the work into the blade. If the blade is dull, the pull is even stronger. If you're cutting short pieces, and have to hold the work in the acute angle the blade makes with the fence, get out the vise if your saw has one. If it doesn't, find a way to clamp the work to the table or fence.   □

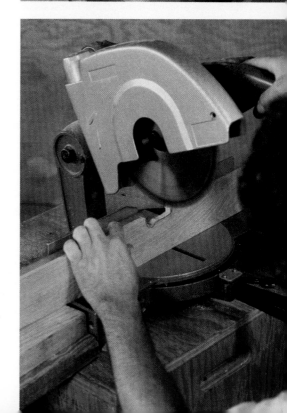

Geoff Alexander is a carpenter and woodworker in Berkeley, Calif.

Plumb Bobs, String and Chalkboxes

Working with string and the tools that hang from it

by Trey Loy

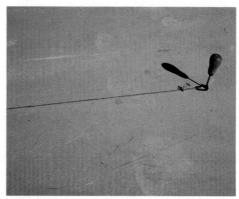

String is pretty basic stuff, but there is no end to the little tricks that make using it easier. If you're cutting gable gypboard or laying out walls, try a small, sharp scratch awl instead of a nail for holding one end of the snapline.

Building, it will surprise no one, is based on geometry and trigonometry. Points are established by measuring, these are connected to form lines, the lines are grouped to form planes, and these planes join to form a solid. Great—in theory. But this whole process relies on establishing straight lines with the right relationship to each other, and keeping them that way as you fill in the outline.

Levels and straightedges are good for short lines. But for long spans, a length of string, whether it is stretched between two nails, suspended by a weight on a plumb line, or coated with chalk and snapped against something, is indispensable. String can be used for many things: to plumb and level your work, to define points in midair, to tell you what's straight and what isn't, to establish grades, to align walls and floors, to make a circle or project its center up or down in space, and to mark your work for cutting or assembly. And if things aren't going well, you can always use the first 100 feet or so to go fishing.

String—String, or twine, is thicker than thread and thinner than cord. Twine is made from natural hard-leaf fibers such as sisal and manila, or from cotton or synthetic fibers like nylon, Dacron and polypropylene. The fibers are drawn into slivers, counted, and spun into threads that are twisted or braided.

Coarse cotton line is ideal for chalkboxes and will do for a plumb bob. Cotton fibers stretch very little, but they rot and mildew around water and cement products, and are easily abraded. Nylon, on the other hand, doesn't absorb water readily and is alkali resistant. Nylon string is also elastic, which is an advantage in stringing a line because you can get it very taut.

Nylon twine is either twisted or braided, and comes in twenty-some sizes ranging from a thin #3 to a thick #120. In carpentry and masonry, you'll be fine with one of two sizes: #15, which measures a skinny 1/16 in. in diameter, and has a breaking strength of 120 lb.; and #18, which measures a fat 1/16 in. and will withstand a 170-lb. force. The twisted version of either of these sizes of nylon line is pretty inexpensive, costing around $3 for a 350-ft. roll, and is adequate for laying out foundations and lining walls. It will stretch up to 8% of its length, and return to normal when released.

Braided nylon twine is a favorite with masons, because it's more durable and easier to work with than twisted line. It costs about twice as much, but it stretches less. One kind of braided string is even heat set for minimum stretch.

Nylon string comes in three colors: white, yellow and green. Green is hard to spot in a background of grass or shrubs, but it is the best color if you've got fishing on your mind. White is very popular, but I like yellow for its high visibility. All three colors are sold in lengths of 250 ft., 350 ft., 500 ft. and 1,000 ft.

One last kind of braided nylon worth mentioning is bonded nylon line. It is woven with an extra thread that is usually a different color and fiber. This bonded thread contributes strength and durability, and gives the twine a flecked appearance that makes it more visible than a solid color. It is treated to make it less slippery, and costs slightly more than braided line.

There are two other synthetics used in construction. For big commercial sites and highway layout, braided Dacron line is often used. It runs about $16 per 500 ft. Dacron has twice the breaking strength of nylon, but isn't elastic. Polypropylene is also very strong, but it's slippery and stiff, and doesn't hold knots well. Like Dacron, poly won't stretch. But it will float.

Securing a line—Knowing how to tie a variety of knots is important to me as a carpenter, yet I notice that many of the people I work with aren't sure how to proceed, and use knots that only hinder their effort. A good knot is not only simple to tie, but also easy to untie. One of the first places you'll need one is when you form a loop at the end of the string. A common knot for this is the *bowline,* which is shown at the top of the facing page (A). The resulting loop can be slipped over a nail or an awl. You can also slip the standing part of the line (the string back down the line from the knot) through the bowline to create a simple slip knot.

String lines aren't much good unless they are taut, and using an elastic line like nylon makes that possible. The knot that holds the tension can't slip, but you should be able to release it without much fumbling and you shouldn't have to cut the string to tie it. A *twist knot* is the knot generally used by carpenters. It's formed by looping the string around your outstretched fingers once, and then twisting the loop three or four times (B). Place this loop over the nail, and stretch the string tightly with one hand, while pulling the excess through the twists with the other. Keep up this routine—heave and pull in the slack—until the line is singing. Sometimes it helps for your partner to pull from the middle of the span. Secure the knot by pulling the free end of the string back toward the nail. This will cause the twists to bunch up next to the nail, overlap themselves and create lots of friction. A lot of carpenters tie a couple of *half hitches* around the nail for security, but these are hard to loosen later. Instead, pass the standing part of the line around the nail once, making sure that it is sitting under the twists—this will provide all the friction needed to keep the string taut.

There's another simple knot that will keep a string tight. I call it a *tension hitch* (C), and I use it when I'm stringing lines between posts or stakes. Just take two wraps around the stake, making sure that the part of the line under tension is laid over the two turns. This combines the elasticity of nylon with the friction between layers of string to hold the hitch.

If you are stringing multiple lines, as you would laying out a foundation, don't cut the string, but take the spool to the next batter board, paying the line out as you go. Use a *clove hitch* (E) to secure it to the next nail, since it will cinch down no matter which end of the line you pull. You tie it by forming two consecutive underhand loops in the line, and laying the second loop on top of the first. The combined loops should then be slipped over the nail, and the ends pulled taut in opposite directions.

Although there isn't any reason to cut the string after making any of these knots, you will sometimes have to join two pieces of string. This is most easily done with a *surgeon's knot* (D). This is merely a square knot with an extra turn taken in the first overhand knot.

Once all of the knots have been untied, a lot of string still gets thrown away, because it's such an effort to wind it up in an orderly way. You

A. Bowline—*The best knot for forming an end loop that won't slip.*

1. Make a loop near end of line.

2. Pass free end up through loop.

3. Pass the free end around the standing string and back down through the loop.

B. Twist knot—*The knotless knot for making lines taut and securing them around a nail.*

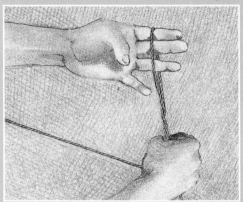

1. Form a loop, twist it around your outstretched fingers three or four times, and lay it over the nail.

2. Tighten the line and retrieve the slack through the knot.

3. Pull the free end of the string back toward the nail and finish off the knot by pulling the free end back under the twists.

C. Tension hitch—*An easy knot to hold an elastic string like nylon taut between posts or stakes.*

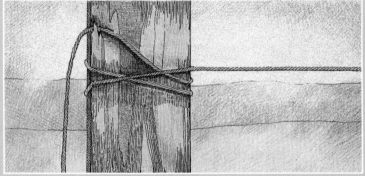

1. Make two wraps around the stake. Recover the slack so the line is taut. Make sure that the end of the line that is under tension crosses over the top of the two wraps.

D. Surgeon's knot—*Used for joining two pieces of string in line. Simply a square knot with an extra turn in the first overhand knot.*

E. Clove or builder's hitch—*Used to fasten the middle of a string so that tension can be brought from either end.*

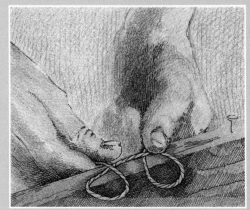

1. Form two consecutive underhand loops in the standing part of the line.

2. Lay the second loop on top of the first.

3. When slipped over the nail and tightened, the first loop should cross over the standing part of the line and the other one pass under.

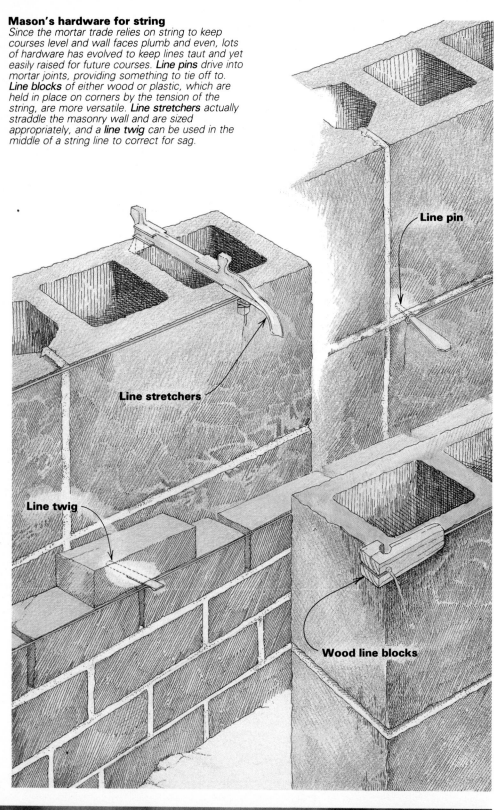

Line pin

Line stretchers

Line twig

Wood line blocks

can buy plastic winders that look like the letter H or make one out of a 1x4, but I was taught to use a 1x2 or a piece of pipe about 10 in. long. With it, you can imitate machine-winding, although your figure-eights aren't going to lay in quite so nicely. After building up a small core of string on the stick, hold it loosely in the middle and twirl the top of the stick in the direction of the string line so that it wraps around once, and then twist the top away so that the bottom gets a wrap. If you rotate the stick in your fingers at the same time you are twirling it in the air, you'll be able to distribute the wraps all along the stick so that you don't get one big ball in the center. When you're finished, secure the loose end by tying it to a nail and sticking it in the string.

Hardware—Masons rely on string to keep their courses straight and level, and they use a variety of fasteners to keep it taut and in place (drawing, left). But unlike most of the string lines that carpenters run, a mason's line has to move up every few courses. One method of securing the line is to use *line pins*. These are steel wedges (tempered ones are best) about 4 in. long that are driven directly into the mortar joint of a built-up corner, and used for tying off the string.

With block and brick, most masons use *line blocks*. These are small blocks of wood or plastic rabbeted on one side to form a heel that fits around a corner block or batter board. They are used in pairs, and the tension of the string between them keeps the blocks in place. A void in the inside corner of the block allows you to secure the string without affecting the block's grip, and a lateral groove on the inside face holds the tensioned string in place. Wood blocks grip the wall best and cost about 50¢ apiece. Plastic ones last longer but cost four times as much.

Line stretchers can also be used for block. These are made of steel or aluminum bar and have knobs on both top and bottom spaced so the stretcher will fit tightly across walls of two widths. The knob that fits over the front face of the wall is notched to hold the string. Line stretchers run about $7 a pair and come in several sizes. Adjustable models cost a bit more.

Even an elastic line like nylon will sag in the middle from its own weight on long walls despite heroic attempts to tighten it. This is where *line twigs* come in. These are flat metal line-supports that clip around the string like hairpins. Once attached to the string, the blade of the twig can then be set on top of a brick or block that has been laid to the working height in the middle of a wall. A loose block or brick can be stacked on top of the twig to hold it in place. Long walls may require more than one twig; they cost about 20¢ apiece.

A lot of masons use corner poles that can be set at the beginning of the job and left until the last course is laid (see *FHB* #15, p. 46). They combine the advantages of line blocks and story

A line level doesn't give a highly accurate reading because of the sag it creates even in the tautest string lines, but it's a useful tool for rough layouts and grading. This aluminum-clad version runs about $3.

poles. The strings are moved up the pole in course increments as the work progresses. Corner poles are available with attachments for both inside and outside corners, and in freestanding models or with telescoping braces. The cost of these aluminum or steel guides is prohibitive unless you do a lot of masonry.

Most of the time, I rely on a builder's level or tubular water level to set my level lines (I make sure that my batter boards are exactly level before I ever get the string out), but for very rough layouts a *line level* (photo facing page) will suffice. It is a lightweight aluminum tube about 3 in. long that is fitted with a level vial. Little hooks project from the tube so that it can be hung from the string. Line levels run about $3. Remember that the line sags somewhat of its own weight, and slightly more with the level, so the reading will only be approximate.

Strings for laying out—I use string the most when I'm laying out foundations. In this process, the strings form a full-scale drawing of the foundation plan. Begin by building batter boards just back from what will be the corners of the building. They should all be level with each other. Then run string lines between the batter boards, and adjust them until the dimensions of the building are correct, and the strings are square to each other (see *FHB* #11, pp. 26-28 for a more detailed explanation of this process). You should set up additional batter boards to line up piers, post brackets or any other hardware that needs accurate placement.

By plumbing down from the perimeter strings, you can establish footing lines with loose chalk. Before removing the strings from the batter boards for the backhoe, clearly mark their final position with a single nail or saw kerf. After the trenching is completed, the lines can easily be replaced for forming.

String is also helpful when you're grading fill for a concrete slab. I run lines 4 ft. or 5 ft. apart just above the ground at finished concrete height, measuring down the thickness of the slab to check the level of the aggregate. A handy gauge for grading this way is to use an eyelet on your workboot that is the correct height to the string from the top of the gravel.

Gauging straightness—Strings that are used to line walls or to check an existing structure require some kind of offset where the string is attached at each end. This will allow the line to run parallel to what's being aligned without actually touching it, so that deviations in the material won't get in the way of the string. It's best to use a standard increment for this offset. Framers usually use a 1x or 2x scrap at each end, and then gauge along the string with a third block of the same thickness.

When lumber comes out of the sawmill it is square and straight, but it's a long way from there to the job site. If you suspect a high joist (you did crown them, didn't you?), use a string before you put the plywood down. This will show you how much to plane off if you have to. The same goes for setting big ridge and purlin beams. What gets sold as "minimum crown" in a huge beam can cause you a lot of problems

once the rafters are up, and it doesn't take long to run a string to find out how bad the hump is, so that you can either ignore it or correct it.

Figuring with string—String is also a good medium for puzzling out how things fit together. Those odd rafters—the ones that aren't in the book or that you can't work out on the square—can be defined with string line. An adjustable T-bevel and a level can then be used to gauge the angles off the string line. If you are doing this, make sure that your string really does represent the top edge of the rafter. It's easy to get in a hurry and stretch the string from the ridge down to the outside edge of the double top plate of a wall.

To get the height of the lower end of the string right, I usually tack a block to the top of the double top plate. It should be the same height as the distance from the seat cut to the top of the rafter. The top end of the string should be stretched over the ridgeboard unless the rafters require a seat cut at the top for a ridge beam or high wall. In that case, cut an appropriately sized block and tack it in place.

Plumb lines—Using a plumb bob is based on the fact that if you suspend a string and weight it at the bottom, it will be vertical and perpendicular to any level plane it passes through. Building projects as impressive as the Pyramids relied solely on plumb bobs to get true vertical. Flooding an irrigation ditch with water gave these builders a gauge of what was level. With these resources they achieved impressive accuracy. Modern builders have more sophisticated tools—transits and spirit levels—for establishing plumb, but there are lots of times when none of these is the right tool, and a plumb bob is.

Plumb bobs can do two basic things: provide a reference for true vertical, and project points up or down. Foundations are a good example of the latter. In this case the string layout hovers above the ground at least a few feet. Superimposing that layout on the ground—first for the backhoe and later for squaring the forms—requires plumbing down from the strings. Provided you didn't do too much celebrating the night before and a gale isn't building, a plumb bob is a very accurate way to do this.

Transferring the corner points to the ground is done by holding the plumb-bob string tight to the 90° intersection of batter-board strings without actually touching them. You should set your legs apart in a secure, comfortable stance, brace your arms against your body and lean directly over the plumb bob. In this case you will have to make sure that you are clear of the batter-board strings, but you'll be using this same position whenever you project a point from above. Hold the bob a fraction of an inch off the ground and concentrate on the intersection of strings above. Once you're satisfied with the position of the string, let it slip from your hand so that the point

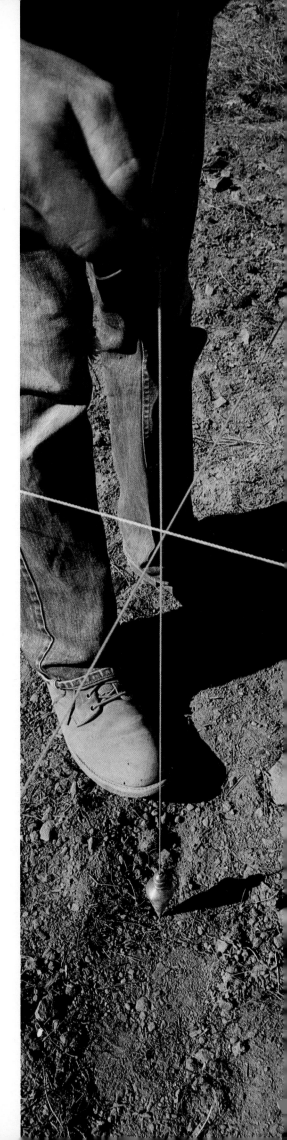

Using a plumb bob to project the corner of a building down to grade or the top of the forms from intersecting batter-board strings requires a surprising degree of steadiness, but it's by far the most accurate and practical method.

hits the ground. If greater accuracy is needed, spot a nail in the dirt where the point hit and plumb it again, making sure that the point centers on the nail head.

To find a point along the line you've established, you will need to use a tape measure in combination with a plumb bob. This requires two people. Your helper should ''burn a foot'' of the tape (hold it on the 1-ft. mark), or use a leather thong to get a good grip while holding the tape accurately over the intersecting strings. A solid stance allows your helper to sight down on the tape and to brace against the tension that you'll be exerting on it. On the smart end of things, lay the plumb-bob string directly over the tape on the correct dimension. Make sure that you are paralleling the batter-board string, which is level and square to the rest of the layout, and then keep the tape taut while you pay attention to the bob itself.

Another common use of a plumb line is projecting an established point up in the air, such as lining up the face of a beam with the layout on the floor. It saves time to use two people here— one on the plumb-bob point at the floor and the other up on the beam.

A plumb bob also makes good sense when you need to establish a true vertical line but the distance is too great to be checked with a level. High walls, posts and very tall door and window jambs are good examples. When you use a plumb line this way, it's necessary to offset the string from the work to allow the plumb bob to hang free. When this offset measurement is constant along the entire height of the string, then the object is plumb.

Because a plumb line is absolutely vertical, it's also indispensable for laying out chimney flues

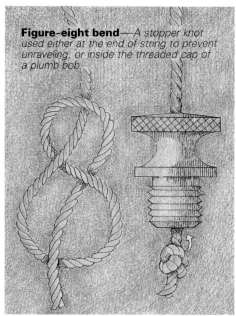

Figure-eight bend—*A stopper knot used either at the end of string to prevent unraveling, or inside the threaded cap of a plumb bob.*

Maybe the most common use of a plumb bob is to project a point up or down in space. Here, the author checks a layout point (represented by the duplex nail driven into the top of the stake) by using the plumb bob together with a tape measure. This requires a braced stance and steady nerves if the tape is going to be stretched tight enough for an accurate measurement from the original benchmark.

and stovepipes, and for figuring out where a light fixture goes in a sloped ceiling.

Almost any type of string will work to suspend a plumb bob, but braided nylon has no natural twist, so it doesn't spin in one direction and then in the other like a confused top. The line should be attached through the hole in the bob's threaded cap and tied off with a figure eight or stopper knot (drawing, facing page). When it's not in use, the string can be wrapped around the plumb bob, reeled into an old chalkbox, or stored on a small winder (see *FHB* #11, p. 14 for a spool that offsets the string when the bob is in use).

Steel plumb bobs are the cheapest. Henry L. Hanson, Inc. (220 Brooks St., Worcester, Mass. 01606) makes a bullet-shaped, hexagonal bob that I see a lot, but there are many other manufacturers. Popular weights are 5 oz. and 8 oz., and they run well under $10. These are fine for short drops, but for anything a story high or more, you should use a heavier version. Solid brass plumb bobs in a teardrop shape are the most popular. General Hardware Mfg. (80 White St., New York, N. Y. 10013) makes them with replaceable steel tips in 6-oz., 8-oz., 10-oz., 12-oz., 16-oz., 24-oz. and 32-oz. weights. They range in price from $8 to $28. Stanley (Stanley Tools, 600 Myrtle St., New Britain, Conn. 06050) makes a painted cast-metal version in 6-oz., 8-oz. and 12-oz. weights for a bit less money.

The heavier the bob, the better chance that you'll be able to get a good reading when there's a breeze. But large plumb bobs are pretty cumbersome to carry around in your nail bags. The ideal setup is a small bob for most work, and a 16-oz. or 24-oz. bob for long drops where the wind is a bigger factor.

There are two other types of plumb bobs that you will occasionally see. The first is the old-fashioned squat type. These are made of iron, but because of their shape it is very difficult to see a mark beneath them. The other kind is the small-diameter, bullet-shaped steel plumb bobs that have been bored and filled with mercury to get a low center of gravity and lots of weight for their size. L. S. Starrett Co. (Athol, Mass. 01331) makes a 12-oz. plumb bob that is only 6 in. long and ⅞ in. in diameter.

Chalklines—Being able to connect two points at considerable distance from each other with a straight line that is highly visible by simply plucking a string is a gift from the gods that a builder couldn't live without. There's no end to the applications. Carpenters of the 19th and early 20th centuries had to chalk their lines by running the string over a hemisphere of solid chalk. A chalkbox is a lot easier.

There are a number of brands on the market priced under $10. The most popular are Stanley's, Evans' (The Evans Rule Co., 768 Freling-

huysen Ave., Newark, N. J. 07114) and Irwin's (The Irwin Co., 92 Grant St., Wilmington, Ohio 45177). Stanley and Evans chalkboxes have aluminum die-cast cases (Stanley also makes a less expensive polypropylene model) that you can fill through a threaded cap. All of these boxes are available in 50-ft. and 100-ft. models (you're better off paying the extra dollar for the extra length). Evans' best model, which runs under $7, has a crank that can be stored completely flush with the case, and will release the string from this position. A slide mechanism allows you to lock down on the string. The Irwin Strait-Line box is an aluminum alloy, but is filled from a nylon sliding window on the side. It costs about half as much as the other two.

There are also geared chalkline reels in ABS plastic (Keson Industries, Inc., 5 South 475 Frontenac Rd., Naperville, Ill. 60540) and aluminum (B & S Patent Developing Corp., Box 1392, Riverside, Calif. 92502) that will recover the string at up to four times the pace of a direct-drive box. This can be a real advantage when you're laying out large spaces. They run about $5 to $7. On the E-Z Fastline, you can disengage the gears, which lets the string free-spool out of the box, by keeping the top of the crank depressed. This is a nice feature, since with some boxes you have to cope with the handle spinning around in your hand as the string is released. Unfortunately, the geared boxes are quite large and take up a lot of room in your nailbags. I also haven't been impressed with their quality—the cranks on the ones I've seen are quite flimsy, and the plastic cover for the filler hole looks like it wouldn't survive an assault by 16d nails in the bottom of a nailbag.

Most chalkboxes are outfitted with cotton line to which a small metal hook/loop combination is attached. Cotton is used because it doesn't stretch, it leaves a crisp line, and the rough natural fiber retains the chalk well. It does, however, abrade easily. Although cotton replacement line is available for less than a dollar, I experimented with braided nylon for its toughness but I didn't have much luck. The line wouldn't hold much chalk, and what chalk did adhere was thrown every which way with the snap. Also, because nylon is elastic, it vibrated after the pluck, leaving a thick line.

Powdered chalk comes in four colors: red, blue, yellow and white. It is packaged in plastic

containers holding 1 oz., 4 oz. or 8 oz. of chalk, and a 1-gal. size weighing 5 lb. I keep an 8-oz. bottle in the toolbox, refilling it from the less expensive gallon jug. It pays to have a couple of different chalkboxes so that you can use contrasting colors when you want to overstrike a mistake or distinguish between two things in a complicated layout.

There aren't too many tricks to using a chalkbox. If it was recently filled, pluck the line in the air a time or two to shake off the excess chalk. But even if the string is really loaded at first, two to four snaps and you won't be able to produce a visible line. You could then rewind the string, but it's faster if you just have another few snaps to do to pull out more string instead.

You will often see carpenters automatically rap their chalkboxes on the floor or against their thighs before attaching the free end of the line. This merely redistributes the remaining chalk in a partially filled box so that the string will come out with a full coating.

The universal hook that comes on the end of all chalkboxes will work on the edge of almost anything but a concrete slab (the edging trowel usually has a larger radius than the hook can accommodate). In this case, you're best off getting a helper. On plywood or lumber, drive an 8d nail and lower the hook onto it or use a scratch awl. For shorter snaps, you can hold one end down with your foot. In a pinch, you can snap a line that is less than 18 in. or so by using your thumbs to hold the ends, and using your little finger to make the snap. This can be useful when you've got to cut a series of jogs in plywood siding, for instance. If you practice this a few times you'll find it faster than taking your combination square apart to connect the lines.

No matter how long the line, it must be quite taut to get a good snap. The pluck must be perpendicular to the surface to be marked, or you'll get a curved line. Lift the line just enough for the ends to clear and let it go. Take time to get it right the first time, because a second snap usually makes a mess. On long runs the line must be lifted so high that you run a real danger of not being able to pull straight up on the string. Solve this by holding the string down in the center and snapping each side individually. □

Trey Loy is a carpenter and contractor in Little River, Calif.

You'll get a clean snap every time if you stretch the string as tightly as you can, raise it straight up off the work with your thumb and forefinger, and then release it quickly in mid-air. For accurate snaps of 12 ft. or more, get a helper to hold the string down in the center and pluck the line separately on both sides.

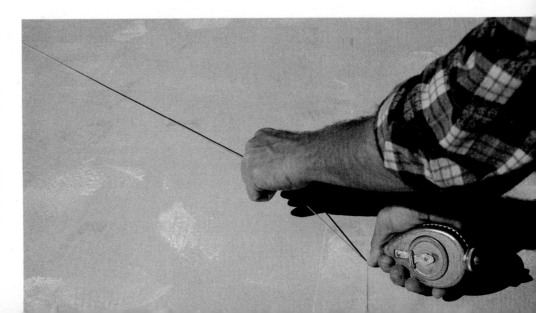

Powder-Actuated Tools

Fastening to concrete or steel without using a drill or a hammer

by Steve Larson

Neglecting to include all the foundation bolts during the insanity of a concrete pour is easy. Adding them later isn't. You can get a roto-hammer and a masonry bit, drill down several inches into the concrete, put in an expansion shield and finally bolt down a sill plate. But if you have to drill more than just a few holes, consider using a stud driver.

Stud gun, stud driver and Ramset (a trade name) are common terms for a type of tool known more formally as a powder-actuated fastening device. They are specialized nail guns, commonly powered by .22-caliber cartridges, that shoot a variety of nail-like fastening pins through wood and steel into concrete. Although it's not a tool that you are likely to carry around every day, if you can beg, borrow, buy or rent one when the need arises, you'll save yourself a lot of time.

Anchoring sill plates to concrete is the most common use for the powder-actuated tool in residential construction. The charge-driven pin is widely substituted for the occasional forgotten foundation bolt, and drive pins have even begun to gain acceptance as primary fasteners for some bearing walls (see the sidebar on p. 88). Stud guns are also commonly used to replace short sections of rotted sill during renovation work, to anchor non-bearing walls to a concrete slab, to attach furring strips to concrete walls, to anchor sleepers, to add plywood subflooring to a slab and to secure braces on concrete formwork.

In addition to nailing wood to concrete, special pins and brackets let you quickly hang a suspended ceiling or anchor electrical conduit. And at the fringe of residential work, steel plates, brackets and anchors can be fastened directly to either concrete or steel.

High and low-velocity tools—Although the operation of powder-actuated tools is simple, there are many brands and models (see the list on p. 90), and each one has special operating and safety procedures, which you must get to know. There are three basic types of powder-actuated tools: high velocity (HV, sometimes called standard velocity), medium velocity and low velocity (LV). But 95% of them are HV or LV.

A high-velocity tool (drawing facing page, left) works like a firearm. As the photos below show, a fastener called a drive pin is inserted into the barrel, then a charge (it looks like a .22-caliber blank) is loaded into the firing chamber. The muzzle of the tool is then placed against the work. By bearing down on the butt of the tool to release the safety, you can fire the charge by pulling the trigger. The resulting explosion in the chamber forces the pin out of the barrel, like a bullet.

A hood called a spall shield fits over the end of the barrel and protects you from flying bits of concrete or a deflected drive pin. The spall shield also holds the tool perpendicular to the work as it's fired. This is very important. A pin fired at an angle can ricochet with dire consequences.

Spall shields are absolutely necessary with HV tools, and manufacturers offer an assortment of specialized shields to allow perpendicular fastenings to irregular materials like corrugated metals and steel channels.

A low-velocity tool (drawing, facing page, right) uses a piston to drive the pin. The charge pushes the piston, which forces the pin out of the barrel at about 330 ft./second—much slower than the almost 500 ft./second of HV tools. The LV tool doesn't scatter bits of concrete at high speeds when it's fired, so it isn't always fitted with a spall shield. As a result, you can poke the barrel right into corners and other tight spots. Even so, I wear goggles when using either kind of stud gun.

Both high and low-velocity tools are commonly available at construction rental yards for a daily fee of about $12 to $20. The cost for pins, washers and charges varies radically—from about $.35 to $.90 for a complete load—so it can pay to shop around. These yards are required by OSHA to show you how to use the tool, and to have you read the operating manual and take a true-false test about the particular tool you are renting. Make sure you get this instruction. The best education would be on the job site, from a certified instructor.

For years, the standard has been the HV tool. Because it takes a stronger charge and will drive a larger pin farther into concrete or steel, it is sometimes the best choice. Concrete hardens with time, and low-velocity tools may not have enough punch to drive a large pin all the way into aged concrete.

Even so, the industry trend is clearly toward the LV tools because they are safer than

Loading sequence for Speed Fastener's high-velocity tool: A drive pin is inserted into the barrel (above) and then pushed into position with a rod made for this purpose (right). The powder load is put in the firing chamber (far right) before the hinged handle is closed.

Photos this page: Steve Larson; Illustrations: Frances Ashforth

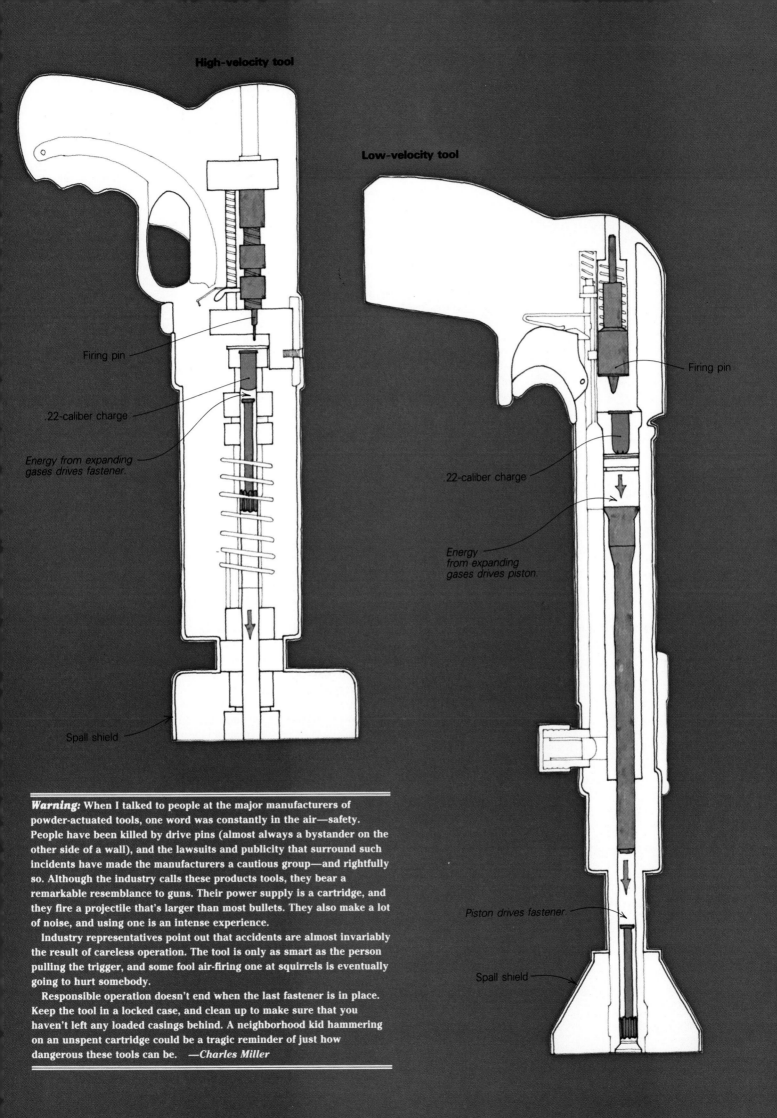

High-velocity tool

Low-velocity tool

Firing pin

.22-caliber charge

Energy from expanding
gases drives fastener.

Spall shield

Firing pin

.22-caliber charge

Energy
from expanding
gases drives piston.

Piston drives fastener.

Spall shield

Warning: When I talked to people at the major manufacturers of
powder-actuated tools, one word was constantly in the air—safety.
People have been killed by drive pins (almost always a bystander on the
other side of a wall), and the lawsuits and publicity that surround such
incidents have made the manufacturers a cautious group—and rightfully
so. Although the industry calls these products tools, they bear a
remarkable resemblance to guns. Their power supply is a cartridge, and
they fire a projectile that's larger than most bullets. They also make a lot
of noise, and using one is an intense experience.

Industry representatives point out that accidents are almost invariably
the result of careless operation. The tool is only as smart as the person
pulling the trigger, and some fool air-firing one at squirrels is eventually
going to hurt somebody.

Responsible operation doesn't end when the last fastener is in place.
Keep the tool in a locked case, and clean up to make sure that you
haven't left any loaded casings behind. A neighborhood kid hammering
on an unspent cartridge could be a tragic reminder of just how
dangerous these tools can be. *—Charles Miller*

Drive pins vs. foundation bolts

The powder-driven pins used for fastening wood sills to concrete look like common framing nails (photo facing page, bottom). The usual range is from 0.143 in. to 0.25 in. in diameter and from 2 in. to 3 in. in length. However, they are made from much harder steel than normal nails, so they can penetrate concrete and resist bending better than the nails in your tool apron. Even so, common sense suggests that a thick, deeply embedded bolt with a hook on its end will hold more tenaciously than a thin, straight, short pin. Research bears this out. For example, a common 3/16-in. dia. drive pin embedded 1¼ in. in concrete has a withdrawal (tension) resistance of about one ton. A ½-in. wedge-bolt anchor embedded 2¼ in. has a resistance of over three tons; and a standard concrete-embedded ½-in. foundation bolt, over 10 tons. This tenfold advantage of foundation bolts over drive pins is the reason they have been preferred for anchoring load-bearing sills, and why they continue to be specified in areas with high winds or seismic activity.

Recently, however, the International Conference of Building Officials (5360 S. Workman Mill Rd., Whittier, Calif. 90601) approved the use of powder-driven pins for anchoring interior and exterior load-bearing walls and shear walls in some circumstances (see ICBO Report No. 1147, March 1982; and check with your building inspector for local acceptance). The researchers found that properly embedded pins driven through ¾-in. washers or 2-in. discs at intervals of 2 ft. to 3 ft. (depending on pin diameter and length, and penetration depth) produced lateral resistance equivalent to using ½-in. anchor bolts 6 ft. o. c. You might not want to hold down an entire house with drive pins, but you don't have to be bashful about using them when they are right for the job and you want to save some time. *—S. L.*

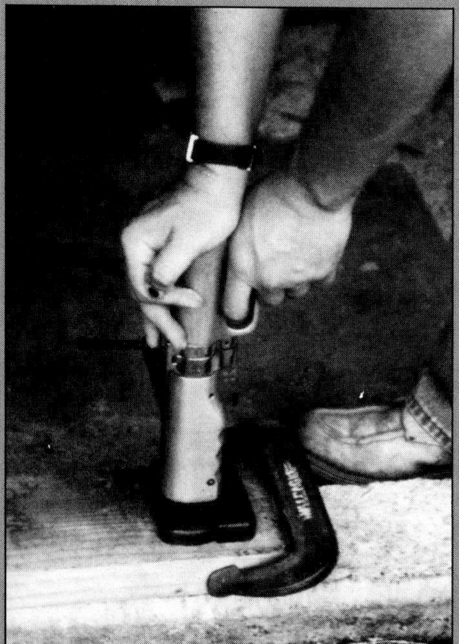

Attaching sill plates to foundations is a common use for the powder-actuated tool. Here a short section is fastened with a high-velocity driver. A C-clamp will keep the wood from splitting when a pin is driven this close to the end of a sill plate.

their high-velocity counterparts. Most of the serious accidents have happened when a drive pin from an HV tool somehow missed or passed through or was deflected by the base material and hit a bystander at high speed. But a pin leaving the muzzle of an LV tool loses velocity so quickly that such an accident is much less likely. Manufacturers estimate that 95% of the fastenings done with HV tools can now be done with the LV devices, and they say in the next few years the LVs will close that gap and take over completely.

Another advantage of the LV tool is the placement of the drive pin within the barrel of the gun before firing. In HV models, the pin is loaded from the firing chamber, and it has to travel all the way down the barrel. In LV tools, the pin sits at the very end of the barrel, and it is loaded from the front, or muzzle, of the tool. Because the pin doesn't travel down the entire barrel of the LV tool, washers or mounting clips can be attached directly to the pin. The HV tool requires that you center the barrel over the washer or clip, and I have missed more than one washer because the spall shield blocked my vision. Although special spall-shield adaptors are available to hold a variety of washers and clips to eliminate this problem, they cost extra and you have to remove them when you aren't using washers or clips. Also, some companies are discontinuing many of their spall-shield adaptors because the LV tools with their pre-assembled clips and washers are clearly more efficient for light to medium-duty uses such as installing conduit outlet boxes and drywall track.

A third advantage of the LV muzzle-loader is apparent during repetitive fastening of furring strips or plywood subflooring to concrete. The pins can first be tacked into the wood by hand and then shot into the concrete by slipping the barrel of the gun over the pin and firing. Low-velocity tools usually cost less than high-velocity tools—$250 to $400 vs. $300 to $500. If you are going to buy one of these tools, shop around. There is a great deal of competition between manufacturers and dealers, and prices vary dramatically.

Some more expensive LV tools feature a semi-automatic magazine that holds up to ten charges at a time, eliminating the time it takes to handle each individual charge. This tool is probably more appropriate for the commercial or large-scale builder who uses a lot of concrete and steel. As a small-time builder and remodeler of wood-frame houses, I wouldn't need one unless I started building a lot more tilt-up or masonry walls. I don't often need ten consecutive identical charges, but I do frequently need a variety of charges of differing strength and pins of different sizes to deal with remodeling situations.

At the low end of the price curve ($30 to $80) are the hammer-detonated tools. Both Remington and Speed Fastener make them. These simple devices use the low-velocity, piston-drive technology. A pin and a charge are inserted into the tool, and the operator fires the charge with a hammer blow. These are handy devices, and I keep one in my toolbox, but some are limited to a 2½-in. pin. Also, since it takes two hands to fire one, I have found them a bit awkward to use at times, with parts of my body uncomfortably close to the detonation point.

Charges—Although there are a few exceptions, charges for the powder-actuated tools are divided into two categories: high and low power. The most visible distinction between the two is that the low-power loads come in brass cases, and the high-power loads are packed in nickel-colored cases. Also, the loads for HV tools are sealed with a cardboard wad, while LV loads are crimped. This is because the cardboard wadding would eventually plug up the piston chamber in the LV tool. Don't pop any old load into your tool. Make sure that the brand and power of charge you are planning to use are appropriate. This information should be printed both on the inside of the tool's case and in the operation manual. If in doubt, check with the tool's distributor. Don't ever use a blank .22 cartridge or bullet in a powder-actuated tool.

Within each category, there are six color-coded levels of strength. If the powder charge is too low, the pin will not penetrate fully; if it's too high, the pin will go in too far. For example, if you're fastening a sill plate and the charge is too low, the pin will be left above the sill and thus won't hold it down. If the charge is too high, you'll blast the pin through the sill, often splitting it along the grain.

You can reduce these problems by shooting the pin through a solid washer. It is more difficult to drive a pin plus washer all the way through a board, and the increased area of compression provided by the washer also holds the wood more securely. When I'm shooting into especially short or brittle lengths of sill, I'll also put a C-clamp on the sill before I fire to keep it from splitting apart (photo left). When the pin doesn't go in all the way, I've found it nearly impossible to bash it home with a sledge or to pull it out with a prybar. If it's in the way, I usually bend it back and forth until it breaks, or just saw it off with a hacksaw or a reciprocating saw.

Pin selection—The first consideration here is safety. Make sure that the pins are designed for your tool. Drive pins are sometimes not interchangeable, and may be dangerous if used in the wrong tool. Consult your supplier or tool manual. The pin-head diameter must match the bore of the tool. Also, some of the specialized low-velocity tools can't handle the long pins needed for anchoring sill plates.

Aside from compatibility with the tool you're using, selecting the proper pin length depends on what you are fastening, and on what you are shooting into. To anchor a mudsill, the pin would have to be at least 1½ in. long to get through the wood, plus enough shank to penetrate the concrete. In average-strength concrete allowed to cure at least 28 days (3,500 to 4,000 psi), the pin should be long enough to penetrate about eight times the shank diameter—about 1½ in. for a 3/16-in.

dia. pin. Thus, you need a 3-in. pin to hold down a 1½-in. sill plate. In soft masonry (2,000 to 2,500 psi), penetration should be nine to ten times the pin diameter. In hard masonry (5,000 to 6,000 psi), penetration of five to six times the pin diameter will ensure a firm anchorage.

Each manufacturer offers a wide variety of pins in various lengths and diameters, and there are charts that give tension and shear values for the various combinations when they are embedded in concrete with different psi ratings. In addition, each company, through the home office or the local distributor, offers technical advice to builders with unusual fastening requirements.

Generally speaking, the holding strength of a powder-actuated pin in concrete increases with the depth of penetration and the compressive strength of the concrete. A pin holds in concrete because the concrete tries to return to its original shape after the pin is driven into it. The resulting compression grips the shank of the pin. Mortar, green concrete and low-strength concrete require even deeper pin penetration for adequate holding power—up to ten times the shank diameter. Don't use a stud gun on concrete that has cured for less than seven days, and even then, the tension resistance will be half as much as for a similar pin in fully cured concrete.

If you're working with brick, never shoot the masonry itself. Instead, aim for a horizontal mortar joint and use a low-power charge.

Testing the base and the fasteners—Although most concrete on construction sites will be suitable, some materials may be too hard or too brittle for firm fastening. Hard materials include hardened steel, welds, cast steel, marble, natural rock and spring steel. Here is a quick test that will help you decide. Take a drive pin and, using it as a center punch, tap it with a hammer against the base material. If the pin is blunted, the base is too hard, and a charge-driven pin will not penetrate. It may even bounce off, inflicting injury.

If the base material cracks or shatters when you hammer the drive pin, it is probably too brittle and may fly to pieces under the pin's impact. Brittle materials include glass block, glazed tile, brick and slate. If you can drive the pin in substantially with a normal hammer blow, the material is too soft and the pin will end up penetrating too far. If the pin makes a clear impression, but isn't blunted, the base material is probably okay and you can try a test fastening with a loaded charge.

If the base material passes this density test, you need to make sure it's thick enough to accept the pin without allowing it to go all the way through or causing bits of masonry to break away on the opposite side. For masonry, the base material should be at least three times as thick as the pin's penetration. This is especially important if someone might be on the other side of a concrete wall or working under a thin-pour concrete floor.

If I don't know how hard the concrete I'm going to fasten to is, I make a test fastening to

Three types of tools. Powder-actuated tools are especially useful when remodeling and adding on. Above, the author attaches a ledger strip to the side of an old foundation, using 3-in. drive pins fired by a low-velocity tool. At right, low-velocity hammer-detonated tools are inexpensive and handy, and they can be used to fasten preassembled clips. However, some are limited to a 2½-in. pin, and because it takes both hands to use one, they can be awkward in situations where you need a free hand to position and hold the work. In the photo below, a high-velocity tool quickly and effectively secures concrete-form bracing to a slab floor.

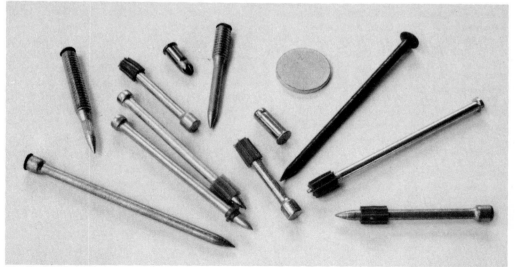

Typical drive pins are shown here next to a 16d nail. Unlike the nail, the pins are case-hardened by heat-treating to produce an outer layer that is tough enough to stand up to concrete or steel. The washer is used along with a drive pin to secure sill plates to concrete.

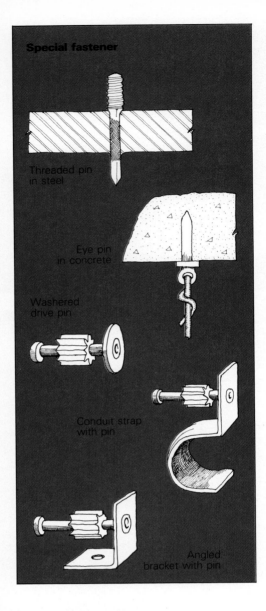

Special fastener

Threaded pin in steel

Eye pin in concrete

Washered drive pin

Conduit strap with pin

Angled bracket with pin

Sources of supply

The following companies are the major manufacturers and distributors of powder-actuated tools in the U. S. They all have catalogs on their tools and manuals on how to operate them. For more on stud-gun fundamentals, consult the "Powder-Actuated Fastening Systems Basic Training Manual," a booklet available free from the Powder-Actuated Tool Manufacturers' Institute (435 N. Michigan Ave., Suite 1717, Chicago, Ill. 60611).

Bostitch, East Greenwich, R. I. 02818.

Gunnebo Corp., 293 Lake Ave., Bristol, Conn. 06010.

Hilti Inc., P.O. Box 45400, Tulsa, Okla. 74147.

Impex/Allied Fastening Systems, 5600 W. Roosevelt Rd., Chicago, Ill. 60650.

Ramset Fastening Systems, Div. of Olin Corp., Shamrock Rd., East Alton, Ill. 62024.

Red Head, ITT, Phillips Drill Div., Construction & Fastening Operation, 5209 SE International Way, Milwaukie, Ore. 97222.

Remington Fastening Systems, AMCA International, P.O. Box 719, Bowling Green, Ky. 42101.

Speed Fastener Corp., 11640 Adie Rd., Maryland Heights, Mo. 63043.

Uniset, P.O. Box 26273, 4130 N. Englewood Dr., Indianapolis, Ind. 46226.

select the proper charge. For safety's sake, I start with a small charge and work up until I get the proper penetration. As you gain experience you'll find that you can usually guess pretty closely what the charge should be.

When I'm working on a base of undetermined compressive strength, I like to use HV tools because I can adjust the effective power of the charge by changing the position of the pin during loading. Increasing the distance between the charge and the pin cushions the charge, reducing the pin's velocity. The trick is useful if you're running short of weaker charges. Several HV models come with a calibrated loading rod for just this purpose. Most of the piston-driven, low-velocity tools are less adaptable; to reduce pin speed, you have to use a smaller charge.

Sometimes a pin will bend on impact, an occurrence called fish-hooking. This may be caused by the pin hitting a dense chunk of aggregate or a piece of rebar, and it will reduce the pin's holding strength. If this happens, drive another pin nearby. Also, the pin's impact may cause the concrete to chip out (spall), making a little crater around the pin. Flying bits of concrete can be dangerous, but spalling, unless it's severe, rarely reduces the holding strength of the pin significantly.

To prevent the drive pin from cracking the concrete, keep it away from the edge of a slab or footing—at least 3 in. if you are firing directly into the concrete, or 2 in. if you are shooting through a sill plate. For much the same reason, don't put successive pins closer together than 6 in. Remember that you are driving a hard steel pin into a brittle substance at high speed. Protect yourself with goggles, always use the spall shield and keep other people away from you. I like to wear heavy boots and stand well balanced with my firing arm fully extended and my head directly above the tool. In a confined space, I use ear protection.

Safety—Safety is a big issue in the use of powder-actuated tools. Some states require special training and certification to operate them. Each tool has its own set of procedures, and you should make sure you know the precautions for the tool you are using.

Maintenance is important. Powder-actuated tools should be cleaned and lubricated after each use. As with any firearm, gunpowder residue will build up in the barrel and firing chamber. On LV tools, disassemble the barrel and remove the drive piston to clean it properly. A lightweight rust-preventive lubricant such as WD-40 should be sprayed on these parts and then wiped dry. Don't use heavier oils because they tend to collect dirt. Most manufacturers also recommend a complete disassembly and cleaning of the tool after firing several thousand rounds or after several months of use. Check your tool's manual.

Special uses—In residential building, stud guns are most often used to fasten a sill or furring strip to concrete. But there are many other kinds of fastening pins and applications for the powder-actuated device, some of which are shown in the drawing at left. These are used more commonly by the big commercial builders, but shouldn't be in their exclusive domain. You can drive pins into steel members like I-beams and metal studs. And you can also fasten soft things like wood or insulation to steel or concrete, steel plates to concrete, steel plates to steel beams and so on.

The pins used for anchoring to steel have knurled shanks and are usually shorter than the pins used for concrete. Since steel is harder than concrete, less penetration is required, but the tip of the pin should penetrate all the way through the steel approximately twice the pin diameter. This is because the pin can be squeezed back out by the steel unless the tip goes all the way through.

Charges need to be selected in relation to the thickness of the steel, and HV tools can shoot through thicker steel than can LV tools: ¾ in. maximum for HV vs. 7⁄16 in. for LV models, depending on the brand of tool.

Both HV and LV tools are ideal for fastening light-gauge metals to concrete. I use 1-in. pins with washers to fasten such things as outlet boxes and masonry ties to concrete walls. And using a drive pin to anchor a light-duty post base is a good alternative to anchoring the base with a lag bolt and shield.

Two other types of pins are also commonly used. One is the threaded stud, which allows you to screw a nut onto it. It is often used to fasten conduit clips to concrete walls, and can serve as an all-purpose anchor anywhere you need to fasten removable hardware. Another is the eye pin for hanging acoustical ceilings from concrete and steel.

There are also pre-assembled clip and pin combinations, which are used only in LV tools. These include washered drive pins for fastening wood, rigid insulation, corrugated-metal roofing and steel siding to concrete or steel structural members. A combination conduit strap and pin is useful for fastening electrical conduit, and angled bracket pins are handy for hanging suspended ceilings or bolting on ductwork or pipe supports.

More than once I've had wall locations changed by clients between the time I've poured the slab and the time to start framing walls. And more than once I've had to cut off a row of misplaced foundation bolts. Now I anchor interior non-bearing partition walls with a stud driver, and save myself aggravation.

Whether you use a stud driver just to get yourself out of a scrape or as a regular part of your tool arsenal will depend on your imagination and your projects. But in any case, time and money are important factors for all builders, and powder-actuated tools can help you save both if you pick the right one for the job and use it safely. The more familiar I have become with these tools, the more I've gone from using them as an emergency backup to using them as a planned part of my construction activities. □

Steve Larson is a building and remodeling contractor in Santa Cruz, Calif.

Reciprocating Saws

These versatile cutting tools, fitted with the right blade, will take on jobs that other saws can't touch

by Craig Stead

A reciprocating saw is rather like a one-man band—certain melodies sound sweeter played by individual instruments, but it deserves applause just for getting through the score. Few tools are shared by as many trades as the reciprocating saw. It's the right hand of plumbers, electricians and sheet-metal workers as well as carpenters. In a remodeling business like mine, reciprocating saws are invaluable time-savers for salvaging door and window jambs, cutting new openings, removing studs, and sawing out holes for pipes, ducts or new lighting. I seldom have a job where demolition work isn't required, and there isn't any tool that can match the raw power and maneuverability of a reciprocating saw. When our crew of three is going full bore on tear-out, it's not unusual for us to have three of them in constant use.

This tool, which is often referred to as a Sawzall or Cut-Saw (brand names of two popular saws), is just a large linear jigsaw with the blade mounted in the nose. It is powered by what amounts to a ½-in. drill motor, averages 16 in. in length and weighs about 7 or 8 lb. To a large degree it has replaced the hacksaw, keyhole saw and handsaw in modern construction. The reciprocating saw has an amazing range because of its shape and the variety of blades available for it. With the blade out front, it can sneak into seemingly inaccessible places and cut almost any material.

The reciprocating saw does not do all cutting jobs well. Its long unsupported blade makes it difficult to get accurate, straight cuts, so a circular saw is often a better choice. A jigsaw is better for cutting curves because it has a larger shoe for its size, and you have a better view of the blade. But a jigsaw can't make a sink cutout flush with the backsplash. And a circular saw can't rough-cut big beams in a single pass. If you've got a lot of rebar to cut, borrow a cutter-bender or a torch, or use a circular saw with an abrasive wheel. But if you need to, you can get by with a reciprocating saw fitted with a metalcutting blade.

In 1975 I bought my first reciprocating saw, a Sawzall, when I started a full-time renovation business. Before, I was a part-time carpenter building my own house. A few months later I was running a crew of six. The Sawzall got such constant use that we picked up several. Now I have reciprocating saws made by Black and Decker, Milwaukee, Makita and Porter-Cable (formerly Rockwell).

Drawing the saw along a line puts more teeth in contact with the work, increasing the vibration. But it also allows the depth of cut to be easily and quickly changed with good control.

Buying a saw—You want a saw that is lightweight, fast cutting, well balanced, low in vibration, and durable. It should have a good view of the blade through the shoe. The saws I own combine most of these features, and all are basically good machines. There are some minor differences between them, and matching your needs to the strengths and weaknesses of the different saws will help you decide which to buy. As with any tool, it pays to try different brands if you can before buying.

Three companies (Milwaukee, Makita, Black and Decker) offer an adjustable shoe on their saws. By loosening one or more screws, the shoe can be moved up to an inch out from the saw body. This feature is convenient when you have dulled the teeth on one part of a blade and you want to use another part of the blade, where the teeth are still sharp. This need is common because the average blade stroke of a reciprocating saw, the amount the blade moves in and out, is 1 in. However, blades range from 3 in. to 12 in. in length. Depending on the thickness of material being

cut, many of the teeth on a blade don't get any use. The Black and Decker saw with a single adjustment screw works most easily, and Makita runs a fair second. I don't use this feature on my Sawzall much because you have to pull the rubber nose boot off the Milwaukee saw to make the adjustment.

All the saws I own have two speeds, controlled by a separate switch on the handle. The higher speed is for cutting wood and the lower speed for cutting metal. You can get the Black and Decker and Milwaukee saws with an optional trigger-controlled, variable-speed drive. This helps if you do a lot of metalcutting because it lets you start your cut slowly before coming to cutting speed. But having variable speed seems to reduce the top cutting speed on the Black and Decker saw, which makes it slower in cutting wood. Variable-speed switches also have a reputation for burning out faster than two-speed switches.

Porter-Cable's Tiger Saw has an action that's orbital as well as reciprocal. Instead of just moving in and out in a linear way, the blade is pushed down on the work on the in-stroke (cutting stroke), and it pulls up slightly on the out-stroke. This elliptical orbit helps clear sawdust from the kerf and digs the teeth of the blade into the wood on the cutting stroke, making the tool more aggressive. This feature, along with a high-end cutting speed of 2,800 strokes per minute, makes it the fastest woodcutting reciprocating saw made. The price you pay for this speed is a heavier tool (8¾ lb.) with more vibration in use. When cutting metal you have to switch off the orbital action on the body of the saw, which leaves you with a straight reciprocating motion.

Although this saw is aggressive, a real advantage in the demolition phases of my renovation business, it vibrates a lot, and has a shoe that limits your view of the blade and makes accurate cutting difficult. The Tiger Saw also has a reputation among people at rental yards and repair shops for wearing out its front-end bearings because of the stresses placed on them by the orbital action.

Milwaukee is justly famous for its durable power tools, and the Sawzall has been the standard in the industry for many years. It is the slowest of all the saws for cutting wood, with a ¾-in. stroke and a top speed of 2,400 strokes per minute, but is a good compromise for many tasks where speed isn't the only requirement. The shoe design provides a very

Woodcutting blades

Scrollcutting blades

Metalcutting blades

good view of the blade as it cuts, important if you do a lot of precision cutting or scrollwork.

Makita makes a lightweight, low-vibration saw that's reasonably priced. It's particularly nice for overhead work because it weighs just 6.4 lb. The stroke on the Makita is the longest of any of my saws (1¾₁₆ in.), but unfortunately the shoe obscures a full view of the blade. I have heard that this design flaw is to be corrected in new models.

My overall favorite is the Black and Decker two-speed Cut-Saw. It's fast cutting because of its 1-in. blade stroke, it doesn't vibrate much in use, and it has an easily adjustable shoe that gives a clear view of the line of cut. This same saw is sold by Sears. It's listed in their industrial line of tools as the Craftsman reciprocating saw.

List price on reciprocating saws runs from $144 to $180, with the Porter-Cable at the upper end, the Black and Decker mid-range, and the Makita the least expensive. All models are commonly available discounted, either through mail-order or lumberyard sales, and they can go for as low as $100 on sale.

I keep each of my saws in a case with the blades, spare allen wrench, and grounded plug adapter. Milwaukee's old case with the saw cradled above the blades was a great design. You could place your boxes of blades in the bottom of the case, and put the dullest blade at the top of the pile. Unfortunately, most of the new cases, including the one offered by Milwaukee, are like suitcases, with no partition between saw and blades. Every time you pick up the case, you end up with a blade stew in the bottom. Instead of buying one of these new cases, pick up an old, used mechanic's toolbox with a top tray to hold your blades. Just make sure the box is long and deep enough for your saw.

Milwaukee makes two accessories for their Sawzall. I haven't found the need for either, but I have friends who do. The first is a cut-off fixture to ensure square cuts through steel and plastic pipe up to 4½ in. o.d. It's a simple blade guide combined with a bicycle-chain grip that will also work on irregular shapes. The second accessory is an offset blade adapter that lets you flush-cut in close quarters without having to put a big bend in the blade.

Blades—A reciprocating saw is only as good as the selection and quality of blades you carry. But deciding which blades to buy can be bewildering. Milwaukee alone lists more than 70 different blades. Most people find a few types that they like, and then try to make these fit every situation.

To make sure you are buying the blades that will be useful to your kind of work, you need to ask yourself a few questions. First,

A versatile assortment of blades. The five blades at the top of the photo are primarily for cutting wood, and range from 3 in. to 12 in. in length, with 3 to 10 teeth per inch. The three metalcutting blades at bottom right have more teeth. The three narrow blades at bottom left are for scrollcutting in wood (7 teeth), composition material (10 teeth) and metal (24 teeth).

what general shape of blade do you need for your cutting task? How long, how wide and how thick does the blade need to be? Cutting a curve calls for a narrow, short blade like a jigsaw blade. If you have thick stock or need to be cutting at a distance, then use a long blade. A straight cut requires a wider blade, which will wander less in the kerf.

Once you have decided on the shape of the blade, then figure out how many teeth you need per inch. This is called tooth pitch. The fewer the teeth per inch, the more rapid and coarse the cut, and the faster the sawdust and chips are cleared from the kerf. The greater the number of teeth, the slower and smoother the cut. Blades are available with as few as 3 and as many as 32 teeth per inch.

What material you are cutting and how smooth the cut needs to be should determine the tooth count of the blade you choose. Soft woods such as pine need a coarse-pitch blade (3 to 7 teeth per inch) because the material is easily penetrated, and will give up a large chip. If you try cutting pine with a fine-pitch blade, the teeth clog and stop cutting. When this happens the blade overheats, and the teeth can lose their temper.

Hard materials, on the other hand, produce smaller chips, so a finer pitch is better. Composition materials like plastic laminate, tempered hardboard and particleboard are best handled with an intermediate number of teeth. I use the coarsest possible blade, so I can cut fast without chipping out the surface material where I want it to look good.

One question frequently asked is whether the blades are interchangeable among different brands of saws. For all of the saws I own, the answer is yes. However, not all blades are the same quality, and this affects the cut and the blade life you can expect. Inexpensive blades are stamped out of coil stock and heat treated. The teeth are formed by the stamping die and the burr left from the stamping process. Better-quality blades have milled or ground teeth—processes that give sharper teeth and more uniform tooth configuration. I prefer milled or ground blades because they cut faster, smoother and hold a line better with less pressure on the blade. You can generally tell these blades by the newly ground metal surfaces of the teeth and gullets.

In my toolbox, I carry ten types of blade, which seem to cover just about all the cutting problems I typically encounter in the remodeling trade. I carry five to ten blades of each type. Most of the blades I buy that aren't bimetal are Milwaukee's because I like their quality and long life.

My basic rough-in blades are 6 in. and 12 in. long with 7 teeth per inch, and work on nail-embedded wood. In nail-free wood, a useful blade is the 6-in. or 12-in. 3-teeth-per-inch blade. These really make sawdust.

Cutting curves in wood calls for scroller blades. I carry 7, 10 and 14-teeth-per-inch blades. With them, I can cut at different speeds and produce finished edges in fiberglass, Formica, and particleboard.

For cutting metal, you need a hacksaw-type

Bimetal blades

Bimetal reciprocating sawblades are relatively new, having been on the market only about five years. Before their introduction, spring steel—the material used to make handsaws—was used exclusively for woodcutting blades. Spring steel cuts wood well, but isn't hard enough to cut any more than the occasional embedded nail.

The other alternative was M2 tool steel, a tungsten-molybdenum alloy commonly called high-speed steel (hss), which is used to make the cutters for metalworking tools. M2 steel can be heat-treated to a high hardness, which makes the teeth last longer when cutting metal. As a result, high-speed steel blades have hard, durable cutting edges. But they are very brittle. They will break when bent, and even shatter when subjected to the shock loading that occurs when a blade is bumped when it's in motion.

A bimetal blade is a marriage of the best qualities of both types. Using space-age technology called electron-beam welding, a thin strip of tool steel is welded to a flexible spring-steel back. The tool-steel strip is 1/16 in. to 1/8 in. wide, and forms the cutting teeth of the blade. After heat treating, the cutting edge is very hard, but the body of the blade is still flexible and tough.

The first bimetal blades I used were Fit-Al, made by Rule Industries (Cape Ann Industrial Park, Gloucester, Mass. 01930). Bimetal blades are now being made by most of the major saw manufacturers as well, and are not hard to find in any of the general types or tooth counts available in conventional blades. They are, however, more expensive. Bimetal blades designed for cutting metal cost about twice as much as regular metalcutting blades. Woodcutting bimetal blades cost about three times their spring-steel equivalents. Depending on the blade and the circumstances, I find that they last anywhere from two to ten times longer than conventional blades.

A couple of my friends—one a builder, the other a tool dealer—had recommended Lenox Hackmaster blades, made by American Saw and Manufacturing Co. (301 Chestnut St., East Longmeadow, Mass. 01028). I found they were manufactured only a few hours from where I live, and visited the plant at the invitation of Marty Kane, Manager of Technical Services, to see first-hand what makes a bimetal reciprocating sawblade different from conventional blades.

American Saw has been making bimetal bandsaw blades since 1960, when the technology became available. While remodeling the company offices, a local contractor broke one metalcutting blade after another on his reciprocating saw while trying to cut away a long run of conduit. Realizing the problem, the machinists in the tool room fabricated a reciprocating sawblade out of bimetal bandsaw-blade stock and gave it to the contractor to try. The blade finished the job, but the contractor had to be told, "No one makes them," when he asked where he could get some more. That changed four months

later, when American Saw introduced the first line of bimetal reciprocating sawblades.

Bill Downing, Technical Services Representative for American Saw, told me that they use an M2 matrix tool steel for their blade edge. This steel is also used for general-purpose lathe tools, milling cutters, taps, dies and reamers. He gave me a bench demonstration of what bimetal blades can do in practice. First, he drove a hardened masonry nail into the end of a 2x4, and then cut off the last inch of the board using a typical spring-steel 6-teeth-per-inch blade. When the blade got to the nail, it stalled and the teeth burned off. Downing then repeated the test using an equivalent bimetal blade. It easily cut through the nail. Some teeth were dulled, but the blade was still usable.

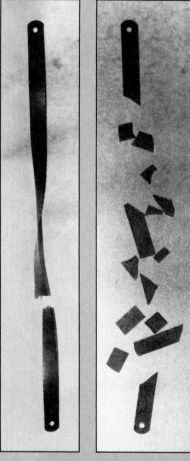

I repeated this test in my shop using a 16d common, a softer nail and one you're more likely to encounter with a reciprocating saw. I was able to cut through the nail with both blades. The teeth on the bimetal blade stayed sharp, while the teeth on the spring-steel blade were rounded over in spots. It required more force on the saw to cut wood as quickly after that.

Next I tried cutting just the exposed shanks of a series of nails driven into a 2x4, still using the two 6-teeth-per-inch woodcutting blades. The teeth on the spring-steel blade were completely worn away after 11 nails. The bimetal blade managed 21 nails. The blade was still cutting, though a number of the teeth had been sheared off by the shock of hitting the nail shanks.

The second of Downing's demonstrations showed the difference in flexibility between blades made from spring steel, high-speed steel and bimetal. Downing took hacksaw blades of all three types and inserted them one at a time into a fixture that would twist the blade until it broke. The spring-steel blade twisted a full revolution and broke in half. The high-speed steel blade made only a one-third revolution before it shattered into 15 pieces. The bimetal blade behaved the same as the spring-steel blade, making a full revolution before snapping.

The third demonstration involved using a reciprocating saw and bumping the end of a metalcutting blade against a steel plate. In this test, the high-speed steel blade broke into three pieces. The bimetal blade was badly bent, but was easily straightened on the flat surface of an anvil with a hammer. —C.S.

Bench tests. Some of the differences between bimetal and conventional reciprocating sawblades are demonstrated in bench tests that simulate job-site situations. In the first test, spring steel (photo, top) is matched against bimetal on a masonry nail driven into a 2x4. The middle photo shows the results of a test for flexibility with hacksaw blades made of spring steel (left), high-speed steel (center) and bimetal (right) that were twisted in a fixture until they broke. The last test (photo, bottom) demonstrated the effects of bumping the end of reciprocating sawblades into a steel plate. The high-speed steel blade shattered, but the bimetal blade only bent and was straightened for further use.

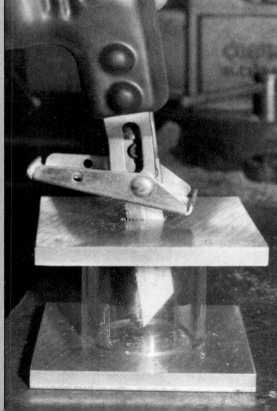

Photos this page: Craig Stead

blade. The rule here is always to have three saw teeth in contact with the metal. Thus, thin tubing and sheet metal require a high tooth-count blade, up to 32 teeth-per-inch. Bolts and rebar, which are thicker, are best cut with a coarser blade. I carry 18 and 24-teeth-per-inch blades in the 6-in. length, and a 32-teeth-per-inch scroller blade for cutting circles in panel boxes and ductwork. The conventional metal-cutting blades of hardened high-speed steel are notoriously brittle, and can shatter if they are pinched or bent. You've got to use them carefully.

In the past few years, bimetal blades have become available. They cut metal easily and yet are flexible, giving them a much longer life. I used to break one or two metalcutting blades a day cutting nail-embedded wood until I stumbled onto my first bimetal blade, which lasted me four months until the teeth wore out. Many companies now distribute them under their private label, so you need to check for the word *bimetal* on the blade.

Installing the blade in a reciprocating saw is a straightforward proposition. The blade is attached to the reciprocating shaft using the clamping action provided by a hex-head cap screw. A conventional allen wrench is used for tightening. All the saws except the Black and Decker have a cross pin that goes through a hole in the blade tang. You should be sure that the blade clamp is free of chips when you insert the blade, and be careful to get the cross pin through the tang hole.

The Black and Decker Cut-Saw relies solely on the compression of the blade clamp to hold the blade. This system has several advantages. First, when you are doing precise scroll cutting, you can flip the blade over so the teeth are in full view, a great aid to staying on the pencil line. It's also easier to blow the sawdust away from the line as you are cutting. When the blade jams in a cut, it will often pull out of the saw instead of breaking at the tang or bending. The third advantage is that you can rework a broken blade on a grinding wheel by taking ⅛ in. off the tooth side of the blade butt, creating a new tang. For how to grind broken blades to fit blade clamps that have cross-pins, see *FHB* #7, p. 10.

It pays to carry a spare cap screw for the blade clamp in case one gets lost or worn. You'll also need an allen wrench to turn the screw. For about $4 you can get one that is extra long, and bent to form a T-handle. It's much harder to lose this key, and you get plenty of torque with just a turn of your wrist. The end of your allen wrench will wear with time and the facets will get rounded, allowing it to spin in the socket. Don't throw it away; just grind it down to an unworn portion of the shank, and it will work like new.

Straight rough cutting with a reciprocating saw is an easy skill to develop, but knowing when to use the tool, what blade to use on it, and how to perform delicate tasks with it calls for some experience. In the photo at left, a pocket cut alongside the rafters and through the drywall is made using a short stiff blade with only a few teeth per inch.

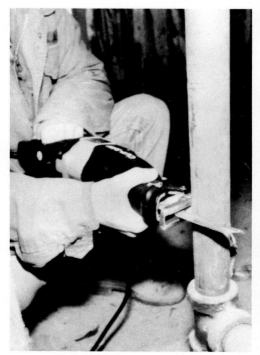

Reciprocating saws are ideal for work in tight places, a constant problem in renovation. To reduce vibration when breaking into old plumbing runs (left), keep the shoe held tightly to the work. When cutting through stud walls (right), make a full-depth circular-saw cut first.

Cutting—Reciprocating saws are are made for a two-handed grip—one hand in the D-handle that houses the trigger, and the other either supporting the nose of the saw from below for control, or bearing down on the nose for faster cutting. It's important to stay loose using any tool. Your grip should be firm but light, letting the saw do the work. If you are cutting stock that hasn't yet been installed, make sure that it's clamped down securely to reduce vibration. This is most important in starting a cut, since you will need both hands to establish a kerf without bouncing the blade all over the work. To begin, press the shoe of the saw against the material and lower the blade. You will be using the teeth farthest back on the blade, which will give you the least vibration and the most control.

How much pressure to apply while cutting is a matter of experience. Too much or too little pressure will produce needless vibration, which increases friction and dulls blades. The blades will also break much sooner, usually at the tang. For faster cutting, rock the saw gently. This action keeps the blade's cutting area smaller, making each tooth in contact with the work penetrate deeper and rake a larger chip.

To make blades last longer, it's common practice to back the saw away from the material once a kerf is established so you can also use the teeth nearer the front of the blade. I usually do this when the increased vibration isn't a problem, but I'm careful not to extend too far or the blade will leave the kerf on the retraction stroke and bend or break when it hits the material on its next forward stroke.

One rule for straight cutting is to match your blade to your material, and make sure that the blade isn't bent and that the teeth are uniformly sharp. Teeth designed for cutting wood are set alternately to either side of the cutting line of the blade. Blades that have been used to cut nail-embedded lumber can develop teeth that are dulled on only one side of the blade, which will cause it to drift to the sharp side when it's cutting. If you get into a cut and you find the blade wandering, you can compensate by twisting or angling the saw (see *FHB* #14, p. 14) You will find that soaping the blade will reduce friction and smoking as you try this maneuver. When cutting metal, use a light oil for lubrication.

In many situations, you should use your circular saw to establish a straight kerf, and finish up with your reciprocating saw. Cutting a window opening into an existing stud wall is a good example. I snap a horizontal chalkline across the exposed studs, and square this mark across each stud face on one side or the other. Using the chalkline for a reference, I cut each stud to the maximum depth (about 2½ in.) using a circular saw with a nail-cutting blade. Then I use my reciprocating saw with a thin woodcutting blade to finish the job.

Holes for large waste pipes, heat-duct floor registers and foundation vents are places where you'll need to make a pocket cut, or plunge cut. Begin by resting the saw on the shoe and blade tip, and then let the point of the blade dig a hole through the wood. This method works best in pine and plywood less than 1-in. thick. Make sure that you are using a short, stiff blade with sharp teeth at the free end. I've found that a variable-speed saw makes it easier to plunge-cut, but a two-speed saw will do. Milwaukee, Black and Decker and Porter-Cable all make a plunge-cutting blade that has teeth on both edges.

For pocket cuts in thicker material, such as cutting register openings through several thicknesses of floor, it's easier to drill holes at the corners. Use a circular saw for the straight runs and then finish off with a reciprocating saw to prevent overcutting your marks.

Safety—Safety glasses or goggles should be worn for overhead work or when using conventional metalcutting blades, which could shatter in use. If you're cutting through siding, floors and interior walls, check before you cut to determine whether any electrical wires are in the blade path. Since this is sometimes impossible, get into the habit of holding the saw only by the plastic nose covering and handle, not by the metal shoe or saw body.

When cutting timbers, in-place studs, pipe or tubing, it is common for the blade to bind in the saw kerf and bend. The immediate impulse is to grab the blade with your bare fingers and straighten it out. You really don't want to do this. It's easy when fighting a kerf to turn blades blue from the heat of the increased friction (about 600°F) and that can mean a nasty second-degree burn. Let the blade cool a few seconds, and then slip on a glove to protect your fingers from getting cut on the teeth or from a blade that snaps when you're straightening it. Use a hammer and a convenient hard surface to take the bend out of a bimetal blade.

Maintenance—Once a year, I do preventive maintenance on all my power tools in the interval between Christmas and New Year's, when I'm typically not working. On my reciprocating saws, I check the brushes for wear. If they're at all short, I replace both of them. I also remove the gear head, clean out the old grease, and repack it two-thirds full of lubricant. To guard against dust, I blow out my power tools once a month with compressed air. This makes the tool run cooler and reduces brush and commutator wear. If I'm cutting a lot of plaster with the saw, then I blow it out daily. □

Craig Stead renovates houses in Putney, Vt.

Photos this page: Craig Stead

Interior Finish and Woodwork

Laying a Tile Floor

Epoxy mastic on a plywood subfloor is a durable alternative to the traditional mortar-bed method

by Michael Byrne

Of all the finish floors available to the builder, none is more permanent than ceramic tile. Tile floors hundreds of years old attest to the durability of the materials, and today ceramic tiles are available in almost limitless colors and patterns. Because ceramic tile can resist the ravages of water (see p. 102), it is often chosen as the finish floor in wet locations like bathrooms, kitchens and mudrooms. But regardless of how durable and pretty the tile is on top, its useful life depends on things that can't be seen: a sturdy subfloor, high-quality mastics or mortar, an accurate layout and sound installation.

This article describes a typical tile installation in a small bathroom. The job was a remodel, but the procedures are the same for new construction. The tools are the same, too (photo below left). Except for the snap cutter, which you can rent from most tool-rental centers, tile-laying equipment isn't expensive.

Preparing the subfloor—The floor in this bathroom had been covered by wall-to-wall carpeting that extended from the neighboring living room. Years of careless splashing in the shower had rotted a corner of the rug, and the subfloor was about to follow. Fortunately the joists weren't damaged, so I didn't have to perform any framing surgery. I just had to remove the toilet and tear out a rotten rug.

The original subfloor was 1x6 T&G planking over joists 16 in. o. c. This is a satisfactory substrate for a finish floor that can flex, like vinyl or carpet, but a tile floor requires a stiffer base, or it will crack. I won't install a tile floor unless the subfloor and the underlayment are at least 1¼ in. thick, so I covered

the planking with ⅝-in. CDX plywood and fastened it down with 2-in. ring-shank nails driven into the joists every 6 in. At the walls, I left a ¼-in. expansion gap.

I never use particleboard as an underlayment because it's not as strong as plywood and it swells if it gets wet. Exterior grades of plywood, when properly fastened, can stand up to the occasional tub overflow without swelling and popping the tiles loose.

I left a gap of about ¹⁄₁₆ in. to ⅛ in. between the pieces of plywood for a glue joint. By edge-gluing the plywood sheets with the epoxy thinset, the underlayment becomes an integral layer that won't move in isolated spots as a result of a water spill. If individual pieces are allowed to move, the result can be cracked tiles or broken grout lines. Unless the subfloor needs a lot of surface preparation, glue the plywood edges as the tiles are set. Adjacent pieces of plywood shouldn't be more than ¹⁄₃₂ in. above or below each other, and for a larger floor, the edges of the plywood sheets should be staggered.

Traditionally, floor tiles are set in a 1-in. thick bed of mortar. This is still the best method for wet locations, but it's also messier and trickier than using epoxy tile adhesives. The epoxy thinset used on this job (Latapoxy 210, manufactured by Laticrete International, 1 Laticrete Park North, Bethany, Conn. 06525) is acceptable for use in wet locations. Epoxy is really the best choice where raising the level of the floor with a layer of mortar is out of the question. Its water resistance, compressive strength and holding power allow it to stand up to daily bathroom use.

I would never use an organic mastic for this

installation for two reasons. (Organic mastics were originally made from rubber-tree extracts. The term now describes a general class of ready-to-use thinset mastics that cure by evaporation.) First, water will eventually get through it to the subfloor and the tiles will loosen. Second, organic mastics don't get hard enough to support floor tiles, so the tiles in the high-traffic area will eventually move, and the grout will crack.

To make sure that the tiles will sit on a flat plane, I knock down any high spots in the subfloor with a disc sander and I fill in the low spots with epoxy mastic. Even slight irregularities can make for wavy grout lines once the tiles are in place. Small tiles look especially bad on uneven surfaces.

To find the low spots, I use a straightedge. I make pencil marks at the edges of the depression just where the light begins to appear under the straightedge. I draw around each depression until an outline defines the low spot. I thoroughly vacuum the floor so that a layer of dust doesn't prevent adhesion between the mastic filler and the subfloor. Then I mix a batch of mastic and trowel it into each low spot with a side-to-side motion of the trowel (this will pick up any remaining dust on the floor), and I screed off excess mastic with a straightedge. Once the mastic has set up, I remove ridges with an abrasive stone.

Laying out the tile—Rather than provide a line for each sheet or piece of tile, it's more convenient to divide a small floor like this one into workable parts. By projecting the location of a few grout lines onto the floor with chalkline or a pencil, I can visualize the alignment

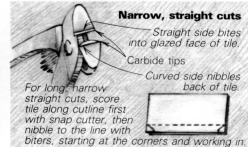

Narrow, straight cuts

Straight side bites into glazed face of tile.

Carbide tips

Curved side nibbles back of tile.

For long, narrow straight cuts, score tile along cutline first with snap cutter, then nibble to the line with biters, starting at the corners and working in.

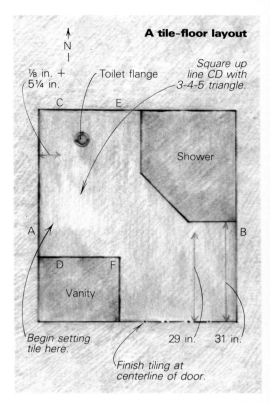

A tile-floor layout

N

⅛ in. +
5¼ in.

Toilet flange

*Square up
line CD with
3-4-5 triangle.*

C   E

Shower

A   B

D   F

Vanity

*Begin setting
tile here.*

29 in.   31 in.

*Finish tiling at
centerline of door.*

Layout. **After the newly laid plywood subfloor has been leveled, it is divided by layout lines into sections that will be set in sequence, beginning with the quadrant farthest from the door and ending with the section around the door. As shown in the drawing above, the layout is square to the door opening, emphasizing the most visible lines in the room.**

of the finished floor. I make each section large enough to give me space for my tools and materials, yet small enough so that I can reach the entire area without straining—it's surprising how easy it is to lose one's balance and fall onto the finished work.

To establish the layout, I usually start at the centerline of the door, directly over the threshold. The most visible portion of the floor is usually just inside the door, so it's important to make sure that the grout lines are straight, and parallel or perpendicular to the threshold. The centerline of the closed door marks the edge of the tile.

The distance between the door centerline and the shower in this bathroom is 31 in. Since the tiles for this job measure 3 in. by 6 in. including the grout lines, there's enough room for 10 full rows of tile plus one row made from pieces about 1 in. wide. But a row of 1-in. wide tiles among 3-in. wide neighbors wouldn't look very good, so I decided to cover the 31-in. width with nine rows of full tiles and two rows of tile trimmed to 2 in. This meant that a grout line would fall at 29 in. from the threshold, so I marked line AB on the subfloor (drawing above).

The distance from A to B is 64½ in. Since 11 rows of tile set lengthwise take up 66 in., I had to trim away 1½ in. from the total. Again, I split the difference by trimming ¾ in. from the tiles that fall at the margins. Whenever possible, plan your layout so that the snap-cut edges of the tiles face the wall. This way,

they'll be covered by the trim tiles. On line AB, I measured out 5¼ in. from the edge of the subfloor to allow for the trimmed margin tiles, and then added another ⅛ in. for a grout line. Then I laid my carpenter's square along line AB, and drew line CD. I checked it with the 3-4-5 triangle method to make sure that line CD was perpendicular to line AB. From line CD, I measured over another 18 in. to mark the grout line for another three rows of tile, and scribed line EF on the plywood. These three lines gave me workable sections that were square to the threshold, and took the most visible tile cuts into consideration. The subfloor was ready (photo above).

Because my epoxy adhesive has such a short pot life (about 30 to 40 minutes, depending on the weather), I like to have all my tools, a sponge and a bucket of clean water nearby before I start setting the tiles. I also cut the first row of tiles to size before I mix up the goo. From experience, I've learned that in spite of everyone's best intentions, walls run out of square and out of parallel a little here and there, and bathroom fixtures like tubs and shower stalls frequently have slightly irregular surfaces. So even though I've allowed 5¼ in. for the first row of tiles, chances are that the actual measurement will vary a bit. Cutting the first row before I mix the adhesive lets me check the tiles in place for fit without having to think about $40 worth of epoxy turning to stone in my best bucket while I'm fiddling with the tile cuts.

The snap cutter—This is the tool you'll need to make straight cuts in tiles. A snap cutter costs about $40, but most tile shops or tool-rental shops rent them out. The cutter (photo facing page, center) consists of a metal base covered with a rubber pad that helps to hold the tile in place while it's scored. An adjustable fence braces the tile for 90° and 45° cuts, and a track over the base holds and guides the cutter wheel.

To make a cut, mark the cutline on the tile (I use a fine-point, felt-tip pen for this) and line it up with the cutter wheel. Hold the tile in place with one hand and the cutter handle in the other. Lift up on the handle so that the wheel touches the glazed surface of the tile, and pull the handle toward you. Try to score the glaze along the cutline in a single pass with the cutters. Making more than one pass will usually result in a bad break.

To break the tile, hold it in place with one hand, position the cutter handle over the tile and hit the handle firmly with the heel of your other hand. If you've done it right, the wings at the base of the handle will push down on the tile, causing it to break along the scored line. If you haven't ever used a snap cutter before, practice on a few scrap pieces of tile to get the technique right.

It's difficult to snap-cut strips narrower than about ¾ in. Usually the tile breaks in the wrong place. Instead, score the tile with the snap cutter, then nibble to the line with a pair of biters (drawing, facing page), specialized

Illustrations: Barbara Smolover

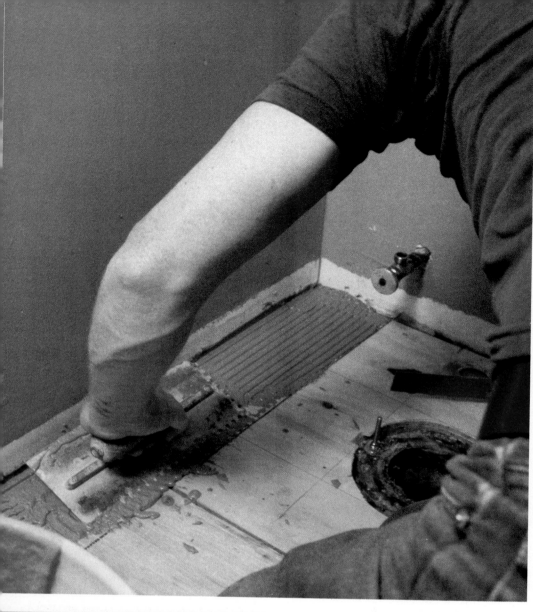

cutters that have a straight cutting edge on one jaw and a curved one on the other. Use the straight jaw on the glazed side, or face, of the tile.

Epoxy mastics—These adhesives were originally developed for industrial applications that required high bonding strength and resistance to chemicals. A tile-setting epoxy is composed of three separate components that are mixed together just before use. Part A is a resin of oxygen, carbon and hydrogen molecules suspended in a water solution. When combined with a hardener (part B) the resin molecules polymerize into long chains that form a resistant skin with high bonding strength. Part C is a mixture of portland cement and very fine sand. It is blended into the concoction of parts A and B to add body and compressive strength.

There are several brands of epoxy mastic, and they all do basically the same thing. They are sold by the unit, half-unit or in bulk. A full unit, enough to do a floor of about 50 sq. ft., costs about $40. I use Latapoxy 210 chiefly because the liquids are packaged in wide-mouthed containers that are easy to use, and they're mixed at a one-to-one ratio.

The two liquids have to be thoroughly mixed in their own containers before they are blended together. Although it's tempting to use a beater mounted on a drill for mixing, don't do so unless you have a very slow drill. High-speed mixing will whip air bubbles into the liquid, and they will weaken the mastic's bonding and compressive strength. I use a stick or a margin trowel for mixing.

Once the individual liquids are homogenized, pour them into a bucket in the proportions specified by the manufacturer, mix thoroughly, and slowly add the cement and sand while you keep stirring. I like to use my margin trowel for this because I can scrape the sides and bottom of the bucket with the trowel's straight sides and broad nose. There should be no lumps in the final mix, and it should be the consistency of very thick syrup. Never add water to thin the blend; it will render the bond useless.

I've found that the ideal temperature for working with epoxy mastics is between 70°F and 85°F. On hot days, the epoxy sets up faster. Conversely, a cold day slows the stuff down. I've laid floors on cold concrete slabs that have taken three days to set up, and others on plywood that are ready to grout three hours later. But on the average, it takes 24 hours before you can walk on the floor.

Setting the tile—Depending on the size of the section I'm about to set, I either pour the adhesive onto the floor or scoop it out of the bucket with my notched trowel. I spread out a skim coat using the flat side of the trowel, and then work it back in the other direction with the notched edge (photo top left). The depth of the notch should be about two-thirds the thickness of the tile. Spreading the mastic in several directions ensures good adhesion.

For this job, I started setting the tiles

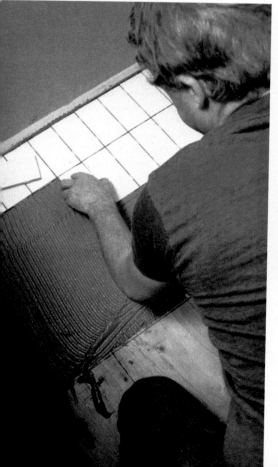

Epoxy mastic mixed to a stiff consistency is spread out along the first layout line (top photo) with a notched trowel. Notch depth should be about two-thirds the thickness of the tile. Frequently checking alignment with a straightedge, Byrne seats each tile with firm hand pressure (left). A light tap with the wooden handle of the trowel will seat the occasional proud edge. Pull up a freshly set tile occasionally to make sure the mastic coverage is correct, as in the photo above.

against the wall beginning at the intersection of lines AB and CD. I pressed each tile down until it met solid resistance (photo facing page, bottom left), and if a tile corner stood a little proud, I tapped it down with the handle of my trowel.

Each tile should be completely embedded in the adhesive, and from time to time I pull up a freshly set tile to make sure the coverage was correct (photo facing page, bottom right). The bottom of the tile should be completely coated with adhesive, and the adhesive has to be wet enough to stick. If the epoxy mortar starts to set up, it has to be thrown out—its pot life can't be extended by adding water. If there isn't enough epoxy on the bottom of the tile I check, I use a trowel with a larger notch.

It's also important to make sure that there isn't too much adhesive because it will squeeze up between the tiles, and fill the gap that has to be occupied by the grout. Trenching out clogged grout lines is a wearisome task. If there's too much epoxy, I turn to a tool with a smaller notch. And if I get some mastic on top of a tile, I wipe it off right away with a wet sponge while it's still easy to remove.

On this job, I continued setting the trimmed tiles along the west wall, carefully aligning them with the layout, and then set in as many full-size tiles as possible around the toilet flange. This left some gaps, and I had to cut some tiles with curves to fit.

Cutting curves—This is where the biters come into play. With them I can shape a tile to fit almost any peculiar gap in the floor, but the tile has to be cut slowly and in the correct sequence (photos and drawing at right). To cut the curve in the tiles around the flange, I first drew the curve on the tiles. Because they will be covered by the toilet, these cuts don't have to be very precise. So I drew freehand guidelines. For radius cuts that are visible, I use a compass. Once the cutlines were drawn, I marked concentric lines about ¼ in. apart and extending to the edge of the tile. The drawing at right shows the sequence for removing the waste bit by bit. Start at the edge of the tile, and always work toward the middle, as shown. This is the best way to shape your cuts and avoid breaks.

Angular cuts—Seven rows in from the north wall of the bathroom, the shower stall takes a 45° jog across the floor, and this change of direction occurs in the center of a tile. To shape this five-sided piece (photos far right), I marked my cutlines with the aid of a sliding bevel and removed most of the waste with the snap cutter, nibbling from the corners toward the center. When I'd finally cut to the line, I smoothed the nibbled edges with my abrasive stone. The stone is also very useful for tile-shaping work that's too minute for the biters or snap cutter.

Trim tiles—The trim tiles where the wall and floor meet can be set as the job proceeds or after all the floor tiles have been placed and the adhesive has set. I used a bullnose base

Cutting curves. Concentric lines leading to the final line mark the sequence of cuts (drawing, below). To trim tiles to fit around the toilet flange, first mark the tiles with the outline of the curve, then nibble away the bits with the biters, starting from the edges and working in (photos below). Two tile shards used as shims under one of the tiles (bottom photo) compensate for an uneven subsurface. The long bolts allow for the thickness of the new floor.

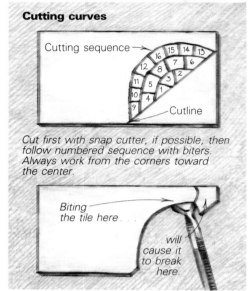

Cutting curves

Cutting sequence →

16	15	14	13
12	8	7	6
11	5	3	2
10	4	1	
9			Cutline

Cut first with snap cutter, if possible, then follow numbered sequence with biters. Always work from the corners toward the center.

Biting the tile here . . .

. . . will cause it to break here.

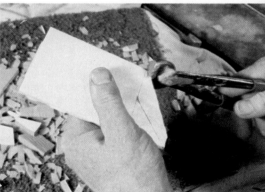

Cutting angles. The shower-stall angle is transferred to the tiles with a bevel gauge (top). To make a pair of angular cuts, begin by removing as much waste as possible with a snap cutter. Then work slowly with the biters from opposite corners to the middle (center). When you reach the line, smooth the raw edges with an abrasive stone (above).

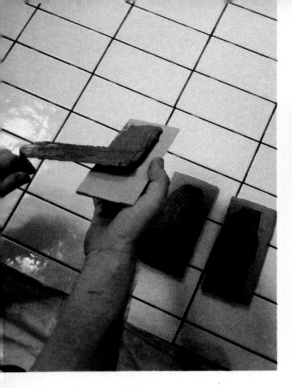

Trim tiles. It's easier to apply the epoxy to individual trim tiles, left, than it is to comb the mastic onto the wall. Trim tiles are temporarily held in place by plastic spacers, below, while the mastic sets up. Notice that the trim tile on the left has been shortened to match the width of the field tiles adjacent to the shower enclosure, so that the grout lines on floor and trim align.

Ceramic tiles

All tiles, even the most elegant, start out as the earthy ooze that lines the bottom of a streambed. As upstream mineral deposits slowly erode, particles are washed downriver where they mingle with organic materials to form sediments known as clay. This is the same soil that expands and contracts as it gets wet and dries out, and it's famous for giving builders fits. But if the mineral particles (primarily aluminum silicates) are removed from the soil, molded into flat pieces, dried and heated to high temperature, the resulting tile is as hard as flint.

Water absorption—The purity of the clay and the temperature at which it's fired determine how much water the tiles will absorb. This is a factor in choosing the right tile for a particular job. Impervious and fully vitrified (made glasslike) tiles are made from highly refined clay (the bisque), and have been fired at temperatures as high as 2,345°F. At this temperature the clay particles begin to fuse, becoming very dense and almost glasslike.

An impervious tile won't absorb more than 0.5% water. This rating means that a tile will absorb no more water than 0.5% of its dry weight after being boiled for five hours. Fully vitrified floor tiles won't absorb more than 3% water; a semivitreous floor tile won't take on more than 7% water while nonvitreous (soft-bodied) tiles absorb more than 7%.

As a rule, 5% and below is considered an acceptable absorption rating for tiles in wet locations; so impervious, vitreous and some semivitreous tiles make good countertops and bathroom floors. Other considerations are acid resistance, resistance to abrasion, slip resistance and glaze hardness. Nonvitreous tiles can also be used in wet places, but they should be secured to a mortar backing over a moisture barrier to prevent water from seeping through and damaging the framing. Never use nonvitreous tiles in wet locations where they might freeze—the moisture in them will expand and pop them loose.

Glazes—Many tiles have color built right into them—terra cotta is a good example. These unglazed tiles range in earth tones from yellow to deep red. Other tiles are finished with a baked-on surface of glass (a glaze) that can take on nearly any color and that's easy to clean.

When you're choosing a floor tile that's glazed, make sure its finish isn't so slick it's dangerous to walk on. And be sure the glazed tile is rated for floor use. Wall tiles are too soft to hold up to foot traffic, their glazes will scratch fairly easily, and they are very slippery.

Some glazed tiles are called "button backed" because of the little feet cast into their backs. These feet allow the tiles to be stacked in the kiln without the use of kiln furniture—tiny, high-fire shims that keep the tiles from fusing together. Integral button shims keep manufacturing costs down without affecting quality. Button-backed tiles must be embedded in a high-compression mastic so foot traffic won't snap off their unsupported corners and edges.

Types of tile—The various types of tile are manufactured in different ways and are suitable for different uses.

Quarry tiles are hard-bodied, with color throughout the clay. They are usually a deep red, but black and off-white are also available. They have a water-absorption rating of 5% or less.

Quarry tiles are made by an extrusion process that squeezes a moist body made from natural clay or shale through a die. This shapes a constant cross section, which is cut at intervals with a wire. Quarry tiles range from 3 in. by 3 in. up to 12 in. by 12 in., their dimensions are reasonably accurate, and they're modestly priced—around $2 to $3 per sq. ft.

Paver tiles can be glazed or unglazed, porcelain or natural clay. The unglazed varieties are less expensive, about $2 per sq. ft.

Most pavers are formed by a process called the dust-press method. Depending on the type or color desired, various clay bodies are mixed together, and the wet mix is squeezed through a filter to remove excess moisture. The resulting crumbly mix is allowed to dry almost completely,

then it's poured into steel molds mounted on hydraulic presses and compressed into tile shapes. This method is especially good for making precisely sized tiles.

Ceramic mosaic tiles are either porcelain or natural clay. They range from ¾ in. by ¾ in. up to 4 in. by 4 in. These tiles come arranged on fabric or paper sheets to make them easier to handle and set. Ceramic mosaics can be either glazed or unglazed, and some contain an abrasive additive to make them a better walking surface. They're made by the dust-press method or by plastic pressing, which uses a die to form wet clay body into tile shapes by direct pressure. Ceramic mosaics come in a mind-boggling variety of colors and designs, and usually cost more than quarry or paver tile—$3 per sq. ft. and up.

Mexican paver tiles are the fat handmade tiles from south of the border. They are rolled out like pastry with a rolling pin. Frequently they are formed directly on the ground, and their backs show the irregular imprint of some faraway courtyard. Their faces sometimes show the tracks of chickens and other barnyard creatures, giving them a quality called "distressed" in the trade. Mexican pavers readily absorb water, and their surfaces will dust with heavy use. They should be coated with a protective sealer (available at any tile store) to create a durable surface. They're not recommended for bathrooms.

Mexican pavers are cut out with big cookie cutters and the fresh tiles are set aside to dry in the sun. They are usually large (12 in. by 12 in.) square tiles, but interlocking shapes like octagons and ogees are also made. These pavers are made of a soft-bodied clay of varied earth tones that is fired at a low temperature and sometimes under primitive conditions. Some small outfits even use old cars as kilns. The sun-dried tiles are arranged inside the car and its trunk. To fire them, old tires are piled around and on top of the car and set ablaze.

For more on tile specifications and installation, see **The Ceramic Tile Manual.** It costs $21 plus $2 shipping (Calif. residents add 6%) from the Ceramic Tile Institute, 700 N. Virgil Ave., Los Angeles, Calif. 90029. —**Charles Miller**

tile for trim here, and made sure that the trim-tile grout lines lined up with the floor-tile grout lines. I back-buttered each tile and pressed it in place, then shimmed each one from below (photos facing page).

Grout—Grouting tiles is both an art and a science, and while there's no substitute for experience, knowing a little of the science can make the art happen a lot easier. Grout doesn't just fill the spaces between the tiles. It should be packed in from the setting bed to the surface of the tile. Sometimes a tile won't be completely supported by mastic, with a gap that may undercut its edge. These gaps have to be filled with grout, so the grout has to be viscous enough to flow into the voids.

Before I mix the grout, I dampen the entire floor with a sponge. This makes it easier to spread the grout, and it keeps water from being sucked out too quickly by a dry setting bed or highly absorbent tiles. Premature drying weakens the grout.

There are many grouts on the market, but for this floor I mixed my own with equal parts of standard portland cement and 30-mesh sand. A few years ago I started using latex additive in my grout, and I wouldn't do a job now without it. The latex makes the grout easier to spread, and it also speeds curing. Latex additives are made by Custom Building Materials (6511 Salt Lake Ave., Bell, Calif. 90201), Upco (3101 Euclid Ave., Cleveland, Ohio 44115) and by Laticrete International; these companies also make epoxy thinset.

Without a latex admix, a freshly grouted floor has to be kept moist and covered for at least 72 hours. With the latex, the floor is ready to use in 24 hours, and it doesn't have to be covered, unless it's especially hot and dry. I haven't had any problems with corner cracks or grout shrinkage. And when used full strength, the latex reduces mold and mildew growth on tiles in damp spots.

Follow the directions on the package if you use a latex admix. The procedure usually amounts to slow machine or hand mixing until the grout is smooth and lump free, and about the consistency of drywall joint compound. After letting it sit about five minutes, you mix the grout again, and it's ready to use.

I remove any standing water from the floor, and if any areas have dried out (a problem on hot days) I remoisten them with my sponge. Then I dump enough grout on the floor to cover about 10 sq. ft. With my rubber trowel at about a 30° angle, I force the grout into the voids (photo above center). I go over each area a few times with a side-to-side motion until the gaps are packed and then I scrape off the excess grout with the edge of the rubber trowel. Next I thoroughly wring out a sponge and I go over the freshly packed area to remove any grout that stands above the level of the tiles (photo top right). The pressure on the sponge determines how quickly the excess is removed.

I avoid bullying the grout, and if it seems a little too soft, I spread and pack another portion of the floor, and then come back. But it's

Grouting. **After excess epoxy is removed from between the tiles, grout is forced into the spaces with the sharp edge of a rubber trowel, below. It takes several passes from different directions to fill the voids. Most of the excess grout is scraped away with the trowel. As the grout begins to set up, the grout lines are brought down and made slightly concave with a sponge, right. The sponge is worked in parallel strokes, and rinsed out frequently to remove the sand and cement that it picks up. The last cleaning, below right, is with a nearly dry sponge that is rinsed out after each pass.**

important not to get too far ahead of the cleaning process—it takes a lot longer than the spreading. During this stage of the job, I remix the grout about every 10 minutes to keep it from setting up in the bucket.

When the joints have been formed and all the sand is gone from the surface, I make parallel sweeps with the sponge to complete the wet cleaning (photo center right). I use one clean side of the sponge per wipe. The rest of the cleanup is done dry.

After about 10 or 15 minutes a cement haze begins to form on top of the tiles. I remove it with cheesecloth, making sure not to distort the grout lines by pressing too hard. If some of the haze won't come off, I rub the area with a moist sponge. Then I remove with a margin trowel any grout that may have found its way into the expansion gap between the plywood underlayment and the wall.

Once the grout has hardened (about 24 hours), I vacuum the dust and grit from the floor, and I apply a bead of silicone caulk to the tile where it abuts the tub or shower stall, and to the trim tiles at the margins of the floor (photo and drawing, right). The caulk keeps the water out and allows the floor to expand and contract without damaging itself. □

Michael Byrne lives in Walnut Creek, Calif.

Caulking and finishing. **Before the grout sets up, it is squared with a margin trowel along the edges of the shower (below), and excess grout is removed from expansion joints. Once it sets, a bead of silicone caulk is applied.**

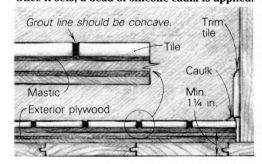

Grout line should be concave. — Trim tile
— Tile
Mastic — Caulk
Exterior plywood — Min. 1¼ in.

Installing a Sheet-Vinyl Floor

Whether flat-lay or coved, use a felt-paper pattern for best results

by Terry Shrode

People still call it linoleum, even though it isn't, and the chances are good that the floors in your bathroom and kitchen are covered with it. It's properly called resilient flooring, and it also turns up in the family room, workshop and utility porch—any place where a durable, low-maintenance and affordable floor is a requirement.

Resilient flooring is available in a wide price range, from as little as $3 a square yard to over $30 a yard, and it comes in colors and designs that will suit just about anybody's taste. Along with the outlay for supplies, you can figure an equal amount for the installation labor. This article is about how to install resilient flooring in two different ways. One is the relatively easy flat-lay method. The other is the more complicated coved style that used to be common with linoleum floors.

Flat-lay is the term for a floor that meets the wall at a 90° angle. This junction is covered by a baseboard or a toe molding. A coved floor requires a little more material than a flat-lay floor, and it wraps right up the wall, where it is finished with a J-section trim piece called cap metal (drawing, facing page). The floor-to-wall intersection is backed with radiused blocking called cove stick. At inside and outside corners, the flooring is mitered—it's the floor mechanic's version of the carpenter's crown-molding problem. A coved floor is more difficult to install than a flat-lay floor, but I think it looks classier, and it's easier to keep clean at the perimeter.

There are two basic types of resilient floors available today (sidebar, facing page): sheet goods, which come in 6-ft., 9-ft., 12-ft. and 15-ft. widths; and tiles, which are usually 1-ft. square. Most sheet goods and tiles are made of either vinyl compositions or pure vinyl. I'm going to concentrate on sheet vinyl because it's the most popular, and it comes in the greatest variety of colors, grades and designs.

Vinyl-tile floors follow the same basic installation procedures as sheet-vinyl floors, and the boxes of tile usually include easy-to-follow instructions. But I think vinyl-tile floors have one serious drawback: they have lots of seams for water or dirt to invade. This can be a serious problem, especially in wet locations with wood subfloors. Don't get me wrong—I've seen properly maintained inexpensive

Terry A. Shrode is a flooring contractor in Richmond, Calif.

vinyl-tile floors over wood substrates that have lasted for many years, but this is the exception and not the usual case.

Estimating materials—The rule of thumb is to lay out the floor with the fewest seams, and with the least amount of waste. There are only two possible directions to run the vinyl, and one is usually better than the other.

In the kitchen plan below, I've shown the layout lines for an installation using 6-ft. wide goods (the most common width available) running the two possible directions. The east-west direction leaves me with both more waste and more seams than the other option, so the north-south orientation is clearly the best choice here. Two pieces, or drops, are needed for this floor. The west-side drop is 14 ft. 6 in. plus an extra 3 in. that should be added to any length to allow for trimming. This floor will be coved, which requires about 4 in. of material per wall, thereby adding another 8 in., for a total of 15 ft. 5 in. If the plan calls for doorways to be covered, this will add to your total—be sure to include them.

The east-side drop is 2 ft. shorter, because of the cabinets, for a sum of 13 ft. 5 in. The tiny gap in the southeast corner will be filled with a piece of scrap from the cabinet cutouts.

The total amount of material you need to order depends on the "pattern match" for the particular piece of flooring in question. If the

piece has no definite design, no extra material is needed. If, on the other hand, there is a recurring design, the distance between repetitions has to be added to each drop after the first one. The design in the material chosen for this kitchen repeats itself every 18 in., so the east-side drop totals 14 ft. 11 in. If there were a third drop, it too would need another 18 in. added to it, and so on. The grand total for this kitchen is 15 ft. 5 in. plus 14 ft. 11 in. With 6-ft. goods, this comes out to 20¼ sq. yd.

Subfloor—It often takes more time to prepare the base under the sheet goods than it takes to install the new flooring, and this is one of the most important steps in the entire process. Every ridge, bump or gouge will telegraph through the new floor, inviting premature wearout. Termite and water damage are the other big problems, and you should take care of any major structural upgrades before you begin work on the new floor. Check to see that the existing floor is securely fastened to its joists. Eliminate squeaky spots or loose floorboards with a few ringshank nails.

A lot of the floors that I install are over existing vinyl or linoleum surfaces. They are adequate subfloors if they are free of wax or grease. I cut back any small imperfections, such as loose seams, gouges or bumps, to solid underlayment, and fill the resulting gaps with a patching compound like Fixall (Dowman Products, Inc., Box 2857, Long Beach, Calif. 90801).

A floor with more extensive damage should be covered with ⅜-in. or thicker plywood or particleboard. This includes old T&G floors that have cupped or twisted. Install the plywood finished side up; if you use particleboard, get the best underlayment grade you can find. Where I work on the West Coast, we don't have the wide humidity swings that are common in other parts of the country. Consequently, I leave about a ³⁄₁₆-in. gap at the walls, but I don't leave much of an expansion gap between the sheets. I just loosely butt them together in a staggered pattern and secure them to the subfloor with ⅞-in. staples. I use staples because they are fast, and because they don't leave dimples in the underlayment the way nails might. If you choose nails, make sure they are ringshanks. Don't use sheetrock nails—they will work their way out over time. In either case, nails or staples should reach ½ in. into the subfloor, and they should be

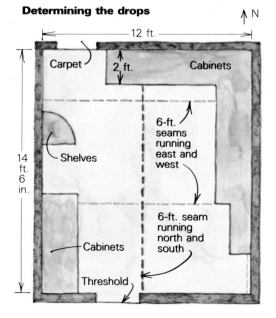

Determining the drops

Two 3-ft. wide pieces of builder's felt are butted together and taped to make a pattern for the first drop in the kitchen shown in the drawing on the facing page. Here the finished pattern has been taped onto the flooring, and Shrode is trimming the material to fit. The corner cut-out at the bottom of the photo indicates that this will be a coved floor.

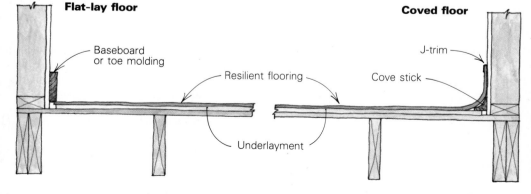

Flat-lay floor

Baseboard or toe molding

Resilient flooring

Underlayment

Coved floor

J-trim

Cove stick

Illustrations: Frances Ashforth

Resilient flooring

A hundred years ago, an Englishman named Walton coined the word linoleum to describe a pliable floor covering made of wood pulp, cork, turpentine, pigments and oxidized linseed oil. This mixture was applied to a burlap or canvas backing, and it was ready to install. Linoleum is still an excellent floor covering, but it requires lots of maintenance and is now made only by the Krommenie Co. in Holland and distributed by the L. D. Brinkman Co. (14425 Clark Ave., Industry, Calif. 91745). People also use it for countertops and drafting boards.

Today's resilient flooring also includes cork and rubber-base products. But by far the most common are vinyl floor coverings. Vinyls are petroleum-base plastics that are made into floor coverings in two basic ways.

The first is the inlaid method, which produces hard-surface goods. Millions of tiny vinyl bits, with color built right into them, are spread out and compressed at high temperature into a thin layer, which is then bonded to a backing material for strength. Inlaid floors are very durable because the color runs deep into the material. They also have a very hard surface and they are quite stiff, which makes them difficult to install.

Rotogravure is the second method. Rotogravure floors make up the bulk of the sheet-goods business, and the cushion layer built into them makes them comfortable to walk on and resistant to dings. To produce them, rolls of blank flooring are run through printing presses. A plate on the press carries the photographic image of the design, which is embossed into the flooring. A thin layer of vinyl above the cushioned core then receives colored dyes from the press to complete the design. A clear layer on top takes the wear.

Once they are installed, cushioned floors have to be sealed at their seams with a special solvent made for each type. It comes in applicator bottles from your supplier, and it welds the two pieces of flooring together to keep out dirt and moisture.

Both inlaid and rotogravure vinyl floorings are available in what manufacturers call wax or no-wax finishes. No-wax means the floor looks waxed, but you don't have to wax it. Wax means you have to wax it. In the past, I avoided cushioned floors with no-wax finishes. They had a thin layer of PVC over the flooring and didn't wear very well. But recently, the manufacturers have come out with new coatings that seem much more durable.

Another change is in the backing material. For years, asbestos backing was the standard. It resisted heat, moisture damage and mildew, but as we now know, is an awesome health hazard. The industry still hasn't settled on its replacement. Some floorings now have vinyl backings, which are an added expense. Others have a fiberglass backing, which can leave you scratching at the end of the day when you install it. But both of these are improvements over asbestos.

In the vinyl-flooring business, new products are being introduced all the time, some of which require new installation procedures. Some call for different adhesives, others shouldn't be back-rolled during installation. It's important to ask your supplier about the specific procedures to follow for your floor.

In general, you get what you pay for in resilient flooring, but after a point your money is spent on style rather than durability. Around here, $20 per yard will buy you a floor that should last for 20 years. Check with your local flooring contractor for the brands and styles that are available. And try this test for durability when you're shopping for a vinyl floor: Scratch the surface of the floor sample with your fingernail. If you can gouge a hole, or come close to it, stay away from it. —*T. S.*

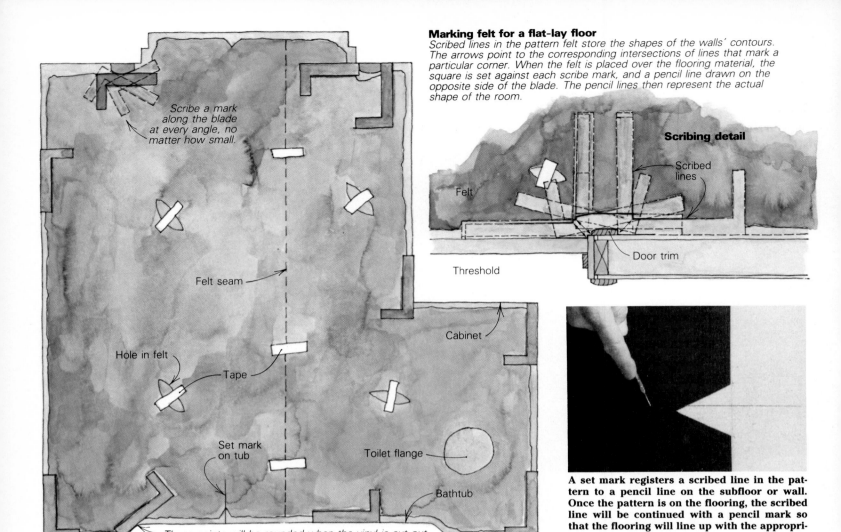

Marking felt for a flat-lay floor
Scribed lines in the pattern felt store the shapes of the walls' contours. The arrows point to the corresponding intersections of lines that mark a particular corner. When the felt is placed over the flooring material, the square is set against each scribe mark, and a pencil line drawn on the opposite side of the blade. The pencil lines then represent the actual shape of the room.

Scribe a mark along the blade at every angle, no matter how small.

Felt seam

Hole in felt

Tape

Set mark on tub

These points will be rounded when the vinyl is cut out.

Cabinet

Toilet flange

Bathtub

Scribing detail

Felt

Scribed lines

Threshold

Door trim

A set mark registers a scribed line in the pattern to a pencil line on the subfloor or wall. Once the pattern is on the flooring, the scribed line will be continued with a pencil mark so that the flooring will line up with the appropriate point in the room.

spaced ½ in. from the seams 3 in. o. c. around the edges and 8 in. o. c. in the field.

Adding a new floor over two or three layers of old flooring is fine. Before you do this, though, check for height problems. Building up the floor this way will sometimes trap the dishwasher or prevent the refrigerator from fitting back into its nook. You may have to remove the old floors. If the old surface is a cushioned-type resilient floor, don't install hard surface goods over it—foot traffic may crack the new material. If the old floor is smooth, you can lay the new one right over it, without new underlayment. Be sure to give it a thorough cleaning with TSP (tri-sodium phosphate) so that the adhesive will stick. Don't sand the old flooring—it might have asbestos in it. If it's a coved floor, cut out the portion that wraps up the wall and remove the old J-trim, but leave the old cove stick. It will work fine for the new floor.

Making a flat-lay pattern—There is nothing mysterious about laying a vinyl floor, even with coving. The key to success is the pattern. The pattern is made of felt paper (15-lb. builder's felt is fine) pieced together with duct tape or masking tape to approximate the shape of the floor. In a room where it takes only one piece of material to cover the floor, I make a pattern (drawing, above left). It begins with a piece of felt butted against the longest wall,

followed by consecutive pieces until the floor is covered.

If the floor is to have more than one drop, I strike a chalkline to mark the first seam. The placement of this line is critical, because the rest of the floor will be affected by its position. When I'm using 6-ft. goods in an area with more than one drop, I use two 3-ft. wide lengths of felt butted together to simulate each drop. One edge of the pattern corresponds with the seam in the vinyl. In either case, I cut eye holes out of the pattern so I can tape it to the work, and I avoid creasing the felt.

Once the pattern paper is secured to the floor, I make set marks, registration points that I mark with a scribed line on the pattern and a pencil line on the underlayment, tub edge or wall (photo above). I notch the pattern with a V at these points to make them easy to find. I like to use at least two set marks for each drop—usually one along the length and one along the width. These marks will later be transferred to the vinyl flooring.

The pattern is cut about an inch shy of the walls all around. This allows me to transfer the shape of the room onto the pattern. I do this with a square with a 12-in. long blade that is 1½ in. wide. The blade spans the gap between the pattern and the wall, and I make a scratch on the room side of the blade (drawing, above right). Then I work my way around the room, marking every angle, no matter how

small, with a scratch on the pattern paper. I make the marks at least 4 in. long. If I have to use the square's short leg, which is 1 in. wide, I make a note on the pattern to remind myself that something out of the ordinary was done. Later, when I use the pattern to mark the correct perimeter on the actual goods, I just reverse this whole process.

I mark curves by moving the square around them at short intervals. These marks will later be connected and rounded off when the material is cut. These scribe marks can look like unintelligible scratches at first, especially around door casings and pipes. Study the drawings to see how the scribe marks capture the room's contours.

Trimming the seams—If the floor will have a seam, the material has to be trimmed. Factory edges must never be used for seams. They will be crooked, gouged or dirty, and all vinyl sheet goods are slightly oversize to allow for trimming.

Sheet goods are installed with reversed seams, unless the manufacturer says otherwise. This means that a roll of sheet goods has a left edge and a right edge. Installed, a right edge butts against a right edge, and a left edge butts against a left edge. Therefore the drops will alternate in direction. I cut the pieces individually, then butt them together at the seam because I've found this to be the quickest and most accurate way for me. Others cut both layers at once.

Roll the flooring out, face up, on a clean, smooth surface. Garage floors are nice if they aren't oily. Don't lay the material in direct sunlight, which will cause it to shrink. Now back-roll the flooring, and roll it out flat again. This also shrinks the material, but since you have to back-roll it during installation, you want the shrinkage to take place before the material is cut.

Where to cut the seam depends on the design of the floor. If it has one with straight lines, they will have a certain width—usually ⅛ in. to ½ in.—to simulate grout lines. Trim to one side or the other; if you leave the line on the first drop, cut it off the next drop. Some floors have extra-wide grout lines at the edges. They have to be trimmed down to match the grout lines in the rest of the design. If the material has no design, take off about ½ in.

Position a long straightedge at the appropriate spot along the border of the material. Kneel on one end, and steady the other end with your hand (photo above left). Use a utility knife with a fresh blade to score the surface. The hand will naturally make a bevel, so turn the knife inward slightly to make sure the blade follows the straightedge. Make a smooth cut approximately 2 or 3 ft. long—whatever length is most comfortable for you. Stop, but don't remove the knife. Slide your hand and knee down the straightedge and repeat the cut. When you get to the end of the straightedge, don't remove the knife. Just slide the straightedge down the border, line it up, and keep cutting.

Use the linoleum knife to finish the seam (photo above right). Place the curved part of the blade in the groove made by the utility knife. Pull gently, letting the knife follow the score mark.

Using the pattern—Align the pattern felt along the edge of the seam, and adjust it so that the design in the vinyl breaks the same way at each end. If the pattern felt covers the entire floor, adjust it to be as square to the design in the flooring as possible, with equal design breaks around the edges. This is usually a matter of compromises, since four walls are rarely square. Using the eye holes in the felt, tape the pattern to the material.

Now transfer the marks on the felt to the flooring. Place the square next to each mark, and make a pencil line on the opposite side of the square onto the material. When you're finished, you have a picture of the floor. Using the utility knife, cut on the waste side of the pencil line. This will make the material a little fat and ensure a tight fit. Round off the corners slightly, and be careful not to cut into the main sheet. For long, straight runs, use the straightedge. Take your time.

Installation—Sweep the floor clean, back-roll the vinyl again to make it easy to carry, and lay it in place to check the fit. Trim any areas that need it, and make sure the set marks are telling the truth. Remove the floor-

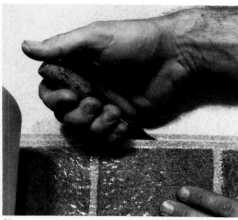

Sheet goods are made extra wide and have to be trimmed at the seams. The author prefers to cut one sheet at a time, using a long straightedge and a utility knife to score the vinyl, left. The cut is finished with a linoleum knife, above.

Shrode uses notches cut in the pattern felt to register the second drop to the first. The notches are placed three design repetitions apart, and let him accurately position the pattern on the material.

ing, and use a ³⁄₃₂-in. notched trowel to spread latex adhesive over the area of the first drop. Stop the adhesive 3 in. shy of the seam line. This area will be spread with adhesive when the second drop goes down. Make sure there are no puddles in the adhesive, and set the flooring into it while it's still wet. One person can do this, but it's a lot easier with two. Position the areas with the set marks first, spread the floor out and use a moist rag to pick up excess adhesive. Immediately use a floor roller, which you can rent, to squeeze the material down into the adhesive. Work from the center toward the edges. After rolling, a few bubbles may persist. They should disappear by morning, because the latex draws the flooring downward as it dries.

The second drop—Matching the design is the challenge of the second drop. For this kitchen floor, I made notches in the pattern felt that corresponded to the grout lines in the flooring design. I lined up these notches with the appropriate grout lines in the second drop. Whether the flooring has a design or not, I make a set mark along each seam to help align the drop (photo above).

This floor is a hard-bodied, inlaid type from Armstrong called Designer Solarian, and it needs an epoxy adhesive under its seams to prevent water penetration. Before the second drop went down, I mixed up a batch of epoxy (Armstrong S-200), lifted up the unglued edge of the first drop, and spread the epoxy under it about 3 in. with a small-notch trowel. I con-

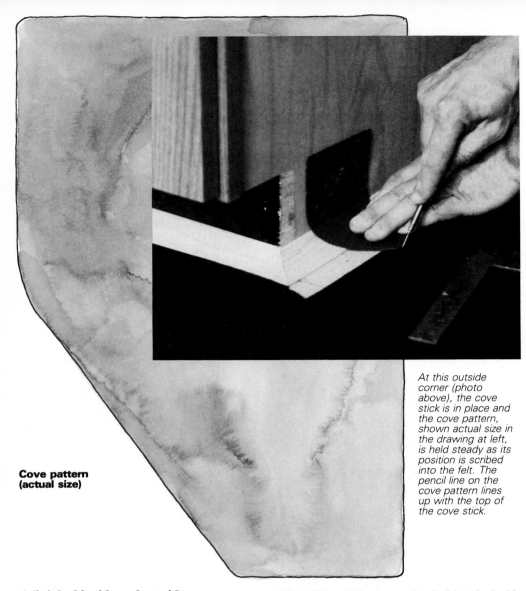

At this outside corner (photo above), the cove stick is in place and the cove pattern, shown actual size in the drawing at left, is held steady as its position is scribed into the felt. The pencil line on the cove pattern lines up with the top of the cove stick.

Cove pattern (actual size)

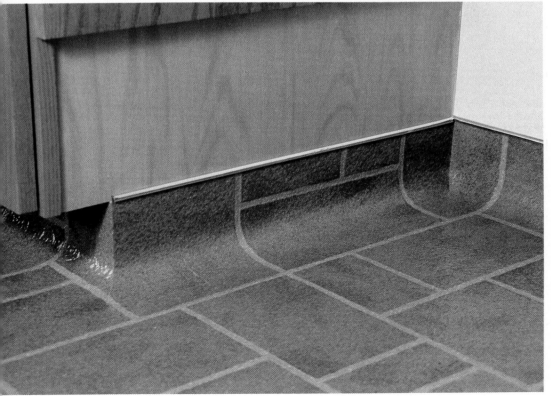

A finished inside and outside corner on a coved floor. Material is removed to fashion the inside corner, while a patch is added to create the outside corner—you can see a faint seam on the right-hand side. Cap metal, held in place by brads, finishes the top edge.

tinued the epoxy another 3 in. out from the seam, and then I troweled latex adhesive over the rest of the drop area. With the second drop in place, I butted the seams together with pressure from my foot, and taped together spots that wanted to separate until the adhesive set up (about an hour).

Coved flooring—A coved floor is essentially the same as a flat-lay floor with a few extra steps. For one, cove stick has to be installed. This looks like a tiny crown molding with a concave face, and it backs the coved flooring. It's mitered at the corners and nailed in place. The pattern felt is laid out the same as in the flat-lay sequence, but the seam line for the drop will be 4 in. closer to the wall, because of the extra material required for the coving.

I trim the felt pattern as closely as possible to the base of the cove stick. Then another pattern comes into play. This one is made from a piece of scrap vinyl, and it duplicates a short section of coving with an inside corner cut on one end (drawing, left). This cove pattern fits over the cove stick just like the finish material. First I decide how high the coving will be—usually 2 in. or 3 in. above the stick. Then I make a pencil mark on the cove pattern to correspond with the top of the stick (photo above left). Starting at the door casing, I align the pencil mark with the stick. Then I note the position of the cove pattern on the felt with scribe marks about every 3 ft. around its base. When the felt is placed over the vinyl to mark the cutlines, I reposition the base of the cove pattern on the marks in the felt. With a pencil, I mark the cutline on the vinyl at the top of the cove pattern.

Inside corner cuts are marked by holding the pattern firmly on the cove stick and fitting the curved side into the corner. Mark the corner of the pattern into the felt. The pencil mark on the pattern won't necessarily line up with the stick. Once the material is cut, the inside corners will look like a square cut-out with a tapered notch in one corner. Installed, the cove flaps fold up and the notch forms a crease in the corner (photo below left).

Outside corners—This is the trickiest part of the job, because it requires cutting a patch of material to fill a gap in the coving. This means matching colors and designs, and making accurate cuts, some with beveled edges. The fill piece usually goes on the less conspicuous side of the corner.

Mark each outside corner on the felt pattern by making a notch in the felt that bisects the angle, then runs parallel with the wall about 1 in. from it (drawing, facing page, top left). This notch indicates which side the fill piece goes on. When all the room's features have been noted on the pattern, place it on the material. Allow room around the edges for the coving, and keep the pattern breaks in mind.

Once the pattern is positioned, tape it down and transfer the scribe marks to the material, using the square around the door trim and the coving pattern along the walls. Connect the cove height marks with the straightedge, and

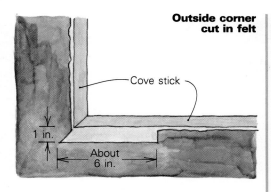

Outside corner cut in felt

Cove stick

1 in.

About 6 in.

Cutting outside corner in flooring

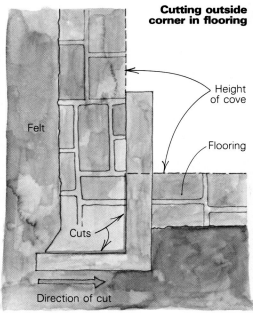

Felt

Height of cove

Flooring

Cuts

Direction of cut

Flat-lay floor around door casing

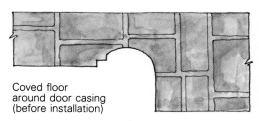

Coved floor around door casing (before installation)

1

2

3

4

Installing an outside corner. When the flooring is in place, the flap protruding at each outside corner will need to be trimmed back. The tape on the vinyl reinforced the unfinished cut while the material was rolled into place. An outside corner scribe (1) is the best tool for transferring the cutline to the finish side of the vinyl. Trim off the flap with a sharp utility knife (2). Start the cut at a 45° bevel, and stay with it through the miter joint in the cove stick and into the field cut (3). Now rub this beveled edge with graphite from a soft pencil, and press the fill piece in place (4). The graphite should leave a cutting pattern to follow (5). Start the cut at the top, about ⅛ in. in from the line, with the knife at 45°. As you reach the curve, bring the knife up to 90°, and finish the cut. The piece should fit well (photo facing page, bottom). If it doesn't, get another scrap and try again.

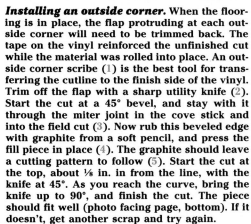

5

cut the material as you would a seam. Stop the cut at an outside corner where the two cove height marks intersect. Begin the outside corner by placing the square along the notch in the felt (drawing, second from top). Cut along the square with the utility knife—these are small seams and must be square and straight. Once this cut is finished, find a piece of scrap that matches the pattern of the 90° corner. This is a fill piece that will be used to complete the corner. Label it, and set it aside. Now reinforce the material around the point of the cut with some tape to keep the vinyl from tearing during installation. The photos above show the installation sequence.

At doorways, you have to deal with a transition from flat-lay to cove. Here the flooring is

cut around the contours of the trim on the door side, then rises in a curve to cove height on the wall side (drawing, above left).

Coved installation—This sequence is the same as the flat-lay process, with an extra step: corner seams have to be spread with epoxy before the field receives its latex adhesive. Spread the epoxy around the door casings, along the tub or shower, in front of the dishwasher, on both sides of an outside corner—wherever water may have a chance to seep in. When both adhesives are spread out, carefully unroll the flooring and tuck it into its corners. If it's cold, hard surface goods may be stiff and reluctant to bend at the cove. Warm it up by passing a hair dryer back and

forth over it, and tack it down with some brads at the top of the coving.

When all the corners are complete, I take out the temporary brads, and install the cap metal with brads 6 in. o. c.

Maintenance—Different types of floors need different types of care, so ask your flooring supplier for warranty information and cleaning instructions. What I have found to work best in my own home is to damp-mop the floor regularly with warm water. Occasionally I add a little mild dish detergent to the mop water, followed by a thorough rinse. I sweep or vacuum the floor frequently, but never use anything abrasive on it. Any vegetable cooking oil will remove scuff marks. □

Raised-Panel Wainscot
Traditional results with table saw and router

by T. F. Smolen

Installing a traditional raised-panel wainscot is a good way to transform a nondescript room into a more formal space. It's also a handsome alternative to replastering old walls that have been damaged over the years by feet, furniture and children. A wood wainscot is more durable than plaster or gypboard, and it relieves the unbroken plane of the wall with the delicate array of shadow lines created by moldings and flat surfaces, as shown in the photo at the top of the facing page.

The term "wainscot" is often loosely applied to various paneling treatments that cover the lower part of a wall. The raised-panel wainscot shown here is a traditional style based on frame-and-panel construction. The frames consist of vertical members, called stiles, and horizontal members, called rails. They support panels whose beveled borders and raised fields give this particular style its name. Each panel rests in its frame with its grain running vertically. Panel width is limited by the width of your stock, unless you edge-join two or more boards together.

Beneath the bottom rail of the frame, a baseboard extends to the floor. At the top of the wainscot, a molding called a chair rail covers the joint between the top rail and the upper section of the wall.

The moldings that I used in making this wainscot can be bought at most lumberyards, and it's possible to make the raised panels and their frames with a table saw. I used a slot cutter and router table to groove the inner edges of my stiles and rails, but you could handle these as well on a table saw with a ¼-in. dado blade.

Panel design—The wainscot that I installed in the dining room of my late Victorian home is traditional in design. I wanted its top to be about 41 in. from the floor. This finished height would include an existing 7-in. baseboard that could be left in place, and a 4-in. wide chair-rail molding that would overlap the top rail of the raised-panel frame. With the stiles and rails 4 in. wide, the panels would show 22½ in. of height. Their actual height would be 23½ in., since ½ in. of the panel edges would be let into the grooved frame all around (drawing, above right).

The width of the raised panels was determined by the distances between corners and the door, window and cabinet frames in the room. In each wainscoted section, I wanted

the panel size to be uniform, so I divided each section to allow equal spacings between stiles. The largest section has four panels, each of which is 12¾ in. wide; the smallest section has a single 18-in. wide panel.

I milled the panel stock from 4/4 roughsawn pine boards about 14 in. wide. The wood, originally intended for flooring, was air dried. I had one side planed down to a thickness of ⅞ in., which allowed for a full ⅛ in. of relief on the raised panel and a sturdy ¼-in. thick tongue around the panel perimeter. Tight knots were acceptable, because I planned to paint the finished wainscot.

Raising panels—I set up a cutting schedule that included all panels for the wainscot, crosscutting the planks to 23½-in. lengths, then ripping them to finished width (the distance between inside edges of the stiles plus a ½-in. tongue allowance on each side). Then I made a template of the panel profile, which I set against the sawblade or dado head when setting up for a cut in order to produce the proper bevel and depth.

Making a raised panel with a profile like the one shown in the drawing below left requires three cuts on each side—one to form the bevel, one to form the tongue on the panel's edge and one to form the shoulder on the edge of the field.

To cut the bevel, I used a carbide-tipped combination blade because it produces a smooth surface that needs little sanding. To set up the saw for the bevel cut, I first set the arbor angle and fence distance to match the bevel on the template. Then I clamped a guide board to the tabletop, parallel to the fence and ⅞ in. away from it. This guide board aligned and steadied the on-edge workpiece as it was fed through the saw (photo facing page, center). Without it, you'd have a troublesome, hazardous time keeping the bevel straight and true. Even with the fence, the blade had to cut through just over 2½ in. of wood, so I held the stock securely and fed with slow, steady pressure. The next time I need a similar setup, I'll use a 4x4 as the auxiliary fence.

Because some boards cup slightly after they are surfaced, I found it best to make cross-grain bevel cuts soon after the boards had been cut to their finished sizes.

When the bevels had been cut on all four sides of the panel, I completed the panel pro-

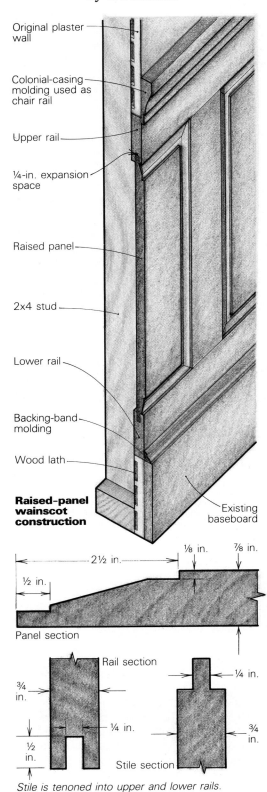

Original plaster wall

Colonial-casing molding used as chair rail

Upper rail

¼-in. expansion space

Raised panel

2x4 stud

Lower rail

Backing-band molding

Wood lath

Raised-panel wainscot construction

Existing baseboard

2½ in.

½ in.

⅛ in. ⅞ in.

Panel section

Rail section

¾ in.

¼ in.

¼ in.

½ in.

¾ in.

Stile section

Stile is tenoned into upper and lower rails.

file using 1-in. wide planer knives (flat across the top) mounted in a Sears molding head.

The second cut (photo bottom left) removes a triangular section of waste at the edge of the panel to create the ¼-in. thick tongue that fits into grooves in the frame. To allow for expansion and contraction of both frames and panels, I trimmed ⅛ in. from the top tongue of each panel and ³⁄₁₆ in. from each side tongue (wood expands more across the grain than along it). The inner edge of the frame would be grooved to a depth of ½ in. to provide expansion space at the top and sides of each panel. Allowing for play in the fit of each panel in its frame is necessary if the wainscot is to survive years of fluctuating humidity. Too tight a fit, and the panels are likely to check or bow out of their frames.

The third cut produces a ⅝-in. wide land at the juncture of bevel and field, and a ⅛-in. high shoulder where the field begins, 2½ in. from the panel edge. No auxiliary fence is required for this cut, but the main fence needs to be set up exactly right. Here again I use the template to get accurate settings for the fence and blade. The blade should just graze the wood surface.

As soon as the panels were cut, I prefin-

In a traditional raised-panel wainscot, a solid wood panel sits in a grooved frame. A chair-rail molding covers the joint between the top rail and the plaster. Overlapping the bottom rail, a baseboard extends to the floor, as shown in the drawing on the facing page.

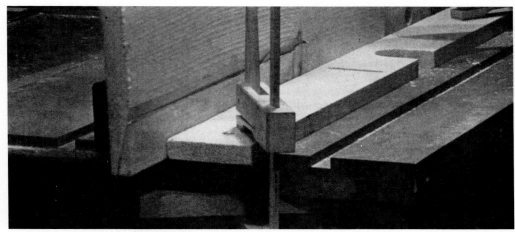

Table-saw setup. Top, a guide board, clamped parallel to the main fence, aligns the panel blank as the bevel is cut. Slow, firm feed and a carbide blade produce a smooth cut. For safety, let the blade stop spinning before the waste piece is removed. Above, a 1-in. wide planer blade cuts the tongue along the edge of the panel. At right, the author uses a tenon-cutting jig, attached to the saw's miter guide, to cut the tenons on a stile.

ished them to keep checking, cupping and wood movement to a minimum while I built the frames. I filled dings and small knots with plastic wood; then I sanded the exposed face with 80-grit paper and sealed knots with shellac to keep sap from bleeding through the finish coats of paint. Finally I gave each panel a coat of oil-base sealer compatible with the enamel finish I planned to use.

Stiles and rails—First I ripped 1x6 pine boards to 4-in. width. I cut all rails about 3 in. longer than their finished length so that after assembly I could scribe them for an exact fit to the walls on either side.

My design called for a ¼-in. wide by ½-in. deep mortise-and-tenon joint between stile and rail, so I grooved the inner edges of all the stiles and rails on my router table, using a ¼-in. slot cutter. Since the slot has to be cut down the exact center of the stock, I tested the setup on ¾-in. thick scrap before running stiles and rails through the machine.

To cut the tenons on the stiles, I used a thin-rim carbide blade on my table saw and a tenon-cutting jig on my miter gauge (photo previous page, bottom right). After a little touchup work with the chisel, I was ready to put frames and panels together.

Assembly and installation—I clamped the bottom rail of each panel section in the end vise of my workbench and fit each section together dry. Small pencil lines on panels and rails served as registration marks for centering each panel in its frame.

In assembling the wainscot, only the stile tenons get coated with glue. The panels are seated firmly in the frame, but not glued. This way, they can respond to changes in humidity and temperature without binding or bowing their frames. I used a small brush to spread glue on the tenons, and then assembled the frames and panels, snugging stile-to-rail joints together with pipe clamps until the glue set.

I wanted to nail the wainscot directly to the studs rather than installing it over the existing plaster, which was sound but presented quite an irregular surface. Leaving the original baseboard intact, I snapped a level line about ¼ in. above the installed level of the top rail and ripped out the plaster and lath. I scribed the plaster along this line with a utility knife, then ripped it off by hand. I cut the lath with a chisel and pried it off the studs with a hammer and small prybar.

Next, I nailed up the paneled sections, which I had purposely built slightly wider than the spaces they would occupy so they could be scribe-fitted to the walls. This left a slight gap between baseboard and bottom rail and between plaster and top rail. I used a 4-in. wide Colonial-casing molding at chair-rail height to cover the joint between the rough plaster and the top rail of the wainscot, and a 1⅝-in. wide backing-band molding at the baseboard and bottom-rail junction.   □

Ted Smolen practices law and does amateur woodworking in Danvers, Mass.

Modified wainscot:
a raised panel with birch-veneer plywood and beveled molding

by Michael Volechenisky

A small ad in the local paper got me interested in building a raised-panel wainscot. It offered for sale "paneled wainscoting from a 150-year-old home" (wainscot is actually the correct term here). This would be just the thing for the dining room in my equally old house, which I was in the middle of redoing. But the age of the wainscot was unfortunately confirmed by its condition, and in any case there wasn't enough of it to go around my 15-ft. by 15-ft. dining room. So I decided to build my own from scratch, enlisting the advice of Luther Martin, a retired builder and woodworker who was sympathetic to the idea of recreating an old look with new materials.

Panel design—I planned to construct a frame-and-panel wainscot along traditional lines, bordering it with a baseboard along the floor line and a chair-rail molding along the top. But Martin didn't want to use solid wood panels because he'd seen long stretches of wood-paneled wainscot push walls out of plumb as a result of normal wood expansion. Cracking and cupping are other risks of solid wood panels. So I made my raised panels from ¾-in. thick lumber-core plywood with birch face veneers, and with slightly modified beveled molding. The molding, which is 2 in. wide, forms the beveled border of the panel, and the lumber-core plywood is the field. A tight tongue-and-groove glue joint between border and field and two coats of white enamel hide the fact that these panels weren't raised in the traditional fashion.

This alternative design has several advantages. First of all, the lumber-core plywood field is far more stable than its solid wood counterpart. Expansion and shrinkage are negligible, as are cracking and checking. And the birch face veneer is better than solid pine or fir if you're planning to paint—as I was—because it contains no knots or resin pockets, which could bleed through the finish.

Third, you can give the panel's bevel a fancier treatment than is possible with conventional techniques, since the border isn't an integral part of the field. The molding I used, for example, has a quirk bead at the inner edge of the bevel—an embellishment that suggests far more intricate work than was actually involved.

Michael Volechenisky lives in Sayre, Pa., and Pompano Beach, Fla. Photo by the author.

Making moldings—From top to bottom, my wainscot contains a chair-rail molding scribed to fit the plastered wall; two smaller moldings (a cove and a beaded stop) that fit over the frame and raised panels; and a baseboard scribe-fitted to the floor, with its top edge covered by a modified scotia molding (drawing, facing page). I made these moldings with my spindle shaper, but could have bought similar ones at the lumber store.

I milled the beveled molding for the panels that make up the wainscot on an old Hebert molder-planer. The Hebert, which is no longer made, is similar to the Williams & Hussey planer (Williams & Hussey Machine Co., Dept. 16, Milford, N. H. 03055), and both machines are shop-size versions of the larger, more powerful planer-molder machines used by lumber mills.

My machine has only one cutterhead, which is mounted horizontally above the table and holds a pair of knives. It is driven by a 1-hp motor, and is powerful enough to complete a molding in a single pass, providing you use knot-free wood that's not too dense. But to be on the safe side, I usually make my first cut to within ¹⁄₁₆ in. of the finished dimension and then run the stock through a second time to get a smooth surface that requires very little sanding.

Molder-planer manufacturers sell a variety of molding cutters to fit their machines, but I've often made my own from precision-ground tool steel (it comes in many sizes and thicknesses, and can be bought wherever metalworking tools are sold). After drilling two holes in each tool-steel blank so they can be bolted to the cutterhead, I transfer my molding outline to the blank and start removing metal. I hacksaw as much as I can, grind the shape to a 30° bevel and hand-file corners and coves where my bench grinder can't reach. After getting both cutters as nearly identical as I can, I mount them in the Hebert and run a trial piece of wood through. This tells me which cutter is doing most of the work by the flecks of wood that adhere to its cutting edges. More filing follows; then I hone the blades and install them. Since most of my molding runs are for 500 ft. or less, it doesn't seem necessary to harden my cutters.

All the moldings, stiles and rails for my wainscot were cut from basswood stock I'd been saving. Basswood works easily, and I've found that you can usually smooth it by hand with a cabinet scraper, with little or no sanding. It also takes a fine coat of paint, because it's knot-free and resin-free.

No matter what type of wood you use on a molder, you'll get better results if you make sure that each board you run through the machine has its grain oriented correctly relative to the cutterhead. To prevent small chips and tears in the molding, feed your boards so that their grain slants down toward the exit side of the machine.

I constructed the panels first. Each completed panel actually consists of five pieces—the birch-veneered lumber-core field and four bordering molding sections. As shown in the drawing, the tongue along the inner edge of the molding is designed to fit in a ¼-in. wide groove cut in the edges of each panel. I grooved the panel edges on my table saw, using dado blades. At the corners of the panel, adjacent molding sections are mitered. Once all the parts for a panel were cut and test-fitted, I glued them up.

After the panels were finished, I built their frames. Top and bottom rails for each 15-ft. side of the room were cut from 16-ft. long basswood boards. Using a hollow-chisel mortiser chucked in my drill press, I cut mortises in the rails to receive stile tenons. Then I grooved the inner edges of the frame on the table saw to receive the ¼-in. by ¼-in. tongue around each panel.

I test-fit the frame-and-panel sections for each side of the room, then glued and clamped them together. Before installing each section, I gave the back of the frames and panels two finish coats—the same number that the front of the wainscot would receive. Thus both sides of the wood can respond equally to temperature and humidity. Perhaps this wasn't necessary, but there hasn't been a single paint crack in the wainscot after eight years on the wall.

Though I could have installed the wainscot directly over the dining room's old plaster walls, they were in such bad shape that I stripped the room down to its studs and nailed up rock lath. Then new plaster was applied down to a temporary ground I nailed just below the height of the chair-rail molding that would top off the wainscot.

The frame-and-panel assembly was the first part of the wainscot to get nailed up. I made sure each wall section was level and used 8d finishing nails, positioning them close to the rail edges so they would be hidden by the covering layer of molding.

Once the four frame-and-panel sections were up, I added moldings to the rails. The baseboard and the scotia-style molding covering its top edge were mitered at the corners, as were the chair rail and its two adjacent moldings. A light sanding, followed by a primer coat and two coats of semi-gloss Kemglow enamel, finished the job. □

The gluing setup used to construct the panel consists of four bar clamps and a Formica-faced base slightly shorter and wider than the panel. At panel corners, the molding joint is a glued miter.

Illustration: Vince Babak

Wainscot construction
Section view

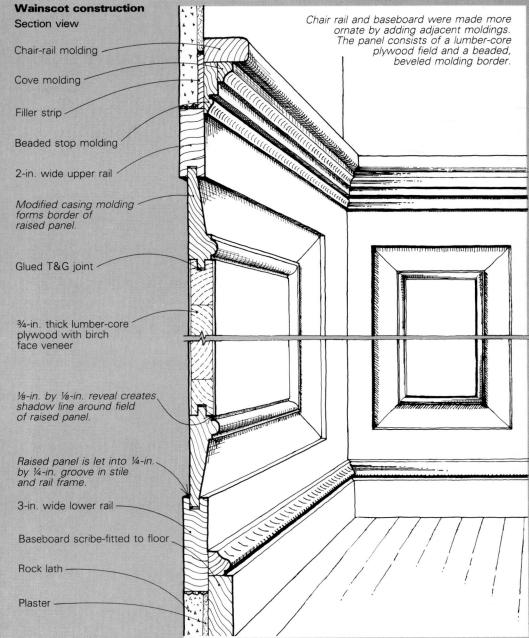

Chair rail and baseboard were made more ornate by adding adjacent moldings. The panel consists of a lumber-core plywood field and a beaded, beveled molding border.

Chair-rail molding

Cove molding

Filler strip

Beaded stop molding

2-in. wide upper rail

Modified casing molding forms border of raised panel.

Glued T&G joint

¾-in. thick lumber-core plywood with birch face veneer

⅛-in. by ⅛-in. reveal creates shadow line around field of raised panel.

Raised panel is let into ¼-in. by ¼-in. groove in stile and rail frame.

3-in. wide lower rail

Baseboard scribe-fitted to floor

Rock lath

Plaster

Making Window Sash

How to do a custom job with ordinary shop tools and a router

by John Leeke

On a historic-restoration project I worked on not long ago, the house's window sash were in poor shape. The original plan for the sash was to repair the worst of them and then replace them all sometime in the future. But before the window work started, the owner decided to have new sash right off. I didn't have enough time to place an order with a custom millworks; so I decided to make the sash in my own small shop, even though it lacks specialized sashmaking machinery. This meant I'd have to match the joinery and molding profile of the originals, and I'd have to work quickly enough to make money on the job without overcharging my customer.

The old sash were hand-made over 150 years ago. One had been without paint for many years, so its joints came apart easily. All I had to do was see how it was made, and reproduce each part. The challenge was to keep track of all those parts, and to make the joints fit properly so the sash would get as much rigidity from its mechanical integrity as from the glue in the joints.

It takes me about 5¾ hours for all the setups needed during a run of sash. The production time for the kind of sash described here on a short run of three or four double units is almost six hours per unit. Considering my time and the cost of materials, the final price was about 20% higher than ordering custom-made sash from the local lumberyard. Not too bad for short-order work that met my specific requirements exactly.

Of course, I could lower these time figures by keeping specialized machinery set up for sash work. If I did, my shop rates would be higher. I'd rather keep my capital expenses low and have more hourly income.

I use white pine for all my sash because it strikes a good balance between machinability and durability. Straight-grain, knot-free wood is essential because the thin, narrow muntins need to be as strong and stable as possible. Also, the outer frame can twist if it's made from wood with unruly grain. I try to use all heartwood, which is stronger and more rot resistant. It's best to cut all the rails and stiles from parts of the board that have vertical grain. These quartersawn lengths of wood are less liable to warp and twist.

John Leeke is an architectural woodworker in Sanford, Maine. Photos by the author, except where noted.

To replace old frames, first remove the exterior casing, as shown above, then remove the interior casing. This exposes the casing nails that hold the jambs to the rough framing. These can either be pulled or cut with a hacksaw blade.

Two kinds of sash—Here I'll describe how I made double-hung sash for jambs that don't have parting strips. So the sash shown in the photos don't have weather stops. *(Parting strips, weather stops* and other sash terms are explained on the facing page.) However, many older sash are made for jambs that have parting strips, and so require meeting rails with weather stops, as shown in the drawing. Meeting rails with weather stops are thicker in section and narrower in elevation than ordinary rails, and are mortised to receive tenons on the stiles, rather than the other way around. If you have to make meeting rails like this, the joinery is the same, except that the stiles are tenoned and the rail is sized to overlap (with

bevel or rabbet) the other rail. You can avoid this trouble altogether, if you wish, by applying the weather stops (with brads and waterproof glue) after the sash are assembled. The instructions that follow are for simple sash.

Sequence of operations—I did all the work on this job with ordinary shop equipment—a table saw, a drill press and a router, which I mounted on the underside of my saw's extension wing. The techniques described in this article can be adapted to produce sash in new construction, casement windows and fixed-glass windows.

I begin by disassembling one of the old sash to determine how it went together, and to get familiar with its decorative and structural details. Then I measure the inner dimensions of the old jamb, and make a drawing of the sash that shows the important features. From the drawing I compile a list that itemizes the parts and tells the dimensions of all the separate pieces.

The sequence of operations in the shop goes as follows: I thickness-plane all the stock (this can be done on the table saw since all the members are fairly narrow), and then cut the tenons and copes on all of the rails and horizontal muntins. Next, I cut the mortises in the stiles and vertical muntins. After this I set up to mold the inner edges of all the frame members on their inside faces, and then I rabbet the same pieces on their outside faces (for glazing). At this point I usually frank the tenons on the rails. Finally, I assemble the sash.

Measurements, drawings, cutting lists—After I take out the old sash, I clean off paint buildup and dirt. If the stiles of the frame are not parallel, I size the sash to the widest measurement and allow a little more time for trimming during installation. If the overall dimensions from sash to sash vary less than ¼ in., I make all the new sash to the largest size and then trim down those that need it after assembly. If the variation is more than ¼ in., I plan to make more than one size of sash. Too much trimming can weaken the frame members.

You can usually make a good guess about the joinery of the original, but if you're doing a precise reconstruction you have to take one of the sash frames apart so you can measure the dimensions of its tenons and mortises.

On my first sash projects I made complete drawings to keep the various parts and joints

Photo: Andrew Edgar

Sash anatomy

A basic sash for a double-hung window consists of an outer frame and an inner framework of smaller members that hold the separate panes of glass. The outer frame is made of vertical members called *stiles* and horizontal members called *rails*. The bottom rail on the upper sash and the top rail on the lower sash are called *meeting rails*, and these are often made to interlock when the windows are closed. This interlock can be a mating pair of bevels or an overlap (see section drawings below), and it helps keep out cold drafts. *Plain-rail sash* have meeting rails that lack the interlock feature; their meeting rails simply abut one another. The lower rail of the bottom sash has to be beveled to fit flush against the sill, which should slope toward the outside to shed water.

In the best construction, frame members are held together by wedged through mortise-and-tenon joints; as a general rule rails get tenoned, stiles get mortised. In some traditional sash, though, the meeting rails are rather narrow and so are mortised to house tenons cut onto the stiles. You can use slip joints, but these lose much of their strength if the glue in them fails, whereas wedged through tenons hold firmly even without glue.

The members of the inner framework or grid that holds the glazing are called by several names. I call them *muntins*, though they're variously known as *mullions, sticks, sticking, glazing bars* or just *bars*. Like the outer-frame pieces, the muntins should be tenoned into the rails and stiles, and into one another.

All the frame members—rails, stiles and muntins—are molded on their inner faces and rabbeted to hold glazing on their outer faces. This arrangement requires that tenon shoulders be made to conform to the molded edge of the mortised member.

In traditional sashmaking, there are two ways to shape the tenon shoulder. The first method involves cutting away the molded wood and shaving the shoulder on the mortised piece flat to receive the flat shoulder of the tenon. This means the beads are mitered on both members. The second, and easier, way is to cope the tenon shoulder. Simply stated, a cope is a negative shape cut to conform precisely to the positive shape that it fits up against. —*J. L.*

Illustrations: Christopher Clapp

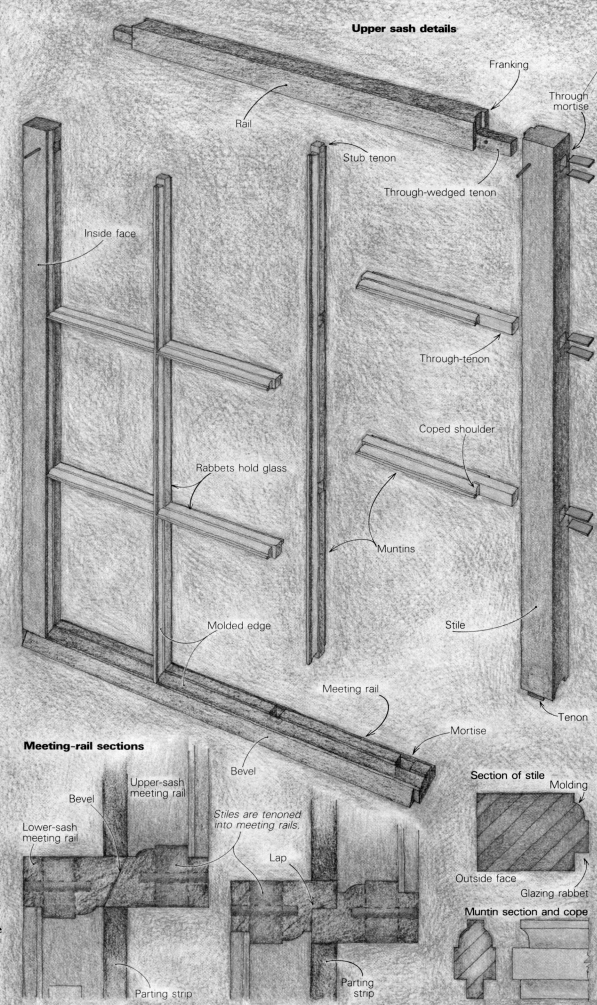

Upper sash details

Rail

Franking

Through mortise

Stub tenon

Through-wedged tenon

Through-tenon

Coped shoulder

Inside face

Rabbets hold glass

Muntins

Molded edge

Stile

Meeting rail

Tenon

Bevel

Mortise

Meeting-rail sections

Bevel

Upper-sash meeting rail

Lower-sash meeting rail

Stiles are tenoned into meeting rails.

Lap

Parting strip

Parting strip

Section of stile

Molding

Outside face

Glazing rabbet

Muntin section and cope

Tenoning with a router. With the router mounted on the underside of his saw's extension wing, Leeke removes the correct amount of wood to produce the cheek of a tenon. The stock is fed into the bit with a push gauge; it squares the workpiece to the cutter and keeps the wood from splintering out on the back side of the cut. A shop-vacuum hose pulls chips through a hole in the fence.

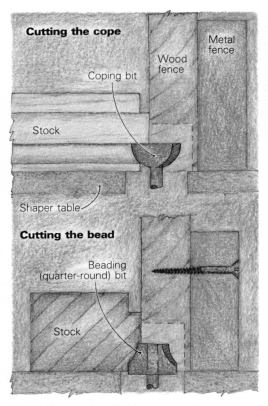

Cutting the cope

Coping bit

Wood fence

Metal fence

Stock

Shaper table

Cutting the bead

Beading (quarter-round) bit

Stock

Ripping muntins to width from stock that has already been tenoned and coped saves time and minimizes splintering and tearout from cross-grain cutting. Because the horizontal muntins are thin and short, a pair of push sticks has to be used.

Rabbeting and molding muntin edges requires the use of two hold-downs. The one attached to the fence holds the stock against the table, and the one clamped to the table holds the work against the fence. Kerfs in the hold-down blocks allow the wood to spring and flex against the stock.

straight in my mind. From these drawings, I made a list of all the pieces I'd need. Each part has its own line, and each line is keyed with a letter to the corresponding part on the drawing. Also on the line is the size, the quantity needed, the name of the part, and its location in the sash.

Router-table joinery—In millwork shops and sash-and-door plants, tenons are cut by single-end or double-end tenoning machines, which cope the shoulder of the joint at the same time they cut the tenon. In smaller custom shops, spindle shapers do the same job. But these are expensive machines, and they take up a lot of space, which I don't have much of. And I was in a hurry. So on this job I improvised a router setup to cut the tenons and copes.

Instead of building an extra table in my already crowded shop, I mounted a router under one of my table-saw extension wings. This arrangement lets me use the saw's rip fence and miter gauge as working parts of the router setup. To keep from drilling needless holes in your saw table, be careful when you lay out and bore the hole in the cast iron for the arbor and cutter and the tapped holes that will let you attach the router base to the underside of the wing with machine screws.

I made a wooden auxiliary fence 2 in. thick and 5 in. high to attach to my saw's metal fence. To suck up chips and dust from around the cut, I bored a hole through the fence and carved a socket to accept the end of my shop vacuum hose (photo top left). This keeps the chips from clogging in the bit during cutting and from building up against the fence.

To guide the workpiece, I use a rectangular push gauge made from a block of pine. By holding the edge of the block against the fence, I stabilize the stock, and square it to the line of cut. Also I notch each corner of the push gauge to each size tenon and to the copes. This way the block backs up the workpiece and keeps the cutter from tearing and splintering the wood as it leaves the cut.

Cutting tenons on rails and muntins—For plain-rail sash there are two tenon lengths—the long through-tenons on the rails and on the stile ends of the horizontal muntins, and the short tenons on the vertical muntins and on horizontal muntins where they are joined to the vertical muntins. It could be that your sash will have a third tenon length for vertical muntins that are joined into the rails, and a fourth tenon length if your stiles are to be tenoned into meeting rails with weather stops. Before setting up to cut tenons, the stock for the rails, stiles and muntins must be surfaced to final thickness and crosscut to finished length. But the stock for the rails and muntins should not at this point be ripped to final width, especially the muntin stock. It's easier and safer to cut the tenons and do the coping on wide boards; it saves time and avoids tearout as well. Remember to mill up some spare pieces for trial fitting, and to be substituted if you ruin good ones. And it's a good idea at this point to set up the hollow-chisel mortiser in your drill press because you'll need to cut some mortises in scrap to test-fit the tenons.

Most tenons are slightly offset from the center of the stock, but because all the framing members are the same thickness, you can set the router bit to cut the tenon cheek on the inside faces of all the pieces, then reset the bit to cut the cheeks on the outside faces. Mark out the dimensions of the two lengths of tenons on a pair of test pieces, and set the bit at the precise height for cutting the face side. This requires careful measuring, for which I use a vernier height gauge.

Calculating tenon length is complicated by the need to cope the tenon shoulder, which in effect lengthens the tenon. This added length equals the depth of the molding profile, and has to be deducted from the length of the tenon. Say your stile width is 1¾ in. and your molding-profile depth is ³⁄₁₆ in.; your through-tenon length before coping will be 1⁹⁄₁₆ in., so you'll set the fence 1⁹⁄₁₆ in. from the farthest point of the bit's cutting arc.

Once the bit is set at the proper height to cut the cheek on the inside-face side of the pieces and the fence is set to cut the longer tenons, you can begin cutting. It's best to make each cut in several passes, even if you're using a large (½-in. or ⅝-in.) carbide-tipped straight bit. You'll get better results without putting an unreasonable demand on the router's motor. Make certain when you make the final pass that the end of the stock is pressed firmly against the fence and at the same time held snugly against your push gauge. Holding the stock this way ensures that the tenon will be the correct length and that the shoulder will be perfectly square.

After the first series of cuts on the rails and on the stile ends of the muntins, you need to set up to make the first cuts on the muntins for the short tenons. To keep from moving the fence and having to set it up again when you cut the cheeks on the outside face of the rails, you can thickness a scrap piece and clamp it to the fence to shim it out from the bit's cutting arc to make a tenon of the correct length.

Once you've made the cuts on the inside face of the muntin stock, you're ready to complete the tenoning by cutting the wood away on the outside face. Leave the shim clamped in place, and reset the router bit to the proper height above the table to make the next cuts. Careful measuring here is critical because your tenons won't fit if the bit is set at the wrong height. So make a cut on one of your spare pieces and trial-fit it in the test mortise. Once you get the bit set correctly, run all your muntin stock through. Then unclamp the spacer from the fence and make the cuts on the outside face of your rail stock and the muntin stock that gets long tenons.

Coping the shoulders—Coping with a router means you have to pattern-grind a matched pair of bits—a concave bit to cut the molding on the inside edges of sash members, and a convex bit to cut the cope on the tenon shoulders (photo above right). The positive and negative shapes of the pair must be perfectly complementary or your joint won't close properly, and will have gaps. The sidebar at right explains how to grind stock high-speed steel bits to get a matched set.

Now you're in a dilemma because you need a molded, rabbeted and mortised stile to test-fit the pieces you've tenoned and are getting ready to cope. The best choice here is to rip a stile to finished width, set up the router to mold the inner edges and rabbet the outer ones according to a full-size drawing of the stile in section. But you're having to perform an operation out of its logical sequence, and that can seem a waste of time. You'll also need a couple of muntin pieces; so rip a couple to width and mold and rabbet them at the same time you do the same to the stile. Whatever you do, don't throw your sample pieces away once you've gone to the trouble to make them. If you ever need them again, you'll save several tedious hours of trial-and-error setup if you have these samples to refer to.

Now that you've got a stile prototype and a couple of muntin samples, chuck the coping bit in the router and set the height so that the top of the cutter just lightly touches the bottom edge of the tenon (stock held inside face down on the table), as shown in the upper drawing on the facing page. Next set the fence to cope the shoulders of the muntins with short tenons. Be conservative when you set up. Make a pass into the cutter, and trial-fit the piece. If the fit is bad, adjust the fence cautiously and try again. Keep making minute adjustments until you get it right. You'll get some tearout on the exit side of the cut because the edge has been molded, but this won't matter for the test piece. Now cope all the shoulders for the short tenons of the muntin stock. Next reset the fence to cope the shoulders of the long tenons.

At this point you should rip the muntins to width (middle photo, facing page). To keep from having to clean up the sawn surface with a plane, I use a sharp planer blade in my bench saw. Because the muntins are thin, I use a pair of push sticks to feed the stock into the blade. Also at this time you should rip the rails and stiles to final width.

Mortising stiles and muntins—I use a ½-in. hollow-chisel mortiser that I keep sharp for this job. You can buy one of these attachments for your drill press at most woodworking-machinery dealers. Be sure to buy a little conical grinding stone that keeps the chisel sharp. A dull chisel will tear the wood on the walls of the mortise and cause nasty splitting out on the back side of the stock. Even a sharp chisel will do some tearing out if you don't back up the cut with a maple block. I use an aluminum plate that's cut out to the precise ½-in. square dimensions of my hollow chisel. The edges of the square hole give positive support to the wood as the chisel exits, and prevent splintering altogether.

To get precise results, each mortise should be laid out with care. It's best if you use a mortising gauge and striking knife, but a sharp

Pattern-grinding router bits

You have to be able to grind your own router-bit profiles if you want to reproduce moldings or make new ones of your own design. For one set of window sash that I was making, I needed a matched pair—a quarter-round bead and a cope. To make such a a pair, I begin by buying "blanks"—common high-speed steel bits, large enough to yield the finished shape I need. For these sash I used a rabbeting bit for the molding cutter and a rounding-over bit for the coping cutter.

To lay out the shape for the cutting edge, I first coat the back face (opposite the beveled face) of each bit with machinist's blue marking dye, and then scratch the profile that I want on the surface. Dye makes the pattern easy to see when grinding. If I am reproducing a molding, I lay a thin cross-sectional piece of the molding on the back face of the bit and scratch around it.

Holding the bit in a hand vise, I grind away the unwanted steel on my bench grinder. As I near the mark I make certain to match or slightly undercut the original bevel. All of this grinding is freehand, steadied a bit by using the grinder's tool rest. For tight inside curves, I use a small, thin grindstone mounted on the work arbor of my lathe. I am careful to remove the same amount of steel from both flutes so the balance of the cutter is maintained.

After I've ground right up to the mark, I refine the shape by hand with medium and fine India slipstones, sharpening the edge in the process. Then I touch up the edge with hard Arkansas slipstones. Finally I mount the cutter in the router to test the shape it will make on a scrap of wood. If the resulting shape is not correct, I grind a little more, resharpen and test again.

To make a coping cutter match another bit, I use the bit itself as a pattern for the layout. When the grinding on the coping cutter is approaching the final shape, I slip both cutters into a simple jig that holds the shanks parallel. The jig is just two ¼-in. holes drilled in a block of hardwood with a drill press. I turn the bits so their cutting edges are next to each other. Holding the assembly up to the light, I can easily see where more steel must be removed to make the edges coincide. —*J. L.*

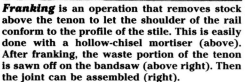

Franking is an operation that removes stock above the tenon to let the shoulder of the rail conform to the profile of the stile. This is easily done with a hollow-chisel mortiser (above). After franking, the waste portion of the tenon is sawn off on the bandsaw (above right). Then the joint can be assembled (right).

Assembly is straightforward. Daub the mortise walls with glue, and tap the frame members together. Then drive in the wedges for the through tenons, and snug up the frame with two bar clamps. The frame shown here is slip-joined because it will be a fixed upper unit and not subjected to the stresses that sliding sash are.

#4 pencil and a try square will do. I clamp an end stop to the drill-press table to ensure that mortises at tops and bottoms of stiles are all the same distance from the ends of the stock.

Molding and rabbeting—The next machining processes mold the inner edges of the frame members on their inside face and rabbet the inner edges on the outside face. First I install the molding bit in the router, adjust its height and set the fence. Having done this already on the test pieces, I use one of them to help with the setup. Then I mold the stiles and rails, after which I clamp hold-downs on the fence and the saw table (bottom photo, p. 116), to run the thin, narrow muntins. Molding done, I cut the rabbets, using the same straight bit that cut the tenons. Again, I clamp hold-downs to the fence and table.

Franking—When you lay out the through mortise-and-tenon joints for the rails and stiles, you'll need to leave at least an inch or more of wood between the end of the mortise and the ends of the stile, or the wood can split out. Therefore, the tenon needs to be an inch or so narrower than the rail is wide. When you cut the tenons with the router, you get a tenon the full width of the rail, so you must cut part of it away to get a tenon the right width and to get the newly exposed shoulder to fit the sectional profile of the stile. Molding and rabbeting the stile produces a proud land (flat ridge) that runs the length of the member, and the new tenon shoulder has to be relieved to accommodate this long, flat ridge.

Relieving the upper shoulder of the tenon in this way is called franking, which I do with the hollow-chisel mortiser (photo top left). What's required is mortising back behind the tenon's shoulder to the width that the land is proud and to a depth that stops at the line of the top of the finished tenon. Then all you do is saw away the waste (photo top right).

Assembly—I really enjoy this part of sashmaking, when all the work finally pays off. Traditionally, joints were put together with thick paint between the parts. I suspect the sashmakers in those days expected the paint to seal moisture out of the joint rather than act as glue to hold it together, but they didn't rely much on the adhesive strength of either paint or glue. They pegged the tenon through the inside face of the stile and wedged it top and bottom from the open end of the mortise.

Most weatherproof glues seem to work well, except for formaldehyde-resorcinol, which can bleed through to the surface after the sash is painted. When all the parts of a sash are fitted together, I snug up the joints with a bar clamp, and wedge the through tenons.

After wedging I check the sash for squareness and then peg the joints. Pegging is especially important for the joints of the meeting rail on the lower sash, because they are subject to a lot of stress during use. If pegs pass through to the outside of the sash, they could let water get into the joint. My pegs stop just short of the outside surface. □

Installing sash

When you're deciding what to do with sash that are in poor shape, you should consider the window as a whole. Examine the frames for deterioration. Windows on the south side of a house are subject to repeated wetting and drying. This can cause large checks or cracks in sills and stiles, so that these members need to be replaced. The north side of the house will be damper because it is in the shade, and you're likely to find rot in the joints of the sill and the jamb stiles.

Jambs and casing—If the sill or jambs need replacing, I usually take the whole window frame out of the wall. I begin by carefully removing the exterior casing and moldings. This exposes the space between the jamb stiles and the structural framing (trimmer studs, header and rough sill), which gives me room to cut the nails that hold the frame in place. I use a reciprocating saw or hacksaw blade for this. There also may be nails that hold the interior casing to the frame that will have to be cut.

Once the frame is loose, it can be pulled out at the top and removed from the wall. I try to use heartwood for replacement parts of the jamb and casing because it is more rot-resistant than sapwood. In any case, the parts should be treated with a water-repellent preservative before assembly and reinstallation. Reusing old casing and molding that are still in good shape helps blend the new work in with the rest of the house. If the old jambs are still good, I scrape and sand their inside surfaces so they are flat and smooth for the sash to slide against.

Fitting the sash—To install double-hung sash with the upper sash fixed in place and the lower sash sliding up and open, you first trim the stiles to fit, then glaze and finally install the sash in the jambs.

Start sizing the top sash by planing the edges of the stiles so the sash will fit into the frame without binding. This should be a free-fit, but not so loose it will rattle side-to-side. Put the sash in place and slide it up to the frame header. If the top rail of the sash doesn't fit uniformly flush along the header, it should be scribed to the header with dividers and trimmed to fit. Then fit in the two outer stops, which are strips of wood that lay flat against the jamb stiles and hold the upper sash in place. If the top rail of the bottom sash and the bottom rail of the top sash are made to overlap and form a weather stop, your jambs must be fitted with parting strips. These are strips of wood that are let into grooves, one in each inner face of the jamb, that run the length of the jamb stile and serve to separate the two sash so they don't slide against one another. Fit the parting strips into the grooves in the stiles so they are held in place by compression only. Don't glue or nail them in place.

Next trim the lower sash to fit by planing its side edges until it runs smoothly up and down in the frame. Set the sash in place with the bottom rail resting on the sill. Then scribe the bottom rail to the sill with dividers set to the distance between the top surfaces of the meeting rails. Plane off the bottom rail to the scribed line (photo above right), forming a bevel that matches the slope of the sill. The weather stops should fit tightly together when the bottom rail of the lower sash is against the sill. If too much is planed off the bottom of the lower sash, this fit is lost. So take some care when trimming for this fit.

When the sash are sized to fit, they should be treated with a water-repellent preservative and primed for painting. Do not prime the side edges of the sash. They should be left bare to slide against the stiles.

Glazing, painting and finishing—I usually take sash to the glass shop to be glazed. The glass should be bedded in a thin layer of glazing compound and set in place with glazing points. When complete, the glazing compound should have a neat beveled appearance and not show from the inside.

The sash should have two top coats of paint. I prefer oil-base paints. Run the paint just slightly onto the glass, thereby sealing the glazing from rainwater. Do not paint the edges of the sash that will slide against the frame. When the paint is dry, wash the glass.

To install the window in its jamb, set the top sash in place and slip the parting strips into their slots. Trim any beads of paint that may have dried on the side edges of the lower sash, and test to see if it still slides freely in the frame. When you are satisfied with the way the sash fits, then secure it in place with the beaded or molded inner stop, taking care to use thin brads so as not to split the wood. These stops should be carefully positioned so the sash is free to move but not so loose that it will rattle in the wind. —*J. L.*

Photo this page: Andrew Edgar

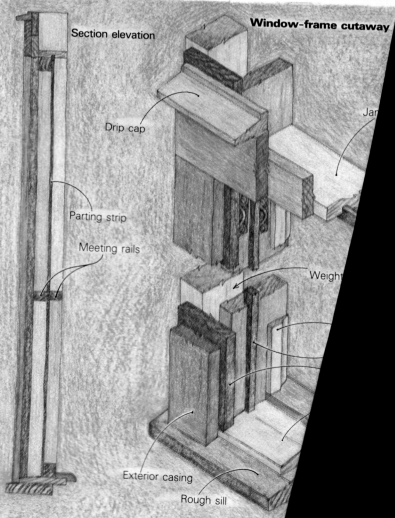

Section elevation

Window-frame cutaway

Drip cap

Parting strip

Meeting rails

Jamb

Weight

Exterior casing

Rough sill

Leaded Glass

The tools, materials and techniques of this craft haven't changed much in 500 years

by Doug Hechter

It wasn't easy to find a large sheet of glass in the Middle Ages, but you could get small, bubble-filled glass discs called crowns. To make them, a glassmaker would gather a blob of molten glass on the end of a blowpipe, and spin it until it formed a circular plate, sometimes as big as 5 ft. in diameter. The resulting glass might be colored or clear, but it was always very expensive.

Once it hardened, the glass was ready to be cut into pieces and assembled [b]y artisans into windows. [The]se craftsmen used lead [cha]nnels called cames to [hold t]he pieces of glass to[gethe]r into a large panel.

[Mak]ing windows this way [was] expensive, time-[consumin]g process, and the [glass w]as just about the [custo]mer who could [afford it.] Church windows often depicted [even]ts, thereby teaching the illiterate [a story with]ing in the light.

[The comin]g of the 15th century brought [a change. Glass] windows were becoming [more common] and a burgeoning middle [class wanted mo]re comfortable, better lit [homes. New re]ctilinear windows, using [diamond] shaped panes linked by [lead, were ma]de to satisfy both mar[kets. Leaded] quarry windows (after [the small panes] of glass, which are called [quarries. Maki]ng one is the topic of this [article.] [These wind]ows remained the stan[dard for cen]turies, and there is a re[newed interest] in today. I think it's a re[action to count]less modern goods that [surround us. T]he individual panes in a [quarry window a]re almost in the same [plane, but th]e tiny differences in [the glass's m]anufacture the reflected [light, which] is the human touch, [missing in] late 20th century.

[The success] of good leaded win[dows as an] object of art sec[tion. To] withstand a rea-

sonable wind load, as well as the sometimes grueling fluctuations of temperature and movement within the building. To achieve this, steel-reinforcement braces are soldered to the inside of the lead frame. Without them the window would eventually sag and fall apart. A well-designed leaded window has lead lines wide enough to hide the braces.

When a building is designed, the architect gives considerable thought to its proportions, its visual mass and the textures and values of the materials. The leaded-glass maker should tune in to these choices. A successful window design reflects the scale of the structure, and if it has colored glass, it should have light and dark areas, just as the building's mass presents highlights and shadows. Like the glazing bars in French doors, the lead lines afford some patterned continuity in the plane of the wall. The larger the window is, the heavier the lines should be.

Strong outdoor light tends to minimize the lead lines once the window is in place. It's especially noticeable in skylights. Colors that might look strong and expressive on the workbench can easily become anemic and indistinct under the glare of natural light.

The cartoon—The cartoon is where construction begins. It is a plan containing all the information pertinent to the building of the window, such as the lead widths, decorative details, the panel size and the client's name. It stays with the window throughout assembly. The cartoon for the window you see on these pages is on 70-weight kraft paper (photo facing page), but any heavyweight paper that will take pencil lines can be used.

I begin a cartoon by marking it with the lines that define the full size of the panel. I get these dimensions by measuring the window opening from the inside of the rabbets, and subtracting ⅛ in. I note them on the cartoon next to the client's name, width first. This is when the window opening has to be checked for square, and any deviations noted.

After the full-size perimeter lines are drawn, I decide how wide the border lead should be. This is a function of how deep the rabbet is at the sash. I like about ¼ in. of lead line around the window, so I add that to the depth of the rabbet for the width of the border lead. For instance, if the rabbet is ⅜ in. deep, the border lead should be ⅝ in. wide. The dimension of the window from the inside edges of the border leads is called the daylight panel size.

Once I've marked the inside edge of the border lead on the cartoon, I draw in the positions of any other lead lines. These are all done at full size, and they are followed by more lines that mark the center, or heart, of the leads.

Patterns—Next, I sandwich a large piece of carbon paper between the cartoon and another piece of kraft paper. Tacks or tape hold both sheets securely to the table. I use a ballpoint pen to transfer the lead centerlines to the lower sheet. These become cutlines for glass patterns. During the copying, I occasionally lift a corner of the cartoon to see if I missed any passages. Each shape represents a separate piece of glass, and each one is designated by location with a letter, as shown in the drawing at right. In a more complicated window, I designate the shapes with numbers. If there are colors in the window, they are also noted on the patterns.

I cut the lower sheet into a pile of patterns with a special pair of shears. Unlike a normal pair of scissors, this tool has two blades opposed by a third in the middle. The pattern shears remove a small fillet of paper as they cut, which compensates for the heart of the lead. There are different shears for different sizes of lead. The cut should be made with the portion of the blade closest to the rivet. The shears should be slid into the cut and only closed completely at the end of a line. Closing them in the middle of a cut will cause a nick that may snag the glass cutter later. The patterns have to be cut accurately, because they will reproduce exactly on the glass. Slight flat spots along a curve can be corrected, but in general the cut should split the carbon line. When the cutting is done, sort the patterns into piles by color and shape.

Cutting glass—Glass can be cut into simple or complex shapes by scoring one side and then applying concentrated pressure opposite the score marks. Glaziers use two types of tools to score glass for a cut: the diamond-tipped cutter and the steel wheel. The steel wheel is the more accessible tool and it's the one I used on this window. Steel wheels are inexpensive, and if they are cared for, they can handle the most complex of cuts. If the wheel is nicked, however, it will make a discontinuous score mark, which will make the break in the glass run wild.

The steel-wheel cutter should be stored in a container with some tissue at the bottom. This tissue should be soaked with a lubricant—mineral spirits, kerosene or any light machine oil. The lubricant is necessary for a successful cut. The oil not only helps the wheel spin freely during the cut, but also seems to affect how long the score mark remains open and breakable. A score is likely to heal over if the glass isn't broken within about 30 seconds.

When I make a cut against a straightedge, I first paint oil along the path of the cutter. This makes for a very clean break. Sometimes cranky old salvaged glass or the more brittle antiques and opals (see the sidebar at right) need this extra help.

The steel wheel feels quite awkward at first. Most beginners either push too hard or are indecisive. The score should be made on the smoother side of the glass. Hold the cutter plumb. The cut should be heard but not easily seen. If the cut is a fuzzy, sputtering line, too much pressure has been applied. Speed is not important but evenness is, so don't stop in mid-cut. You can let the wheel roll off the edge of the sheet to complete the cut without worrying about breaking the glass.

To make the break, hold the glass with a thumb on each side of the score at one end, and your index fingers bent and touching

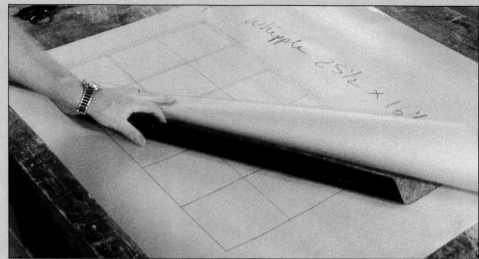

The cartoon is a full-size drawing of the window with all the information pertinent to its construction. For this window, the border cames will be ⅝ in. wide, and the field cames ⅜ in. The lead lines are transferred, via carbon paper, to the bottom sheet. The shapes inside these lines represent the glass, and they are labeled, cut out and used as patterns.

Window layout
Even though the panes appear to be the same size in the finished window, those that intersect with the border cames need to be a little larger to fit into the deeper channel. The window at right has panes of four different sizes.

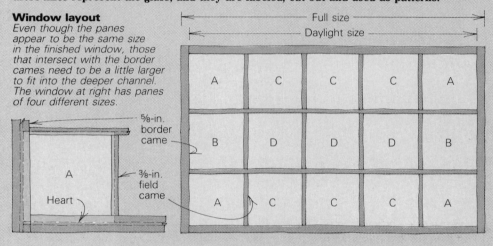

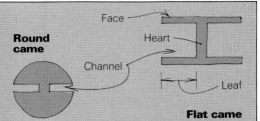

Round came · Face · Heart · Channel · Leaf · Flat came

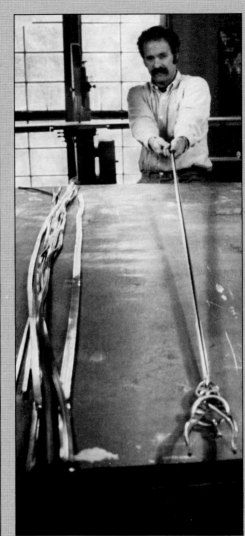

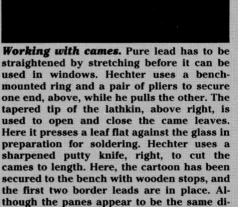

Working with cames. Pure lead has to be straightened by stretching before it can be used in windows. Hechter uses a bench-mounted ring and a pair of pliers to secure one end, above, while he pulls the other. The tapered tip of the lathkin, above right, is used to open and close the came leaves. Here it presses a leaf flat against the glass in preparation for soldering. Hechter uses a sharpened putty knife, right, to cut the cames to length. Here, the cartoon has been secured to the bench with wooden stops, and the first two border leads are in place. Although the panes appear to be the same dimension, there are actually four different sizes in this panel.

wheel. This method works for both curved and straight-sided shapes. If I'm going to make repetitive 90° cuts, such as the ones for the glass in this window, I set up a jig using a T-square and a stop, and I occasionally check the cuts against the pattern.

Working with cames—I'm often asked how I pour the lead between the panes. The lead is not, and never was, poured around the glass. It starts out cold as H-shaped strips called cames. They are about 6 ft. long, bend easily by hand, and can be cut with a knife. Cames may be less than ⅛ in. wide, or exceed 1 in. in width. The faces are usually round or flat (drawing, above left). Flat lead has a flexible leaf that closes down snug against the glass. The round style is usually heavier in section, with a stiff leaf not intended to be closed down. Round lead looks more delicate than flat lead, and it can be bent into tight, intricate curves. Most Victorian work uses round came.

Not all brands of came are pure lead. Antimony and other metals are being used more and more. These mixes are economical, and are not as prone to tarnish as pure lead. Many craftsmen prefer this new product because it solders so well, but its drawbacks make me hesitant to use it. It is harder and more brittle than pure lead, which makes it more difficult to cut and shortens its lifespan.

Lead cames must be stretched before they're used (photo far left). This hardens the lead, so I'm careful to pull just enough to straighten the came. Over-stretching also gives the surface a scaly texture. Alloy leads usually don't need stretching.

I used flat lead in this window, and the channels had to be opened with a lathkin. It looks like a fat knife blade (photo top center), and I use it constantly. Mine is 2 in. by 6 in. by ⅜ in. thick, and it's made of Teflon. Traditional lathkins are made of waxed hardwood.

Once the came is prepared for glazing, it should be set carefully aside. Its pliability causes it to kink readily, and if it's bent more than once through careless handling, it will lose its crisp look in the finished window.

In addition to the lathkin, I use a small hammer, a lead knife, a stopping knife and horseshoe nails. The lead knife I use is a hand-forged, high-carbon steel putty knife made in England under the brand name of Footprint. I have shortened the blade a bit and honed a sharp edge onto it. Many window makers use the German-style lead knife. It has a rounded cutting edge that ends in a hook. Its point is also useful for lifting the leads and poking around in tight places. I make a cut by setting the knife on the face of the lead, applying light pressure and rocking the knife back and forth (photo bottom left). After the blade penetrates the top leaves, the lead can be quickly chopped without crushing it. If many leads of a single length are needed, you can use a bandsaw or a radial-arm saw. Still, my everyday cutting is done by hand. The tools used in glazing have worked well for centuries, and there's little room for improvement.

Partner to my lead knife is my stopping

each other at both knuckles beneath the glass. As the thumbs press down, the index fingers become a fulcrum, and the glass will break along the weakened score line. On a stubborn piece or a tight curve, you can tap the score from beneath to start a crack. Use either the handle of the cutter or the grozing teeth found on the back side of the cutting end. This crack can then be chased along the whole cut, or used only to encourage a break. Breaking pliers can also be used to assist a stubborn cut. They have jaws up to 1 in. wide that meet only at the end of the tool. They are used to provide more leverage. But a tool-assisted break is never as clean as one done by hand.

Grozing is the term for nibbling away at a

piece of glass. A notch on the cutter handle slips over the edge of the glass, and when the handle is levered upward, a bit of glass breaks off. This is how glass was shaped during the Middle Ages. Most grozing is done with a pair of grozing pliers. This blunt-nosed, untempered, parallel-jawed tool can be very useful at the cutting bench. It has jaws that meet only at the business end, and it is used as the name implies—when a cut can't be completed by hand. There is no inside curve that cannot be shaped by patient grozing, and I have won beer proving this.

To cut glass with the pattern, hold the pattern against the glass with one hand, and follow the edge of the paper with the cutter knife. Made from a wood-handled oyster knife, it is used for probing tight corners, lifting came leaves and prying. I cast a lead plug into its handle for tacking nails or tapping glass into place.

Glazing—This is the term used for the assembly of glass and lead into a window panel, and it takes place on a sturdy, flat workspace. The top of my glazing bench is a series of tightly fitting 2x planks, and it's 36 in. high. The cartoon is placed at one end, with ½-in. by 2-in. stops nailed along the bottom and the left-side lines (photo right). Check the stops with a framing square. If they're too long for the square, use the 3-4-5 method. After cutting and placing the border leads, I insert the first glass into the channels. The open edges of the glass should align precisely with the inside edge of the penline that transferred the design to the pattern. These lines will be the references throughout the glazing stage. If the glass is allowed to deviate from this line, the result will be distortions that will multiply as the panel progresses.

Flat came is tucked wherever it intersects. The flexible leaf is gently pried up to accept the intersecting lead. The hump that results is then carefully hammered flat. This tucking strengthens the window and enables the glazier to adjust the leadwork without recutting any of it. The tucked end should not extend all the way to the heart but should be held back slightly to prevent the panel from glazing full (expanding beyond the reference lines).

I tack horseshoe nails into the bench alongside each piece of glass as it goes into place. They pin the loose pieces against the stops until the cames are soldered, which happens when the entire panel is in place. As the window grows, I pull them up and stick them around the working perimeter.

Round-faced lead can not be tucked. It is always butted or mitered, and for it to be done right, the cuts have to be straight. If any undercutting occurs, what looks tight on one side will have gaps on the other.

Working from left to right, the quarry pattern is glazed one course at a time with one long lead capping each course. These leads can be kept straight by placing a stop alongside and tapping it with a hammer.

Occasionally, handmade glass runs too thick to fit into the channel. Many people avoid using it for precisely this reason. Seen from the outside, leaded glass has a certain faceted quality, and glass of random thickness contributes to this effect. Thicker glass also bends the light as it passes through the window, casting wavy patterns of sunlight around the room. Much new leaded glass is flat and inanimate because the craftsman holds flawless lead as the highest priority, using only glass that cuts and fits easily.

To modify the channel to accommodate thick glass, I sever the leaf from the heart with a pair of front-cutting nippers, the type that carpenters use for clipping nails. The leaf should be cut only on the side where the glass is uneven, leaving the other half intact. If necessary, I make several cuts ⅛ in. apart. This segmented leaf yields to the glass. It is then pressed flat against the glass and reconnected with solder. The soldering iron should be on the cool side. A scar remains, but the overall product is improved.

Glazing. **A window builds from left to right, with a long lead came capping each course. The short leads are tucked under the leaves of the intersecting cames, strengthening the panel. At the top of each course, the leads are adjusted with a few taps from a hammer against a stop. Note how the leads and the panes align with the cartoon.**

One further note on glazing: a neat, uncluttered bench shows up in the finished product. It is better to have unbroken leads wherever possible, for this reason. Leads partially woven into the unfinished panel will trail out onto the bench. They scratch easily, so don't let them mingle with tools, and can your lead scraps frequently.

Placing the last two border leads along the upper and right-hand edges completes the panel. Nail two more wooden stops to the bench, recheck your full-size dimensions and adjust the border leads accordingly. With the flat came there is a bit of give and take, and I slide the tucked leads in and out to adjust the size of the panel.

Once all of the joints have been hammered flat, the leads can be pressed down against the glass with the lathkin. The tool rides over the heart of the cames, and you have to be careful not to stress the glass. Individual leads can be straightened or aligned by placing the tip of the lead knife on the came at about a 45° angle, and then tapping its handle with a hammer.

Leads down and aligned, I affix a small pa-

Soldering. A special tip solders the cames, but before it can be used it must be heated, filed and tinned. Hechter's tinning tray is an old coffee can, above, and it contains a small amount of solder and powdered rosin or flux. The hot soldering tip is held slightly above the joint and a piece of wire solder placed between the two. The three are brought together for a few seconds, and the heat transferred through the solder to the came. When the solder melts, the tip is lifted and the finished joint looks like the one at right.

Finishing. Pigmented cement is forced into the cracks on the weather side of the window using a stiff bristle brush, left. In addition to sealing the window, the pigment tones down the brightness of the fresh lead. Excess cement is removed with brushes, rags and sawdust. Steel braces keep a leaded window from sagging or giving in to the wind. They are soldered to the interior side of the window. The ends of the braces are pounded flat into 'paddles' and soldered to the came with a chisel-tipped iron, above.

per label to the glass in the upper right-hand corner. This identifies the panel, and marks the interior side.

Soldering—A glazed panel should be soldered as soon as possible. If you wait, you get oxidation, which hampers the soldering. Minor oxidation can be cleaned away with steel wool or a wire brush. When the lead is badly oxidized, as is often the case with pure lead came exposed to moisture, brush the panel with flux and allow it to sit awhile before you scrub it. Be sure the panel is free of dust and debris before soldering. Any bits of lead or loose solder caught under a leaf will break the glass during cementing.

Oleic acid, a derivative of sheep fat, is the best flux for lead. With a brush, I coat each joint with flux just before soldering. Use the flux sparingly, because any excess will seep into the panel, form a puddle and then ooze out after the window is finished.

A soldering gun is of little use for all but the smallest of lead. The tool worth owning is a 250-watt American Beauty soldering iron with a 5⁄8-in. tip. Hexicon also makes a good iron.

The iron doesn't need a thermostat. Its temperature can be controlled by switching the power on and off. My iron is connected to a switched socket, and I know it's on when I see the red light out of the corner of my eye. The temperature demands for soldering leaded glass differ greatly from moment to moment, and the switch gives me the most flexibility.

For soldering the leads, I use a homemade tip made from a 5⁄8-in. piece of copper rod. It's bent at a 90° angle, and it has a blunt end. As soon as the iron starts to heat up, I file the tip smooth of the burrs and pits caused by the acid in the flux. I then tin the clean tip (coat it with a thin layer of solder) on the lid of an ordinary can (photo facing page, top left) on which is placed a little flux or powdered rosin and some solder. The tinning tray will draw excess heat off the iron, which makes it another means of temperature control. Before I start in on the panel, I test the iron on a piece of scrap lead.

Soldering done well requires a light touch, proper temperature and pacing. The tip hovers above the joint as a piece of 50/50 wire solder is placed between the two (photo facing page, top right). The solder transmits the heat as the tip is rolled slightly. When the lead has taken on sufficient heat, it will accept the solder and the joint will flatten. If the iron is too hot, the lead (which melts at a slightly higher temperature than the solder) will mingle with the puddled solder and make a lumpy joint. These can be flattened, but messing with them doesn't help them to look their best. When every joint has been soldered, I go back and check the panel for the joint that has inevitably been passed over. Then I wipe off the flux with a rag, pull the stops and dress down the unused channels on the outside of the border lead with the lathkin. The panel is now ready to turn over.

A half-soldered panel is still flimsy, and must be turned carefully. I pull it halfway off the bench and pivot it to vertical using the bench edge as a support. Then I lay it back down using the same technique in reverse. After truing any misaligned leads, I hammer the joints flat. This is the exterior side of the window, and the leads will not be closed down until the panel is cemented (explained below). Once the soldering is finished, I let the panel stand for a few days. This allows the flux to dry out. It should be tilted slightly, backed by boards for support. Larger panels should be stacked with a board in front as well to prevent them from folding over on themselves under their own weight.

Cementing—At this stage, the panel still rattles when it is moved, and it would leak on a dewy morning. It becomes a solid, impermeable unit when a waterproofing compound called cement is forced into the spaces between lead and glass on the outside of the panel. Cement is pigmented, and it turns the new metal shine of the panel to a dull grey. Quick cement can be had by mixing white gas and lampblack into a portion of steel-sash glazing putty. Stirred with an electric drill and paint-mixing attachment, it will flow freely, yet support a popsicle stick upright when its consistency is right. The best cement takes longer to concoct, but it is tenacious. The formula, in volume measurement, is: 12 units powdered whiting (powdered calcium carbonate), 1⁄2 unit powdered lampblack, 1½ units Japan or cobalt drier, 1⁄2 unit grey floor and deck enamel (alkyd resin type), 1¼ units boiled linseed oil, 1⁄2 unit turpentine. This compound is best when it's fresh, but it can be stirred up and used for a week if it's kept covered.

Gloves and an apron are a good idea for messy jobs like mixing and applying cement. The tools I use for cementing include one large and one small natural-bristle scrub brush, two good rags, sawdust, a stiff bent putty knife and a sharpened bamboo stick.

I ladle some cement to the open-lead side of the window, and work it into the cracks with the small brush (photo facing page, bottom left). Only one side is cemented because air pockets deep in the window will heat up under sun and blow out the seal of one side or the other if both are cemented. The one-sided approach works only with flat lead.

I brush the cement under the leads in a circular motion. Every space, every corner must be filled. A heavy stroke discharges cement from the brush, a light stroke pulls it off the window. After the entire side is covered, I lift excess cement off the panel with light strokes and scrape the brush on the side of the can. Then I lay the leads down with the bent putty knife, trapping the cement under the leaf. Ride the heart of the lead to avoid cracking the glass.

When the leads are down, I rub a handful of sawdust over the surface with a clean rag. Ideally, the rag itself should not touch the panel—just the sawdust, if you can manage it. This cleans the panel nicely.

Though the interior side of the panel does not get cemented, it needs a little color to break the shine of the lead. I rub the cement brush over this side, then clean up with the sawdust. To further clean the panel, I scrub both sides with the clean brush, working in the direction of the lead. Then I polish the panel with sawdust, applied with the clean rag, and I finish by removing excess cement along the cames with the bamboo stick. A window with round-faced lead should be cleaned with whiting instead of sawdust, which would embed itself in the open channel.

The cement takes at least a week to cure, so I let the panels rest before installation. I lean them up, and brace them so they don't bow. Straightening a bowed panel after the cement has set up will break the seal.

Bracing—Horizontal braces should be used every 18 in. to 24 in. on panels that are 12 in. wide or wider. Bracing stock is usually 3⁄8-in. or ½-in. by 1⁄8-in. galvanized flat bar. It is soldered on edge along a lead line on the inside surface of the window. Both ends of the bar should fall on a solder joint along the edge of the panel. Braces are cut with a hacksaw or bench shear to a length 3⁄8 in. short of the full-size dimension. The ends should be gently tapered, then flared out into hammered paddles (photo facing page, bottom right). The paddles enable the bar to make contact with the sash; this braces the window without having to modify the rabbet. The bar can be bent to conform to a curved lead, but the greater the bend, the weaker the brace will be.

Sometimes passing a bar over a section of glass can't be avoided. But the bars, which can look obtrusive during construction, usually become a lot less obvious when the window is finally in place.

To secure a bar, I first center it over the lead line I want it to follow. Then I paint tinner's fluid (flux for galvanized metal) on each side of the bar over every solder joint the bar intersects. The soldering iron must be fitted with a flat chisel-tip, and it must be very hot. The tip is loaded with solder and held to one side of the bar without touching the lead. I hold a solder wire to the other side of the bar, which will conduct enough heat to melt it. At this point the iron is punched down onto the lead, pulling the solder down and around the bar with it. The iron is hot enough to fry the lead and should not be in contact with it for long. The bar is held still until the solder is completely cooled. Next, I tin the open cuts to keep them from corroding, and then solder the paddles and bring them flush with the border lead. I remove any residual flux, first with a rag and then with newspaper and window cleaner. If a lot of flux is left, I spray glass cleaner on the affected areas and scrub them with a brush and sawdust. A little gun bluing (sold at most sporting-goods stores) brushed on the brace bar will eliminate its shine.

The window is now ready to install. If you built it right and install it properly, it should last for 500 years.   □

Doug Hechter is a licensed glass contractor working in Santa Barbara, Calif.

Drywall Finishing

One contractor's techniques, using a taping banjo and stilts

by Craig Stead

It was a hot July in 1977, and I was working on a run-down two-story house. Much of the work involved hiding ugly cracked and crumbling plaster walls and ceilings with sheetrock. I was over budget and late on schedule. Large-scale drywall finishing was rumored to be an esoteric art practiced only by initiates, so I decided to subcontract it out. I started phoning finishers listed in the Yellow Pages. Fifteen calls later, I had not a single bid on the work. At that point, I was getting a sinking feeling in my stomach because I could see the whole project coming unglued. My client was eager to move in. The crew was standing around waiting to start the painting. Desperate action was needed. I popped down to the local lumberyard, bought everything that looked related to drywall finishing, grabbed a stack of sandpaper, and set to work. The going was slow, the sanding endless, the dust every-

where. I found that with enough time and sanding, you can achieve an acceptable finish. I also found that I hated sanding drywall seams. I got the job done, and vowed that I would never get caught in that bind again.

Since then, I have finished the drywall on all my jobs; and I have talked to every pro who would give me the time. I even worked one house alongside a drywall contractor to learn the tricks that are so important to speed and final quality. I found that professional rockers don't sand much. In fact, they don't sand at all until the last coat has gone on. The professional approach is to use the finishing knife or trowel to smooth the drywall mud while it's wet so the coat is free of ridges and high spots.

There are specialized tools for drywall work. In some parts of the country, sheetrock contractors use bazookas (they look a little

like anti-tank weapons). In the time it takes to walk the length of a joint, a bazooka can apply the tape, smooth it down and cut it off. Bazookas are typically used with "boxing" equipment that lays down properly feathered fill and finish coats without using a trowel. For years, bazookas could only be leased, and the charge was well over $1,000 a month. Even though they can now be purchased, they cost a lot and only high-volume drywall contractors would want them. If you're a professional who builds on a large scale or a novice with limited time and expertise, you may find it cheaper and quicker to call one of these well-equipped specialists. But if you're out in the boonies, or you are a small contractor with time between jobs, you're probably better off doing the work yourself.

I do a lot of renovation, which most drywall contractors don't like to touch. For me, aluminum stilts and a taping banjo like the one I'm using in the photo, facing page—a metal box that holds mud and 500 ft. of tape—were investments worth making as a general contractor. A banjo isn't as slick and quick as a bazooka, but it costs a lot less, and it helps me do a good job with some speed. I'll be talking more about the banjo and stilts below.

Hanging the rock—The first step to a quality drywall finishing job is hanging the sheets right. You've got to have flat walls, solid nailing and correctly installed outside corner beads. Most of this is discussed in Bob Syvanen's "Drywall" (*FHB* #8, pp. 52-57), but a brief review is probably in order.

Drywall comes with its long edges tapered, and its short edges the full thickness of the rock. The sheets are hung so that the full-thickness edges butt on studs. The tapered edges span between studs. Butt seams should be staggered from one course to the next so they don't all line up. If you have a tapered edge meeting a butt edge, you have a mismatch that is difficult to finish.

For drywall, walls should be framed with surfaced, kiln-dried lumber. In remodeling, where you often uncover walls built of rough cut lumber, you have to shim the studs to get them all to align. I always check the walls and ceilings on my remodel jobs with a 6-ft. aluminum level or a string line. I correct small variations with the 1/16-in. cardboard drywall shims that are commonly available around Boulder, Colo., where I used to work, or with 4-ft. long shims that I rip on my table saw in 1/16-in. increments from 1/16 in. to 3/4 in. thick. For a ceiling that's in bad shape, I either strap the whole thing with 1x4s, leveling them with shim shingles, or I frame a drop ceiling using 2-in. lumber. This is a tedious job, but if you don't take the time to do it, your final results will suffer.

Once you have flush framing, the hanging hints in the box at right can save you time.

Materials—To get a high-quality finishing job, you will need paper joint tape and premixed drywall finishing compound. I also use a filler called Durabond 90, made by U. S. Gypsum Corp. (101 S. Wacker Dr., Chicago, Ill. 60606). The paper tape bridges the joints between the sheets. Fiberglass mesh tape with an adhesive backing is also available, but it's much more expensive than the paper tape and is more typically used with veneer plasterboard (blueboard).

Drywall mud, or joint compound, is both an adhesive and a filling material. You use it to glue the tape over the seam, and then you fill the seam with it. It sets up as the water in it evaporates. After it dries, it can be redissolved to a certain extent by water. This means that you can smooth and feather a joint by washing it with a sponge if you need to. Tool cleanup is simple, because dried compound soaks off with water.

Don't dilute premixed drywall mud unless you're using it in the taping banjo. I mix each new pail of mud with a power paint mixer in a

---

1/2-in. drill, just enough to give the mud a smooth, even texture. I keep the pail covered at all times so the mud won't dry out. As I use the mud, I continually clean down the sides of the pail with a drywall knife to keep the thin side layers of mud from drying out and falling into the bucket. These lumps create dragouts (trenches) in the fill coat when you are smoothing the seam. If you get a dragout, clean your knife on your mudpan and continue smoothing the seam.

I keep a 6-in. drywall knife stuck in my mud pail to clean the bucket sides and to fill the mudpans and banjo. Each evening I empty my mudpans into the pail, then smooth the top of the mud. If I'm storing a partially filled bucket of mud for a long period of time, I lay a sheet of plastic over the surface or pour 1/4 in. of water over it to keep the mud from drying out. Keeping your mud pail and tools clean goes a

---

long way toward a satisfactory finishing job.

Durabond 90 is similar to patching plaster. It sets up by chemical reaction (hydration) rather than by evaporation. Once it's dry, it is not water soluble, does not shrink, and is hard to sand. It comes as a powder, and you mix only as much of it on the job site as you can use before it hardens (it has a 90-minute working time). You must also clean your tools immediately, or the stuff will stick hard and be difficult to get off.

I learned about Durabond 90 in Colorado from a fellow on my crew who had spent two years doing nothing but drywall. His outfit used it a lot to fill cracks, and sometimes as the fill coat, although most drywallers use only joint compound. I think the stuff is well worth the trouble of mixing it. I use it to fill all large gaps in joints after the rock has been hung. It's good for filling gapped corners, holes, and the crevice formed where a knee-wall intersects a cathedral ceiling. On my jobs, any gap greater than 1/4 in. gets a Durabond fill. Otherwise, the mud can suck the tape into the crack. This is particularly bothersome on inside corners. When I'm trying to move quickly on a job, I often use Durabond 90 for the first fill coat on outside corners and tapered joints because I can go back and get a coat of mud over it in an hour and a half instead of having to wait until the next day. The important thing to remember when you use the material this way is not to overfill a seam. It's hard to sand down high spots.

A compound called Gypsolite (Gold Bond Building Products, a Division of National Gypsum, Charlotte, N.C.) is useful for very heavy fills, like patching a large hole in a plaster wall. This is brown-coat gypsum plaster with perlite added. It resists cracking in thick sections, and it's inexpensive. But you have to overcoat it because it is grainy. It is not often used for drywall finishing, but it can be useful in remodeling work when the rock is running to an old plaster wall. In these situations, you can have voids as deep as several inches, and Gypsolite is just the ticket for filling them.

The taping banjo—For the kind of work I do, the taping banjo (photo facing page) is a terrific tool. It puts a layer of thinned drywall mud on the tape as you lay the tape on the joint, so you just smooth it down with your knife, saving the step of mudding the joint before applying the tape. The nose of the banjo has a toothed edge for cutting the tape. It's a great time-saver, and with it I can tape off a three-bedroom house in a little over a day, about a third the time it takes with taping knives alone (with a bazooka, the same job might take four to six hours).

Two types of banjos are available: dry and wet. In the dry type, the roll of tape is kept separate from the mudbox, which holds the thinned mud. In the wet type, the roll of tape is mounted inside the mudbox. Mine is a dry type made by Marshalltown (Marshalltown Trowel Co., Box 738, Marshalltown, Iowa 50158). It works well. I have never tried the wet type. Banjos are made of aluminum or

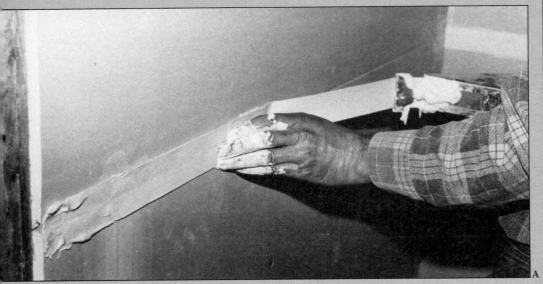

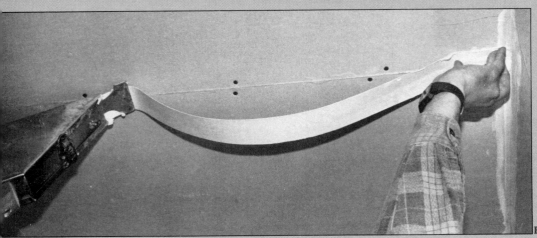

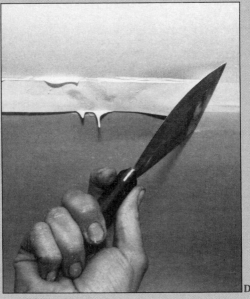

Taping seams. A taping banjo (A) lays the mudded tape along the joints between sheets of drywall. You pull tape out of the tool's throat to get it started, then press the tape to the wall every foot or so with your hand to hold it in place. On ceilings, tape is liable to fall off, so once Stead gets it started, he lets a bit sag (B). Then he tugs the banjo sharply and snaps the tape up against the drywall. The force of the snap holds the tape in place until it's smoothed down with a knife. At corners, it's important to keep the fold centered, and press the tape snug every 5 in. or so (C). Once the tape is applied, it's smoothed into place with a 6-in. knife (D).

stainless steel. I have used both types and prefer the aluminum model because it weighs a lot less.

To set up a dry banjo, you first install the roll of tape in the holder. I use 500-ft. rolls, which save reloading time. A typical three-bedroom house requires five to six rolls of tape. Next, you thin your mud to the consistency of condensed tomato soup—about ½ gal. of water per 5-gal. pail. If the mud is too thin, it will drip out of the nose of the banjo; if the mud is too thick, the joint tape will be hard to pull through, and may break.

Next, set the adjustable slide in the nose of the banjo so that it deposits a $\frac{1}{16}$-in. thick layer of mud on the tape. This minimizes squeeze-out when you smooth the tape down with your knife. Now loop the tape through the mudbox of the banjo, as shown in the photo, facing page, left.

Load the mudbox with your thinned mud and you are ready to go. Before you begin, grease your hands with Vaseline. You'll be getting mud all over them, and this keeps them from drying out.

The order in which you tape the seams is important. Do all the butt seams first, the tapered seams second, and the inside corners last. Lap your tape. The butt-seam tapes should go from, say, a corner to the center of a tapered seam. The tape on the tapered seams and corner tape will then lap over the ends of the tape on the butt seams. This way, no tape ends will be torn loose by the finish knife, and the corners will come up smoothly when troweled.

Pull the tape out of the banjo and press it to the seam with your free hand (photo A, top left). You can press it every foot or so just to hold the tape to the seam until you smooth it down with your knife. Once you have run the tape over the seam, set the banjo down and smooth the tape down with a 6-in. knife (D). You want that tape tight, particularly on the butt seams. Clean the excess mud on your knife into the mudpan and return this mud to the banjo for reuse. Wipe both sides of the joint clean of ridges so you have a well-bedded tape with no streaks of mud on either side. On tapered seams, the tape will not lie as tight because you are smoothing it into the dip between the two sheets. Don't worry about this—a heavy fill coat will cover up the tape later.

Slightly mismatched butt seams where one side is higher than the other can produce tape bubbles. To avoid these, smooth the low half of the tape first, and then smooth the high side, creating a step effect with the tape. Don't worry if it looks funny; a crown trowel—a 12-in. trowel that is slightly curved along its length—will take care of the problem when the fill coat is applied.

If you forget to smooth down a tape and it dries to a wrinkled mess on the seam, wet the tape with a sponge periodically to soften the mud. Then scrape off the tape with a drywall knife and try again.

If I'm working alone, I tape four or five seams with the banjo before doing the

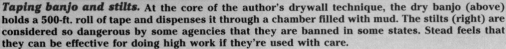
Taping banjo and stilts. At the core of the author's drywall technique, the dry banjo (above) holds a 500-ft. roll of tape and dispenses it through a chamber filled with mud. The stilts (right) are considered so dangerous by some agencies that they are banned in some states. Stead feels that they can be effective for doing high work if they're used with care.

smooth-down. This reduces the number of times I have to switch from the banjo to the smoothing knife and mudpan. For large jobs, the most efficient procedure is to have one person running the banjo and another smoothing the tape. On ceilings, pull out about 3 ft. of tape, allow it to sag 3 in. or so from the ceiling, and then quickly pull the banjo to snap the loop of tape to the ceiling (B, facing page). This tug slaps the mud side of the tape to the ceiling and holds it for its full length. Otherwise, the tape may peel off and drop in a sticky mess on the floor.

Inside corners are done last and require some fussing to get good results. I tape the vertical corners first, from ceiling to the floor, and then do the horizontal ceiling corners. What you want is a smooth, continuous tape that exactly meets the corner at either end. If the tape is too short, it leaves a hole in the corner. If it's too long, it piles up in the corner and must be trimmed to fit. If you do cut the tape short, splice in another piece. Any overrun can be cut back with a utility knife.

As you are pulling the tape from the banjo, tuck it into the corner centered on the fold, and press it in about every 5 in. (C, facing page). Smooth the corner tape carefully with your 6-in. knife, and try to form a wrinkle-free 90° angle. Your corner trowel is going to ride on this tape during the fill coat—some care here will make quality easier to achieve with your final coat.

The banjo is difficult to clean. If you are going to be using it for several days, store it nose down in a pail of water at night. This keeps the tape from drying out and sticking to the nose of the banjo. When the job is over, hose out the banjo and use the scrub brush to get into the corners of the mudbox.

Tools for finishing—I use a mudpan and drywall knife once the tape is on the wall. I also use an inside-corner trowel, and a crown trowel for finishing butt seams and problem seams. I prefer a mudpan to a hawk because it's easier to clean knives on the edge of it, and because it exposes less of the mud to the air so the mud doesn't dry out so quickly.

I have finishing knives from 6 in. wide to 12 in. wide in 2-in. increments. The 12-incher is the largest knife that will fit my mudpan. I like Harrington knives (Harrington Tools, Box 39879, Los Angeles, Calif. 90039) because they are more flexible than other brands, and I can feather the final mud coat better with them than with a stiffer knife. Every once in a while, I want to feather a seam real wide, or coat two neighboring seams at once. For this special occasion, I have an 18-in. wide knife.

I carry a hammer on my belt when laying the first coat of mud to set high nails. Tool cleanup is always needed on a drywall finishing job, so I put a mud pail filled with water in the center of the floor. In the pail floats a stiff tire scrub brush for brushing off hardened mud. All my equipment, the water pail plus the mud I'm using, sits on a sheet of plywood or tin to protect the floor from spills.

Stilts—In some states, safety regulations don't allow the use of stilts and will not award Workman's Compensation to anyone injured using them on the job. Many people, including a lot of professional tapers, feel that stilts are among the most dangerous tools around. A fall can pop your knee or snap your leg. I feel, however, that stilts are an effective way to handle high work if you're reasonably careful. I use stilts to finish ceilings and upper walls. With them, you can get smooth ceiling seams

without sanding because you can smooth the seams with one continuous motion of your finishing knife. They do require some practice to use comfortably and safely.

Before you start your drywall finishing, clean up the area thoroughly. Get rid of all debris, scrap lumber, and leftover rock. You're liable to trip over these things when you are doing overhead finishing on stilts. Besides cleaning up the area, I stack up drywall pails so I can refill my mudpan conveniently when I'm up in the air.

Various brands of stilts are available; the ones I use are made by Dura-Stilt Corp. (8316 S.W. 8th, Oklahoma City, Okla. 73128) and cost around $180. When you order a pair of stilts, get the accessory that lets you strap them to your boots. Otherwise you have to mount your boots permanently to the stilts with screws. Make sure your stilts are in good condition and inspect them before every use. Adjust the stilts to your weight and body size as recommended in the manufacturer's literature. Stilts that are out of adjustment can tire you out, and make you more liable to fall. If you are new to stilts, or haven't worked with them for a while, take some time to get the feel of them by walking around the room near the walls so if you start to fall you can catch yourself on the wall. Mount your stilts leaning against a wall, and attach the upper leg strap first before you fasten the boot straps (photo, above right). Going up and down stairs on stilts is not recommended, though I know people who do it. I scaffold stair halls and don't use my stilts on the stairs.

The fill coat—Finishing drywall is like a striptease—you put it on and then you take it off. You apply a coat of mud, smooth it out

129

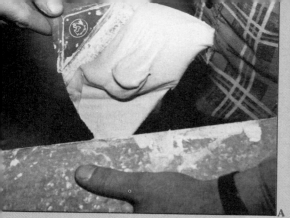

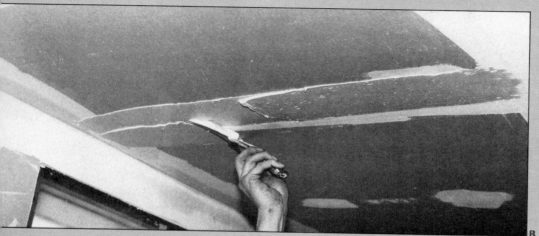

Applying the fill coat. To cut down on squeeze-out when he's applying mud to a seam, the author uses the sides of his mudpan to cut the compound back on the corner of his knife (A). Stead tries to avoid butt seams, but it's virtually impossible to do a drywall job without having to deal with a few. To fill butt seams, a separate band of mud is laid on each side of the joint (B), then smoothed down with a crown trowel (C). Filling inside corners is tricky, too. Stead lays a band of mud on each side of the corner, then smooths them down with a corner trowel (D) before smoothing the remaining mud ridges on each side of the corner out with a 6-in. knife (E).

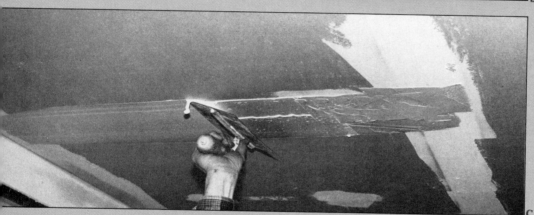

with the finishing knife by removing some of the layer, and clean your knife on the edge of the mudpan with each stroke. Using the knife to smooth and feather the wet mud is easier and faster than sanding the stuff when it dries. This approach avoids a lot of dust, and you don't have problems with fuzzing the paper covering of the drywall. Any small ridges or blips of dry mud that need removing between mud coats are knocked off with the edge of a finishing knife.

I apply two or three coats of mud after I lay the joint tape on with the banjo. A systematic approach to the application of each coat of mud is the key. You must do things in order and not bounce around the room, wasting time and energy. The principles are the same in covering all those nails or screws—three coats of mud, generously applied and smoothed off level with the surface of the sheetrock (assuming you want smooth, not textured, walls).

The first finishing coat of mud is called the fill coat. It is thick, so to make sure it dries in a reasonable time, keep the temperature of the room between 70°F and 75°F. As the fill coat dries, it shrinks, and looks pretty bad. Novices are tempted to start sanding immediately. Don't.

When you apply mud with a knife to a seam, the mud tends to ooze off its ends as you are laying it on and then drips onto the floor. Cutting back the mud on the corners of the knife before applying it substantially reduces this problem (A).

Fill the tapered seams first, the butts second, the outside corners third, and the inside corners last. The tapered seams are filled first because you want to get the surface flat before dealing with the other types of seams. Lay on your mud with a 6-in. knife and smooth the joint with an 8-in. knife. Be sure you completely fill the depression to the surface. In smoothing the joint, keep the knife almost perpendicular to the wall, and use light pressure on the knife. Too sharp a knife angle coupled with a lot of pressure can curve the blade and dish the joint.

In Colorado, I often used drywall shims under the rock one stud back in each direction from butt joints. This created a $\frac{1}{16}$-in. dip at the joints, which I treated as if they were factory-tapered. Since the cardboard shims are not available in Vermont, I lay two bands of mud on butt seams, one to either side of the tape, and smooth the seam with a crown trowel to cover the tape with mud and to establish a knife guide for the next coat (B and C). Here I want a good layer of mud over the tape. Don't worry if the edges of mud farthest from the seam are a little ragged. The next coat will take care of them.

Outside corners take a lot of mud on the fill coat. If I'm filling a lot of them, I lay on the mud with a 12-in. knife and apply the mud with the handle of the knife held parallel to the floor. Then I smooth down each side of the corner using an 8-in. knife. Make sure you smooth the mud that goes around the corner and gets on the other side. If you forget to do

Finishing. To finish inside corners, Stead lays a thin band of mud with a 6-in. knife over the remaining ridges (above), then smooths them down with an 8-in. knife (right).

this, you can knock the dry globs off with your knife before you do the next coat.

Inside corners are the trickiest to fill. Lay a band of mud to each side of the corner using a 6-in. knife, making sure that you get the mud fully into the corner. Then smooth the corner down using a corner trowel (D, facing page). You may have to smooth it several times to get it good and clean. Make sure you start the trowel tight into your starting point so the intersections of the two walls and the ceiling come up clean. After running your corner trowel, smooth the mud ridges left on the wall with your 6-in. knife (E, facing page).

If you pull your knife off the seam as you are smoothing the joint, it will leave a ridge of mud that may need sanding when it dries. Try to trowel your joints in one smooth, continuous motion. If you have to break the knife off of the joint, do so where another joint crosses the seam. If you are smoothing a tapered seam, lift your knife as you come to a butt seam that intersects it. This way your next coat on the butt seam will cover the ridge on the tapered seam.

The finish coats—The finishing coats are thin. They dry in three or four hours, and shrink very little. The final coat is almost ironed on and is thinner than a dime.

The order of seams is not important on this coat; just make sure that you do all of one type of seam before jumping to the next type. That way, you are unlikely to cross a wet seam as you proceed around the room. I use the same order of seams as in the fill coat, or put one worker on one type of seam while another does the corners.

For tapered seams, I lay on the mud with an 8-in. knife and smooth it down with a 10-in. knife. Butt seams are done the same way, only two bands of mud are laid down, on either side of the tape, and then smoothed down in two passes. On outside corners, I use a 12-in. knife to lay on the mud as before and then smooth it down with a 10-in. knife. Inside corners are fun at this point because they merely need a thin band of mud laid on with a 6-in.

knife where the ridge was left by the corner trowel. Then this is smoothed down with an 8-in. knife (photos above).

Smoothing a seam requires that your knife have two smooth guides to run on. For example, an outside corner is covered with metal bead not only to give it strength and protection from impact, but also to provide a guide for the finishing knife. In smoothing such a corner, your knife rides on the corner bead on one side, and the smooth surface of the drywall on the other. On any seam I'm finishing, I'm looking for my smooth guides. On tapered seams, the smooth surfaces of the drywall on each side of the seam are the guides. On inside corners, the corner tape and the drywall surface are the guides. That crisp, straight corner tape is one of your knife guides for feathering the coats of mud.

Butt seams are the greatest problem because the center where you lay your tape is a high point and the guides for the knife thus become the center of the tape and the smooth drywall to either side of the seam. Butt-seam tapes are often wrinkled from the nails underneath them, and the smoothing knife chatters on the lumpy tape. One approach to finishing a butt seam is to lay two bands of mud on it, with the tape in the center of the two bands, leaving a ridge to sand later. This works, but you must have a lot of knife control as you are working with only one knife guide and floating the center of the seam. Commonly, you get either too thick a coat, which shows as a bulge over the seam, or you starve the seam for mud, get a very thin mud coat over the tape, and break through the mud to the tape when you sand out the ridge. This leaves a fuzzy tape, which shows after you paint. My method is to use the crowned bed of mud as the guide for the knife, and subsequent coats of mud completely hide the seam.

The second and final finish coat is very thin. It conceals minor imperfections. It is smoothed off with a lot of pressure on the knife, and if it's done right, it feathers out to almost no ridge on the seam. I lay on and smooth off the mud by running the knife with

the seam at a 45° angle, or crosswise. On the final coat, I lay on my mud with a 10-in. knife, and smooth with my 12-in. knife. Keep smoothing the joint until you have no ridges of mud left, just an even, well-ironed, thin coat of mud. This coat dries in about an hour and goes on quite quickly. The order of the seams is not important. Some seams give you a chatter or washboard effect. If you encounter this, lay your mud on crosswise to the chatter, and smooth it off in the same direction.

Some light sanding and sponging of the seams are all you need to complete the finishing job. Inspect the seams as you work, and circle any areas needing touchup with a pencil so you can correct them before painting. A strong light on an extension cord is useful for finding problem areas in hallways and dark corners. Also inspect your work after the first coat of paint is applied; sometimes previously unseen imperfections become noticeable when the surface is all one color.

If I haven't done any drywall finishing for a while or if I'm training someone new, I start in the closets to warm up on techniques. Work your back rooms first and save the living room and dining room for last. That way your best work will be where people notice it the most. The new worker gets the nails as a starter to get the feel of the 6-in. knife, the drywall mud and the mudpan. My rule is that you always sand your own work so you can get feedback on how you are smoothing the joints. A common problem with novice workers is oversanding tapered seams, leaving them slightly depressed.

There isn't one way to finish a drywall job, just a final result that must look right. Your corners must be clean and crisp, especially where the ceiling meets the intersection of the two walls; the tapered seams must be fully filled and flush with the surface of the rock, and the butt seams must be feathered out far enough to each side so they don't show as a bulge under raking rays of light. That is your final test and grade on your finishing work. □

Craig Stead is a builder in Putney, Vt.

Tiling a Mortar-Bed Counter

How one tile setter builds the classic kitchen work surface

by Michael Byrne

I think the best part of being a tile setter is that my work doesn't get covered up by someone else's labors. On the other hand, setting tile is tough, physical work—especially large floors, where my knees cry out for a desk job and my back creaks from all the bending. So it's no wonder that I enjoy tiling countertops.

Ceramic tile offers many advantages as a finish material in the kitchen. A hot pot won't damage a tile surface, and a properly waterproofed installation can stand up to all of the splashes and spills that cooks can dish out.

The best tile countertops are done on a thick bed of mortar called a *float*. The float is usually ¾ in. to 1 in. thick, and the solid base it provides for the tile isn't affected by moisture. I'll be describing the most common type of counter that I do. It has V-cap face trim and a single row of tiles for the backsplash. To make cleanup easier for the cook, the sink is recessed beneath the surface of the counter and is trimmed with quarter-round tiles (drawing, facing page).

Choosing the tile—Kitchen-counter tiles should be either impervious or fully vitrified (see p. 102). These ratings mean that the tile will absorb almost no water, a property that increases the life expectancy of the installation.

Many tiles are designed to decorate rather than protect. You should be able to find a tile that does both. But you need to be careful even with heavy glazes since some of these are easily marked by metal cooking utensils. I urge my customers to get samples of their favorite tiles, and to rub them with a stainless-steel pan, an aluminum pot and a copper penny. Some tiles can be cleaned up after this kind of abuse—others can't. The surface is important in another way, too. Because most appliances need a flat surface to work efficiently, tiles with irregular faces, such as Mexican pavers, make beautiful backsplashes but lousy work surfaces.

Layout—The goal here is to keep tile cuts to a minimum, to locate them in the least conspicuous places, and to eliminate tiles that are less than half-size. On a straight-run counter, this usually means beginning the layout halfway

Tile layout. **Byrne has used a ³⁄₁₆-in. trowel to spread thinset mortar over the float, and he now aligns the rows of tile with a straightedge. The chalklines at the inside corner of the counter mark the position of the V-cap trim, and the starting point for the first full sheet of tile.**

along its length or at the centerline of the sink. If you take a close look at sink installations, you'll find that there is often a trimmed tile in the center of the front edge, in line with the spout. This trimmed tile keeps things symmetrical, and it allows full tiles along the edges of the sink. I begin the layout on an L-shaped counter at the intersection of the two wings (photo facing page).

Once I have the tiles in hand, I use the direct method of measurement to help lay out the job. I unpack some of the tiles and move them around the cabinet top. Sometimes, shifting the tiles an inch this way or that can make a substantial improvement in the finished appearance. Small tiles are more forgiving, allowing you to adjust the width of the grout lines to make things fit. But unless the counter has been meticulously designed with the tiles as modules, cuts are inevitable.

The substrate—The substrate should be at least ¾-in. plywood rated for exterior use. Particleboard won't do. The waterproofing on most counters consists of a layer of 15-lb. asphalt-saturated felt or a similar protective paper. I take this a step further and laminate the felt to the plywood with wet-patch fibered roofing cement.

Tiling a counter is a messy, gritty project. To keep the asphalt (and later the mortar and grout) from soiling the cabinets, I first drape kraft paper or plastic film over the face of the cabinets, and staple it to the counter plywood. I also protect the floors with canvas dropcloths.

I use a ⅛-in. V-notched trowel to spread a thin layer of roofing cement onto the plywood around the sink and about 3 ft. to each side. This helps to protect the vulnerable areas under the dish drainers. Ideally, the asphalt should cover the exposed plywood end grain in the sink cutout (drawing, below). If the sink is already in place, I squeeze the asphalt into the junction between the sink and the plywood. At the backsplash I make a tight crease in the felt and lap it up the wall about 3 in. Later this flap will be trimmed to about ¾ in. above the finished counter. Combined with the backsplash tiles, it makes an effective water barrier, keeping moisture out of the rear of the cabinets.

Screeds and rails—Once the waterproofing is completed, I set the sink rail. This rigid galvanized sheet-metal channel reinforces the mortar bed down the entire front edge of the counter and makes it easier to level the bed. The rail has narrow vertical slots every 3 in. for nails or screws (photo below left). Along its top edge are ⅝-in. dia. holes. Mortar will ooze through these holes, linking the mortar that faces the counter edge with the countertop float. This helps to anchor the V-cap finish on the edge.

I start screwing the rail in front of the sink, adjusting it to suit the height of the quarter-round trim in relation to the top of the sink. Once all the rail is in place, I use ½-in. thick pine to box in any openings in the substrate that have been cut for the cooktop, chopping block or other built-ins. These are installed at the exact height of the finished float. Attaching the sink rail and boxing the openings is a lot like setting up the forms for a slab floor.

Metal reinforcing—My experience has taught me that metal reinforcing in a counter float reduces or eliminates cracked tiles and grout. Consequently, I use plenty of it. First, I cut 20-ga. 1-in. wire mesh (chicken wire) and secure it to the plywood substrate with ½-in. staples, overlapping neighboring pieces at least 4 in. The mesh (photo below right) extends from the sink rail to the back wall and covers the entire substrate. Rather than cut the mesh a little short to make an easy fit between the sink rail and the wall, I cut it a bit long and bend the excess back over itself.

Finally, I use 9-ga. galvanized wire like rebar to strengthen those parts of the mortar bed that will be narrow in cross section. This prevents the cracks that often appear in the tiles close to the front or back corners of sinks and cooktops. I center the wire and run it parallel to these narrow sections, and I anchor it with ¼-in. or ⅜-in. furring nails. Then I bend it at about a 45° angle where the counter broadens, and extend it at least 6 in. toward the center of the field. At first, I used the wire rather sparingly. But now I use it all over the countertop—at inside and outside corners and across peninsulas—and I've found cracking problems a thing of the past. When all the reinforcing is in place, I check to make sure none of it protrudes above the top level of the sink rail. This can be done either with a 2-ft. level or by sighting the top of the sink rail.

Deck mud—Most of my jobs are in the San Francisco Bay Area, where the adobe soil swells during the winter rains and shrinks in the long

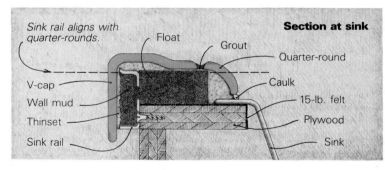

Section at sink

Sink rail aligns with quarter-rounds.
Float — Grout — Quarter-round — Caulk — 15-lb. felt — Plywood — Sink
V-cap — Wall mud — Thinset — Sink rail

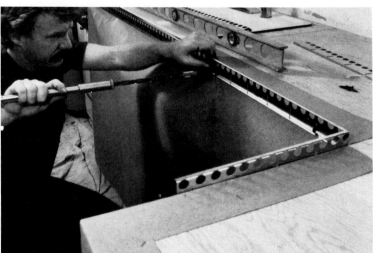

Sink rail. **Before floating the mortar bed, Byrne screws a galvanized strip called a sink rail to the edge of the plywood substrate. It serves as a screed, edge reinforcement and a framework to anchor the thin layer of mortar that will face the edge of the countertop.**

Reinforcement. **Chicken wire stapled to the plywood substrate covers the area to be tiled. If a mortar counter cracks, it usually does so in the narrows around the sink, or at inside corners. These areas are reinforced with 9-ga. galvanized wire held in place by furring nails.**

A. *Float strips* are the key to controlling the thickness of the mortar countertop. Here the author beds a float strip in a mortar pad, getting the strip's relationship to the sink rail right with a level. The float strips will guide the screed board while the mortar is leveled. Later, the strips will be removed and the resulting voids filled with mortar.

B. *Wall mud*—a special mortar blend that includes masonry lime and latex additive, which help the mortar to cling to vertical surfaces—fills the sink rail before the counter is floated. Some of this wall mud squeezes through the holes in the rail, keying it into the deck mortar.

C. *Deck mud* is loosely spread across the counter with a wood float after the sink rail is filled with mortar. Because the mortar in the rail is still fragile, it has to be supported by a steel trowel so it won't break away.

D. *Leveling the deck mud.* Byrne uses an aluminum straightedge as a screed board, and he moves it in a side-to-side motion as he gradually pulls it forward. Here both ends of the screed are resting on float strips. Note the difference in texture between the crumbly deck mud and the smoother wall mud used on the rail and around the sink.

E. *Edging.* Before it hardens, any mortar that overlaps the sink rim is trimmed away. This allows the sink to be removed later if necessary, without damaging the mortar bed. The open edge will be trimmed with quarter-round tiles.

hot summer. Add to this the occasional earthquake tremors, and you have mortar beds that tend to move around quite a bit. I use 3701, a mortar-and-grout admix made by Laticrete International (1 Laticrete Park North, Bethany, Conn. 06525) that allows my floats to flex a little without cracking. Other companies, like Custom Building Materials (6511 Salt Lake Ave., Bell, Calif. 90201) and Upco (3101 Euclid Ave., Cleveland, Ohio 44115), make similar products.

The amount of admix I need depends on the weather and how wet the sand is (see the sidebar, facing page). This deck mud, as it is called, is considerably drier than brick-type mortars—it has just enough moisture to bind the ingredients and no more. This means that the mix can be compacted into a uniformly dense slab.

Floating the counter—Before I can start spreading the mortar around the countertop, I have to install float strips along any edges that aren't boxed or that don't have a sink rail. Float strips are ¼ in. thick and 1¼ in. wide pine or fir rippings that will sit temporarily atop a layer of mud as I level the mortar. Each float strip begins on a mound of loose deck mud piled slightly higher (about 1 in.) than the height of the finished float. Then I take a level and, placing one end of it on the sink rail for a reference and the rest of it on the float strip, I tap the strip with a hammer until the strip is leveled, as shown in photo A, top left.

Floating begins with filling the front edge of the counter. The channel formed by the sink rail must be filled with what's called wall mud. To make the wall mud, I take a small portion of deck mud (for this job about 2½ gallons), add about a quart (dry measure) of masonry lime, and enough Laticrete 3701 to make a thick, heavy paste. Then, using a flat trowel, I press the mix onto the face of the sink rail until the mud is forced through the ⅝-in. holes (photo B). The resulting extruded lumps of mortar will key into the deck mud.

Once the sink-rail face is filled, I use the remaining wall mud to surround the sink. Then I dump the deck mud onto the countertop and spread it around with a wooden float while I keep the rail and its wall mud steady with my steel trowel (photo C). By this time, the mud in the sink rail has begun to harden, but if it is not supported, it will be pushed off the rail when the nearby deck mud is compacted.

To level the deck mud, I use a straightedge as a screed board (photo D). Using a side-to-side motion, I pull it toward me, gradually removing the excess mud until the straightedge makes contact with the float strips or sink rail. I apply a horizontal rather than vertical pressure on the straightedge to avoid mashing the strips out of position. Smoothing out one area at a time, I gradually work my way around the countertop until the screeding is done. The surface is now flat, but not all the mud is compacted. The float strips also have to be removed, and the resulting voids have to be filled.

I take a lot of pride in my finished floats—they are my pieces of sculpture, and the wood float is my finishing tool. First, I scoop some deck mud onto the flattened top and ram it into the voids where the strips sat with the float. Then with the edge of the tool, I gradually slice off the excess. Experience allows me to "feel" my way across the surface by the way the float sits in my hand. I scour the top until it feels right. With all the voids filled, the top is an unbroken expanse of grey.

The last two areas of mortar to clean up are the sink rail and the sink perimeter. The top and bottom edges of the metal sink rail provide a good surface for the float to trim off the excess mud. Around the sink, I square up the mortar with a trowel, trimming it back far enough to expose the edges of the sink (photo E, facing page). Although it is no picnic, this makes it possible to remove the sink without having to rip up the field tiles. These edges are covered later with quarter-rounds.

Inevitably, some mud will fall away from the rail, or the screed will knock a float strip out of position. Fortunately, the material is very forgiving, and problems are easy to fix. I skip over these minor accidents until the initial work is done, then I go back and fill in dings with fresh mortar before everything sets up.

Setting the tiles—Instead of laying the tiles as soon as I finish the float, I let it harden overnight. This way most of the shrinkage likely to occur will happen before the tiles are in place and grouted, and I can be less concerned about deforming the float as I set the tiles.

The next day, the first order of business is to vacuum loose sand and cement particles from the top to increase the grip of the thinset mortar that bonds the tiles to the float. Then I snap chalklines along the edges to mark the layout for the V-cap trim. I usually spread a few sheets of tile around to confirm my earlier layout; then I mix up enough thinset mortar to last through a couple of hours of setting.

Thinset is a portland-cement based mortar that contains very fine sand. The bond it forms is unaffected by moisture, and it is ideal for applying ceramic tiles to a mortar base. On this job I used Bon-Don (Garland-White & Co., P.O. Box 365, Union City, Calif. 94587). I mix the stuff with water to the consistency of toothpaste, using a drill and a mixing paddle.

The sheet-mounted tiles going on this counter are a little less than ¼ in. thick, so I used a ³⁄₁₆-in. V-notched trowel to comb out the thinset. Spreading too thick a layer will cause the adhesive to ooze up between the tiles. On the other hand, the backs of the tiles must be completely covered. These 12-in. by 12-in. sheets covered the top quickly. I used a short straightedge to help align them.

Everything went smoothly on this job until I reached the open side of the L. There I realized that the tiles were falling short of the V-cap layout line by about ⁷⁄₁₆ in. Checking back, I found that the sheets in one box were all undersized. Adding a narrow row of tiles that have been trimmed to make up for a mistake like this never looks right, so before the thinset dried I quickly widened the grout lines between the rows of tile. The string backing prevents the tiles from being spread apart, so I cut through it with a utility knife, and used a long straightedge to

Mixing the mortar

Of all the skills necessary to produce durable tile installations, none is more perplexing to the novice than mixing mud. There is no substitute for experience, but having a good recipe, the right tools, and knowing a few good mixing techniques can produce workable deck mud. The recipe I use comes from instructions printed on the bucket of latex admix (when using various mortar additives, always follow the manufacturer's recommendations). With Laticrete 3701, the mix is 1 part portland cement, 3 parts mason's sand, and about 4½ to 5 gal. of the admix per sack of cement. To help keep the batches consistent, I measure the dry ingredients in 3-gal. or 5-gal. buckets instead of counting shovelfuls. A full 5-gal. bucket holds ¾ cu. ft. of sand, and when I calculate the volume of mortar for a job, I disregard the cement. It fills the spaces between the sand particles. The sand I use comes damp from the yard, although occasionally I use dry sand shipped in paper sacks. With the dry sand, I measure out the amount I need and mix it with just enough water to dampen it.

I use a steel mixing box and a slotted mason's hoe rather than a rotary mixer, which can cause the mix to form marble-sized lumps. I layer the sand and water evenly in the box and chop them three times back and forth with the hoe. Each time, I take lots of small bites with the hoe, and I pull the ingredients toward me to form a pile at one end of the box. Before any liquid can be added, the sand and cement must be thoroughly blended to prevent lumps from forming.

Next I level the dry ingredients and use the handle of my hoe to punch holes in the mix (photo above right). This allows the liquid to distribute itself more evenly instead of just sitting on top. Then I repeat the mixing procedure, chopping back and forth three times. At this point, I pick up a handful and squeeze it. If the moisture content is right, the deck mud will form a tight ball that sticks together without cracking apart (photo below right). If it oozes through my fingers, the mix is too wet and must be adjusted by adding some dry sand with the right proportions of cement. If the ball falls apart, I need to add more liquid.

The direct rays of the sun can ruin the mud at this point, so I pack it into buckets and get it inside the house. If it's above 90°F, I'll have only about a half-hour to work the mortar. If it's 65°F to 75°F, I may have as long as two hours. —*M. B.*

A steel mixing box (top) is the place to prepare a batch of deck mud. Byrne blends the dry ingredients with his hoe, then pokes holes in the mix with its handle to help spread the latex admix. Properly blended mud is fairly dry, but it will cling together in a ball when you squeeze a handful of it (above).

open up the joints, as shown in the photo at left. For getting out of a jam, nothing beats a good set of straightedges.

Cutting the tiles—The narrow tiles in front of and behind the sink can be cut with a snap cutter (see p. 98), but I prefer to use a diamond-bladed wet saw for the accuracy and smoothness of cut I get in one step. The saw is set up outside the house, and running back and forth for each cut eats up time, so I accumulate a stack of tiles to be cut for each trip. You can use a ruler to take measurements and then set the saw fence to these, but that leaves a lot more room for error than just marking the tile directly. The water jet on the saw can sometimes blast away a pencil mark while cutting, so I cover the tile with masking tape and make my mark on the tape.

V-cap, backsplash and quarter-rounds—After all the field tiles are positioned, I set the V-cap. Complicated trim tiles like these often distort a bit in the kiln, so they must be set with extra care. I usually butter each piece with thinset and then tap it into place, controlling the amount of thinset I use to suit the alignment (photo facing page, top left). At inside and outside corners, the V-cap tiles are mitered, and I cut them a bit short to allow for a grout line.

Before I can set the single row of backsplash tiles, I trim the excess tar paper down to about ½ in. to ¾ in. above the deck tiles (drawing, facing page). The joint between the backsplash and the deck must allow for free movement, so later, when the grout is dry, I seal it with a bead of silicone caulk. I allow a full-width joint here rather than have the splash tiles rest directly on the deck tiles. Bon-Don is especially sticky thinset, allowing me to hang these relatively light tiles on the wall without any support from below. Heavier tiles usually require wood or plastic shims between the last course of deck tiles and the bottom edge of the backsplash tiles.

The last tiles to go down are the small radiused tiles that trim the sink. Unlike the other tiles, these quarter-round trim pieces are set on a bed of grout. This grout is the same used to pack the joints, only it is mixed stiffer. To make sure that the quarter-rounds adhere to the float mud around the sink, I coat both the float and the back of each quarter-round with thinset for a stronger bond between the tile and the grout.

Factory-made inside corner pieces look and feel better than the miter cuts you can make on a tile saw. They are set before the straight sections of quarter-round. With quarter-rounds, it's important to apply more grout than is actually needed to set each piece. As the tile is slowly pushed home (photo facing page, below left), the excess grout is squeezed out of the joint. Once the piece is in the right position, I support it with my fingers for a few seconds to prevent it from moving. When all the pieces are set, I

Adjusting the courses. **Instead of adding an unsightly row of narrow tiles, the distance between courses can be slightly increased. This strategy can spread out a discrepancy so it can't be seen, and save tedious tile-trimming.**

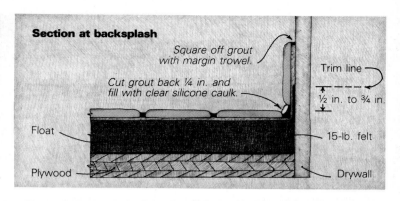

Section at backsplash

Square off grout with margin trowel.

Cut grout back ¼ in. and fill with clear silicone caulk.

Trim line

½ in. to ¾ in.

Float

15-lb. felt

Plywood

Drywall

V-cap tiles, which trim the leading edge of the counter, receive a lot of contact. It's important that they be securely anchored to the mortar bed— any voids between them are unacceptable. Byrne butters the back of each trim piece with a generous helping of thinset, and presses it in place until it's in the same plane as its neighbor.

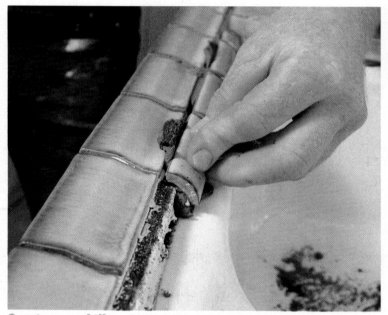

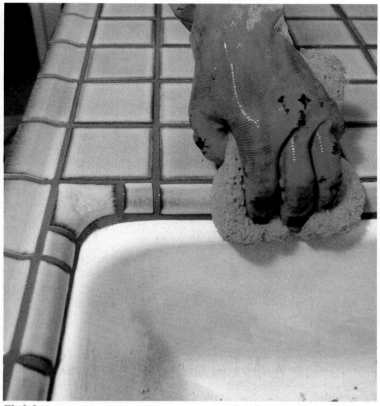

Quarter-round tiles, which trim the edge of the sink, are the last tiles to be placed. They are set on a bed of stiff grout. Before bedding them, Byrne applies a thin layer of thinset mortar to the tiles and to the float. The thinset mortar helps to strengthen the bond between the two.

Finish. The entire counter has been grouted and sponged. Residual cement is cleaned up with a damp sponge. The corner of the sink is trimmed with a factory-made inside-corner piece. Next to it, the quarter-rounds have been trimmed to align with the V-cap edge trim and the field tiles. The last step will be to undercut the grout around the sink, and fill it with silicone caulk once the grout dries.

leave them alone for about a half hour or so to allow the grout to set up. Meanwhile, I prepare another batch of grout.

Grouting—I prefer a grout made with a latex admix because it is a lot stickier than regular grout. This allows it to adhere tenaciously to the slick edges of glazed tiles—an important advantage on a tile work surface that gets constant use. Also, grout with admix is far more resistant to liquids, and to the erosion they can cause.

To prepare the grout, I follow the directions on the sack, which usually recommend combining the dry ingredients with water or a latex admix to the proper consistency, and then allowing the mix to sit for five or ten minutes. The grout is then mixed again and it's ready for use. During this wait, I trim the excess grout from the sink quarter-rounds.

There is no single method for grouting, and the techniques for grouting floor tiles (see p. 103) also apply to a counter. The porosity of the tiles, the moisture content of the setting bed, the addition of admixes, temperature and humidity levels are all factors that determine how much grout can be spread before it's time to clean off the excess. Usually, I begin by spreading about 8 to 10 square feet. I hold the rubber trowel at an angle between 30° to 40° as I force the grout into the joints. I work the grout from different directions until I'm satisfied that the joints are packed solid.

I start the cleaning by scraping away loose grout with the edge of my rubber trowel. Then I take a wet sponge and wring out as much water as possible. This is important because any excess moisture will weaken the grout. I work the sponge across the counter, gradually lowering the level of the grout until it is slightly below the plane of the tile, with a concave surface. During this process, the pores of the sponge quickly fill with grout and must be flushed constantly. Once

the entire counter has undergone this step, I go back over it with a clean sponge to remove most of the cement haze (photo above right).

The last step is to trim the grout in a few places. At the sink, I undercut the grout below the quarter-rounds about ⅛ in. so the joint can be caulked with clear silicone. This allows the sink to move a little, without breaking the waterproof seal, and lets the color of the grout show through the caulk. Because the counter and the wall will move slightly in relation to one another, I use the same technique to seal the joint between the deck tiles and the backsplash. At the top of the backsplash, I square up the grout line with my margin trowel. This makes it easier to paint or paper the wall.

Finally, I remove any grout haze with cheesecloth or fine steel wool, followed by a thorough vacuuming to take away the loose particles. □

Michael Byrne lives in Walnut Creek, Calif.

Stripping Trim

Why and how to take old paint off interior woodwork

by Mark Alvarez

You don't always have to strip moldings, baseboards and trim. Sometimes, the best approach is to clean and paint over what's already there or—in extreme cases—to rip everything off the wall and install new stuff. But stripping trim is a job all renovators face at one time or another. It's never neat or easy, but there are efficient and effective ways to approach it, and the results can be spectacular. Your house may be a Victorian, for example, with fine hardwood trim that was meant to be seen in its natural state but has been painted and repainted over the years. It's a prime candidate for stripping. Even if you have simple pine or fir trim that was meant to be painted, it may be peeling and cracking or have so many layers of paint globbed over it that its details have been obscured, and its visual effect ruined. If the wood's still in good shape, it makes sense to strip it down to a clean, smooth surface before you repaint.

There are three basic ways to remove paint from wood: abrasives, heat and chemicals. Abrading paint off with sandpaper or those spinning wire wheels you chuck up in your electric drill is either very slow (orbital and hand sanding), very hard to control (belt sanding), or both (wire wheels). It's tough to get into crevices and cracks with these tools, too, and they all kick up a lot of paint dust. If your house is 30 years old or more, this dust will probably contain lead, a hazardous material you don't want to breathe. So always wear a dust mask if you decide to sand.

The heat plate that is so commonly used on house exteriors works only on flat surfaces. It won't do the job on moldings, in corners or along the hard-to-reach edges of your trim. Heat guns—those giant electric hair driers that generate astonishingly high temperatures—are better for trim, but on curved surfaces especially, inexperienced operators almost invariably end up scorching some wood. The heat these machines put out can crack glass, too, so you shouldn't use them around windows. With any heat method, there's always some danger of fire. Never use a propane torch or any other open flame.

Chemical paint removers—strippers—are either caustic or solvent. Caustic strippers, which are usually used in commercial dip-stripping operations, are lye-based—usually sodium-hydroxide or potassium-hydroxide formulations—and the residue they leave on

the wood must be neutralized with an acid after the finish has been removed. They are sold either as powders to be mixed with water, or as pastes.

Caustic strippers are cheap, but they darken wood and raise its grain, so you shouldn't use them if you intend to leave your trim clear. They are often used by professionals when the trim is going to be repainted, but many clients specify that the contractor run pH tests to be sure of adequate neutralization. Many restorationists feel that caustic strippers can never be satisfactorily neutralized, and that they will eventually begin to bleed out of the wood and cause paint failure.

Most experts agree that solvent removers, which have a methylene-chloride base, are best for stripping interior trim, especially if you've got a lot to do. They are more expensive than caustic strippers, and they present hazards of their own, which will be discussed below, but they do a thorough job without harming the wood.

Tom Schmuecker's family owns Bix Process Systems, Inc., which manufactures stripping products for the general public and also trains and licenses dealers who use the products professionally. Schmuecker strips wood professionally himself, and over the last few years, he's seen his business, once almost entirely centered on stripping and refinishing furniture, expand strongly into renovation and restoration work. He's stripped a lot of trim, and has developed techniques that keep the job straightforward and manageable.

Getting ready—Solvent removers aren't as dangerous as lye, but you should always protect your eyes with goggles. Schmuecker also

wears rubber gloves, coveralls and old shoes. And he makes sure that pets are kept outside.

The fumes of solvent strippers can make you dizzy and sick, so you need good ventilation when you work. A couple of open windows are sometimes enough, but you might want to set up a fan. Strippers work best at room temperature, but it was cold the day we took these photos, and while Schmuecker had a couple of windows open, he didn't want to turn on the fan and lower the room's temperature too much, so he wore a respirator with an organic-solvent filter. If you do the same, remember that the room should be thoroughly ventilated once you're done. Don't heat the area directly. This is a good way to make the remover dry out quickly, limiting its effectiveness. Schmuecker also had a bucket of cold water and a mop handy in case of a spill. On larger jobs he relies on a wet/dry vacuum.

It's easier to strip wood in the workshop than on the walls, so on big jobs Schmuecker likes to take down as much of the trim as possible—crown molding almost always, because it's so hard to get at—and haul the pieces to his own shop, where he can strip it in a tray. Then he returns it to the house and nails it back up. This is an especially good idea if other parts of the room have already been restored or refinished, because it keeps the mess in the shop. But before you do it, consider the damage you may do to the surrounding wall, and how easily it can be patched.

Schmuecker has found that if he's careful, he can pry trim off old plaster walls without harming them. But if you've got original plaster walls in perfect condition, you might decide to work on the trim in place. To strip the window shown here, he removed the window guides and scratch-marked them (stripper removes ink and crayon) so they could be returned to their original places. He stripped these pieces and the lower sash in his tray. On other jobs, he's removed every foot of molding and baseboard.

Schmuecker advises planning remodeling or restoration work so that you can proceed from the top of the room down. Scheduling subcontractors is always a problem, though, and he's had to strip trim in a number of rooms whose floors had already been expensively refinished. Even if your floor has not recently been returned to glory, you'll want to protect it from stripper and paint residue. Schmuecker first lays down a few layers of newspaper, then covers them with plastic sheeting, firmly duct-taped around its edges, snug up to the baseboard. As an added precaution, he lays down a painter's dropcloth over the other layers. The newspaper acts as a final line of defense in case the dropcloth and plastic are punctured, and also keeps the plastic from slipping. The wall near the trim also has to be protected. Schmuecker uses plastic here, too, taping it down first with masking tape and then with duct tape. He also covers radiators and the like with plastic sheeting (photo right).

Testing—There are different kinds of solvent strippers. Some are flammable and some are not. Some require a solvent or a water rinse, and some require no rinse at all. For most of his work, Schmuecker prefers non-flammable removers that need to be rinsed. Flammable removers are cheaper, but not quite as effective on paint, because they contain between 30% and 50% methylene chloride as opposed to 50% to 85% in the non-flammable paint removers. No-rinse strippers are fine for overhead work, for example, where rinsing liquids would drip on you and the floor, but they just don't work as well as rinse-removers against either paints or applied stains.

Once he's set up, Schmuecker always does a test patch, applying remover to a small portion of the wood to be stripped. If the molding is intricate, he tests its most ornate section, which is usually the most troublesome area. The patch verifies that he's using the right stripper, and gives him some idea of how many layers of paint he's dealing with, and how long it will take to remove them all.

At this point, if the wood is undistinguished, its location is awkward, and the finish promises to be tough to remove, Schmuecker sometimes recommends simply replacing it. Anything can be stripped, he says, but there are times when it just isn't worth the trouble.

Some paint removers work faster than others. Strippers that raise quick blisters are usually breaking bonds between layers of paint, or between the first coat of paint and the wood surface. These fast strippers work fine if there are only a few coats of paint, especially if they were applied over a varnish that sealed the wood. Slow-acting removers, on the other hand, dissolve paint rather than breaking bonds. If left on long enough, they can go through six to eight layers at a time, and they also get paint out of unsealed wood better.

If he's working in several rooms, Schmuecker does a test patch in each, because the woodwork in different rooms will probably have different tales to tell. Kitchen trim, he says, has usually been repainted many times and will probably require the use of a slow-acting remover, but bedroom trim will probably have only a few coats of paint on it, and a faster stripper might be the best choice. This information is especially important to a pro, for whom time is money.

Professionals know which strippers will harm which materials. If you aren't sure, and

Applying the stripper. **Schmuecker wears goggles, rubber gloves and coveralls when he works with paint remover. The room isn't well ventilated, so he's wearing a respirator, too. Plastic sheeting protects the walls. He's run a test patch at the junction of sill and casing. Each application consists of two coats of remover; a light one followed two to five minutes later by a heavy one.**

In areas where the paint remover might drip off the surface, it can be covered with cellophane. A cellophane cover also lengthens the stripper's wet life and keeps it working longer.

After the paint remover has been applied, it can be worked into corners and gaps with a knife blade to get as much paint as possible out of these tough spots.

Before starting to scrape, Schmuecker makes sure that the paint has been dissolved down to the bare wood. The stripper he uses may be renewed with another light application if it dries out.

can't find the information on the can, you should run your own tests. Dab some paint remover on the plastic sheeting before you tape it in place. Wipe a touch on your weatherstripping, and anything else you don't want destroyed that will come into contact with the stripper. Wait a few hours and see what happens. Don't rush this process if you want a reliable reading.

Applying the remover—On a small job, Schmuecker simply pours stripper into a coffee can and works with a 2-in. to 3-in. brush. For heavy paint, he applies a light coat first, letting it stand for two to five minutes, and then covers it with a much heavier coat. He brushes the remover on in one direction only, so as not to thin it out and encourage evaporation. To lay on the thickest possible coat, he dabs, rather than sweeps, with the brush.

On larger jobs, he often sprays the stripper on. Sprayers are professional tools, and cost anywhere from about $200 to about $800. Don't try to use a common paint sprayer. It's important that the tank, gaskets and hoses be compatible with the remover you're using. You shouldn't, for example, use an aluminum

tank with a methylene-chloride stripper. The metal will react with the chemical, and the tank could burst. If you have questions about the compatibility of a sprayer and a stripper, write to the paint remover's manufacturer, and request its material safety data sheet.

Whether he's brushed or sprayed the remover on, Schmuecker covers it with Saran Wrap to hold it against the surface in spots like the underside of the window rails, where the stripper might drip off (photo top left). The plastic wrap also retards evaporative drying and increases the stripper's wet life by up to an hour, so Schmuecker covers his remover whenever he's working on a small area with an especially thick paint layer.

After 10 or 15 minutes, he uses a knife or an ice pick to work remover into cracks, corners and the most deeply carved parts of ornate trim (photo above left). Then he lightly recoats the area.

It may take as long as four hours for the surface to be ready for scraping. Many people begin to scrape the remover off much too soon, before it has had a chance to penetrate and dissolve the layers of paint. By doing this they work a lot harder than they have to, and use

much more stripper than necessary. Schmuecker usually coats more windows or trim while he's waiting, but he cautions that it's very easy to get ahead of yourself. Scraping takes much longer than applying remover, so by the time you've come back and scraped a few windows or feet of trim clean, the last areas you coated could have dried out. Work only a window or a few feet of molding ahead until you can judge your speed.

The instructions on some strippers tell you not to let them dry out. This is because they'll harden if they're allowed to dry, and then you've got to start all over again. Others can be reactivated with another light layer of remover. The strippers Schmuecker works with can be reactivated.

Scrape, rinse and clean—It's time to scrape off the remover with its paint residue when a test scrape shows you're either down to bare wood (photo above right) or that the lowest layer of stripper is drying out. This is the most tedious and time-consuming part of the job. It invariably takes a lot longer than an inexperienced person thinks it will. The first time you strip molding, give yourself twice as much time as you think you'll need. Begin by gently scraping the flat areas with a putty knife (photo top right). Drop the sludge into a pan or bucket. Schmuecker uses a disposable aluminum baking pan, which is big, light and cheap. In corners, a soft brass brush (photo center right) or a stiff plastic brush works well. Old dental tools are best for the most ornate areas. They come in all sorts of odd and helpful shapes, and they're best for stripping just at the point when they're no longer good enough to be ground down again for use by your dentist. A shavehook (photo center left) is especially good for getting paint out of dents and dings in the wood.

After you've scraped and brushed, there will still be a few stubborn spots of paint left. If you are going to repaint, you can stop here and apply the rinse, but if you're after a clear surface, recoat and scrub the wood with a plastic steel-wool substitute (the real thing can leave particles embedded in the wood to oxidize, change color and leave a pebbled finish that looks like black-eyed peas).

Tool catalogs offer scrapers shaped to fit some standard moldings. It's worth buying one if it fits the trim you're working on. When he has to strip hundreds of feet of a single, unusual molding, Schmuecker takes a section of it to a machine shop and has a scraper ground to fit the pattern. He says the frustration and time this saves is well worth the $30 to $40 for the custom grinding.

Sometimes, if the paint is very thick, another complete application of remover is necessary. Just repeat the whole procedure, using both the light and heavy applications. The five or six coats on the window shown here came off with a single application, and Schmuecker proceeded to rinse. You can use either water or solvent—lacquer thinner or paint thinner—to rinse the wood down. Schmuecker prefers solvent, even though it's flammable. It dries much more quickly, and it doesn't raise the grain or make the wood swell. He rubs on a light coat with the steel-wool substitute, then wipes the surface down with a damp cloth and lets it dry (photo bottom right). The trim is ready to refinish in a day, after a light sanding with 220-grit paper.

As anyone who's done any stripping knows, it's next to impossible to get paint out of joints. This is no problem if you're going to repaint, but is trouble if you want to leave the finish natural. Schmuecker doesn't even try to remove all the paint. Instead, he scrapes it back, much like a mason raking a mortar joint, and fills the joints in either with putty, which he later stains, or with colored lacquer sticks or wax crayons.

Another common problem is finding heat or water marks, mottled wood or stain residue on wood that you want to leave clear. Schmuecker recommends applying—right after the final rinse, if possible—an oxalic acid solution (½ cup to 1 gal. warm water). Let it stand for two or three minutes, then rub it down with lacquer thinner, paint thinner or water. This will lighten the marks, mottles and residue so that the wood can take a stain. □

Paint-stripping tools. It sometimes takes as long as four hours for a heavy coating of remover to dissolve paint down to the bare wood. Once it's time to scrape, a putty knife (top) will lift globs of paint and stripper off the surface. A shavehook (above left) and a plastic or brass brush (above right) come in handy for corners and edges.

A good scrub with synthetic steel wool gets off the last specks of paint. If you're going to repaint, you don't need to do this. The last step is to wipe the surface down with solvent, as below.

Designing and Building Stairs

Stairways can be minimal or very elaborate, but they're all based on simple geometry and accurate finish work

by Bob Syvanen

Today, few people can afford the time and expense of building classical 19th-century stairways. Tastes have changed too, and as a result, stairs have become simpler. But the principles of stair design and construction are the same as they've always been, and so are the skills that the builder must bring to the task. If modern stairs aren't ornate, they still remain a focal point of a house—your work is out there for all to see.

One of the problems builders face in cutting stairs is getting rusty. Unlike the master stairbuilders of the past, on-site carpenters typically build only a few stairways a year. Fortunately, designing a stair and laying out the stringers uses the same language and concepts as roof framing. Fitting treads and risers, on the other hand, is nothing more than simple, if demanding, finish work.

Basic stair types—The simplest stairway is a cleated stairway, which relies on wood or metal cleats fastened to the carriages to support the treads. You could use a cleated stairway for a back porch or cellar, but count on repairing the ever-loosening cleats. Wood cleats are typically 1x4s, and screwing them in is a big improvement over nails. But angle-iron cleats will last longer.

Another open-tread stairway—one without risers—uses dadoed carriages. The treads have ½ in. or more bearing on the inside face of the carriages, and they are either nailed or screwed in place. I use a circular saw, and set the depth of cut to half the thickness of the carriage, to make the parallel cuts. Then I clean up the bottom of the dado with a chisel. A router and a simple fixture built to the stair pitch and clamped to the carriage will also do a quick, neat job.

In the past, dadoed carriages were used mostly for utility stairs for porches, decks and the like, but more and more I'm asked to build open-tread oak stairs that have to be nearly furniture quality. One type, the stepladder stair, can be used when limited space doesn't allow any other solution (for more, see p. 148).

On finished stairs, the dado is usually stopped (that is, its length is limited to the

Stairbuilding in its highest forms requires the conceptual skills of a roof framer, and the fitting talents of a cabinetmaker. But simple open tread runs like the basement stair at left only require understanding the basics.

width of the tread), and squared up at the end with a chisel. If you begin the dado on the front of the carriage, leave the tread nosing protruding slightly. If you start the dado at the back of the carriage, the treads will be a little inset. They can look nice either way.

Other stairways use cut-out carriages. The simplest of these is a typical basement stair where the rough carriages are exposed, and risers are optional. The most complicated is a housed-stringer stairway. It is based primarily on patient and accurate work with a router. The stringer is mortised out along the outline of the treads and risers with a graduated allowance behind the riser and tread locations for driving in wedges. Adjustable commercial fixtures or wooden shop-made templates are used to guide the router. Although the first stairway that I built by myself on my own had a housed stringer, I won't try to give complete instructions here.

A finished stairway that still requires patient finish work, but is much less tedious to build, uses cut-out carriages hidden below the treads, and stringboards or skirtboards as the finish against the stairwall. The treads and risers are scribed to the skirts. The example I'll be using to describe final assembly is one of these that also has an open side, which uses a mitered stringer. This side of the stair requires a balustrade—handrail, balusters, and newel posts—but that's a separate topic.

Designing a stairway—No matter what style stairway you want to build, the design factors that you'll need to consider are comfort, code, safety and cost. Comfort gives the ideal conditions for good walking, and code dictates what you can and cannot do. Safety is largely a matter of common sense, and cost limits your grand ideas.

Although comfort is very much a subjective notion, there are three objective factors to consider in designing a stair—stair width, headroom and the relationship between the height of the riser and the width of the tread. Of the three, stair width is the easiest to deal with. Most building codes require utility stairs to be at least 2 ft. 6 in. wide and house stairs to be 3 ft. from wall finish to wall finish, but 3 ft. 6 in. feels a lot less restrictive. However, if the stairway gets beyond 44 in. in width, most building codes require a handrail on each stair wall.

Headroom is not quite as simple, although most codes agree that basement stairs need a minimum of 6 ft. 6 in., and house stairs need at least 6 ft. 8 in. This measurement is made from the nosing line (for a definition, see *headroom* in the glossary on the next page) to the lowest point on the ceiling or beam above. A lack of headroom is most noticeable when you're going down the stairs, because you are walking erect and bouncing off the balls of your feet. Ideal headroom allows you to swing an arm overhead going downstairs, but this requires a clearance of nearly 7 ft. 4 in. Headroom can be increased by enlarging the size of the stairwell, decreasing the riser height, or increasing the width of the treads (tread

width is the distance from the front to the rear of the tread).

Certain combinations of riser and tread are more comfortable than others. Most codes set limits—a maximum rise of 8¼ in. and a minimum tread width of 9 in.—but these are based on safety, not comfort. A 7-in. rise is just about ideal, but it has to be coupled with the right tread width.

The timeworn formula for getting the tread width and riser height in the right relationship is: riser + tread = 17½ (a 7-in. riser + a 10½-in. tread = 17½ in.). Another rule of thumb is: riser × tread = 75 (7 × 10½ = 73.5, which is close enough). Still another formula is: two risers + one tread = 24 in. I've always found the first formula the easiest. All of them establish an incline between 33° and 37°. This creates a stairway that is comfortable for most people.

More considerations—Each tread should project over the riser below it. This projection, or nosing, should be no more than 1¼ in. and no less than 1 in. In open-tread stairways, a tread shouldn't overlap the tread beneath it by more than ½ in. The nosing adds to the area where your foot falls, but doesn't affect the rise-run dimensions. The top of a handrail should be between 30 in. and 34 in. above this nosing, with a 1½-in. clearance between the handrail and the wall. If the top of the stairway has a door, use a landing at least as long as the door is wide.

Keep in mind that people aren't the only things moved up and down stairs. I once lived in an old Cape Cod house that had a stairway with 10½-in. risers, 6½-in. treads, and not too much headroom. You could negotiate it if you exercised a little caution, but moving heavy furniture up and down was another story.

The size of a stairwell is based on the riser and tread dimensions, and on how much headroom you need. A typical basement stair can be gotten into a rough opening 9 ft. 6 in. long by 32 in. wide. A main stairwell should be a minimum of 10 ft. by 3 ft.

Framing a stairwell isn't complicated and is usually defined in the local building code. If the long dimension is parallel to the joists, the trimmer joists on each side are doubled, as are the headers. Similar framing is required if the long dimension is perpendicular to the joists (drawings, above right). If the header isn't carried by a partition below, it will have to be designed for the load.

Making the calculations—Careless measuring can get you in a lot of trouble when you're building stairs. Although the initial figuring may seem a little theoretical, you'll soon be doing some fussy finish work based on these calculations and the resulting carriages. First, check both the stair opening and the floor below for level. If either is out of level, determine how much. You'll have to compensate for it later. This problem occurs most often with basement slabs.

There are many ways to lay out stairs; the following system teamed with a pocket calcu-

Framing stair openings

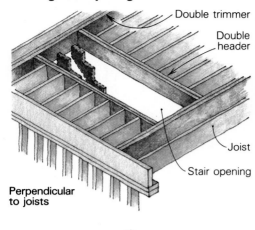

Perpendicular to joists

Double trimmer

Double header

Joist

Stair opening

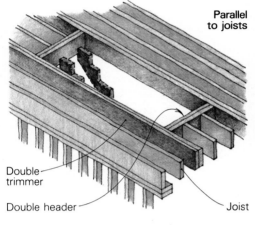

Parallel to joists

Double trimmer

Double header

Joist

lator works well for me, even though I have to scratch my head to recall what I did last time. Start with the measurement that has already been determined, the finished floor-to-floor height. I'll use 108 in. in this case. Then, just for a starting point to get you close, divide by 14, the average number of risers used in residential stairs. This gives a riser of 7.71 in., which is a little high. Adding another riser will reduce this measurement some: 108 in. ÷ 15 = 7.20 in. Sounds good. You now know how many risers you'll be using and how high they are.

To get the width of the treads (remember, this means the front-to-back measurement of each step), use the rise-plus-tread formula in reverse: 17.5 in. − 7.2 in. = 10.3 in.

All stairs have one more riser than treads. This is because the floors above and below act as initial treads, but aren't a part of the stair carriage calculations. You'll have to keep reminding yourself of this when you lay out the carriages. I don't know a good way of remembering this, and I have resorted to drawing a sketch of a couple of steps and using it to count the difference. In the example we're using, then, there are 14 treads and 15 risers.

The last calculation is the total run—the length of the stairway from the face of the first riser to the face of the last riser. This is simply the total number of treads multiplied by their width: 14 × 10.3 in., or 144.2 in. With this figure you can check to see whether the stairway will fit in the space available. Although I now use a calculator for the math, my main tool used to be a stair table like the

A Glossary of Stair Terms

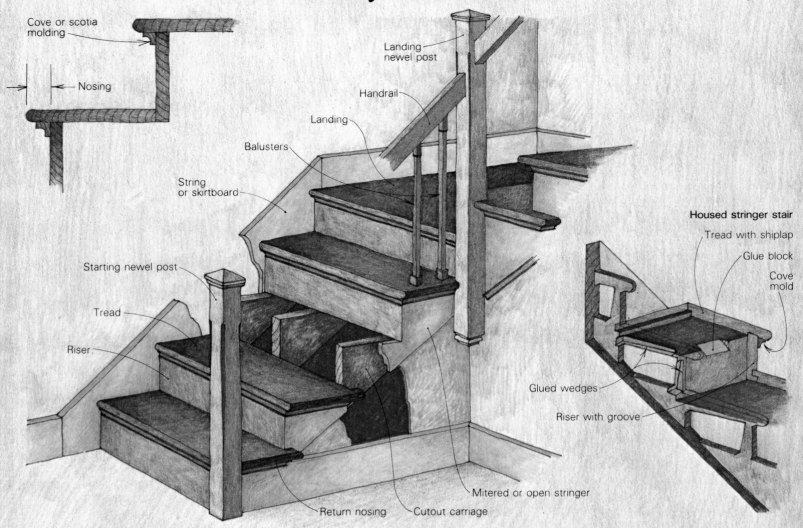

Cove or scotia molding

Nosing

Landing newel post

Handrail

Landing

Balusters

String or skirtboard

Starting newel post

Tread

Riser

Housed stringer stair

Tread with shiplap

Glue block

Cove mold

Glued wedges

Riser with groove

Mitered or open stringer

Return nosing

Cutout carriage

Balusters—The posts or other vertical members that hold up the handrail, usually two per tread.

Balustrade—The complete railing, including newel posts, balusters and a handrail. Most of these parts are available as stock finished items at lumberyards.

Carriages—Also called *stair stringers, stair horses* or *stair jacks*. They are the diagonal members that support the treads. Carriages can either be *finish stringers* or rough stringers—for an outside stairway, or for an inside stairway hidden from view. *Rough carriages*, whether they are *cut-out carriages* or just *dadoed* or *cleated stairs*, are made of 2x10 or 2x12 softwood lumber. Finish stringers are usually made of ¾-in. or 1⅛-in. stock. They either can be cut out (an *open or mitered stringer*) or routed (a *housed stringer*).

Closed stairway—Stairs with walls on both sides. In this case a *wall stringer*, whether it is a *housed stringer* or just a *stringboard*, is nailed to each wall. Closed stairways use handrails, not a balustrade.

Finished stairway—Any of several interior stair types that have risers, treads, stringers and a handrail or balustrade.

Handrail—This rail runs parallel to the pitch of the stairs. It's held by balusters or brackets.

Headroom—The vertical distance from the lowest point of the ceiling or soffit directly above the stair to the *nosing line*, an imaginary diagonal connecting the top outside corners of treads. Most codes require at least 6 ft. 8 in. for stairs in living areas, and 6 ft. 6 in. for basement utility stairs.

Housed stringer—The profile of the treads, nosing and risers is routed into a finish stringer. Extra room is left for wedges to be driven and glued in between the stringer and the treads and risers. Rabbeted and grooved risers and treads are also used.

Landing—A platform separating two sets of stairs.

Newel post—The large post at the end of the handrail. There is a *starting newel* at the base of the stairs, and a *landing newel* at turns.

Nosing—The rounded front of the tread that projects beyond the face of the riser 1 in. to 1¼ in. In the case of *open-tread stairways*, it shouldn't exceed ½ in. In most cases, the nosing is milled on the tread stock. On open stairways, a half-round molding called *return nosing* is nailed to the end of the tread.

Open or mitered stringer—This is a cut-out finish stringer used in open stairways. The treads carry over the stringer, but the vertical cut-outs on the carriage are mitered with the risers at 45°.

Open stairway—This can be open on one or both sides, requiring a balustrade. In finished stairways, the open sides will use a *mitered* or *open stringer*.

Rise—The height of each step from the surface of one tread to the next. Just as in roof framing, this measurement is sometimes called the *unit rise*. Many codes call for a maximum rise of 8¼ in. The height of the entire stair, from finished floor to finished floor, is the *total rise*.

Riser—Describes the rise of one step. It is also a stair part—the vertical board of each step that is fastened to the carriages. Risers for a *housed stringer stair* are rabbeted at the top to fit the tread above, and grooved near the bottom for the tread below. Other stairs use 1x square-edged stock. *Open-tread stairs* don't have risers.

Run—Also called *unit run*, this is the horizontal distance traveled by a single tread. A 9-in. run is the code minimum for main stairs. *Total run* is the measured distance from the beginning of the first tread to the end of the last tread—the horizontal length of the entire stairway.

Stairwell—The framed opening in the floor that incorporates the stairs. Its long dimension affects how much headroom the stair has.

Stringboard—Diagonal trim, not used to support the treads, that is nailed to the stair walls. Finished treads and risers butt these. Often called *skirtboards, backing stringers*, or *plain stringers*.

Tread—It is both the horizontal distance from the face of one riser to the next, and the board nailed to the carriages that takes the weight of your foot. Exterior stairs typically use 2x softwood treads. Interior stairs use either 1⅛-in. hardwood stock milled with a rabbet and groove to join it to the risers, or 1³⁄₁₆-in. square-edged stock. Both are usually nosed.

Winder—Wedge-shaped treads used in place of a landing when space is cramped, and a turn is required in the stairway. Many building codes state that treads should be at least the full width of the non-winder treads, 12 in. in from their narrow end; or that the narrow end be no less than 6 in. wide.

—Paul Spring

Figuring carriage length and layout

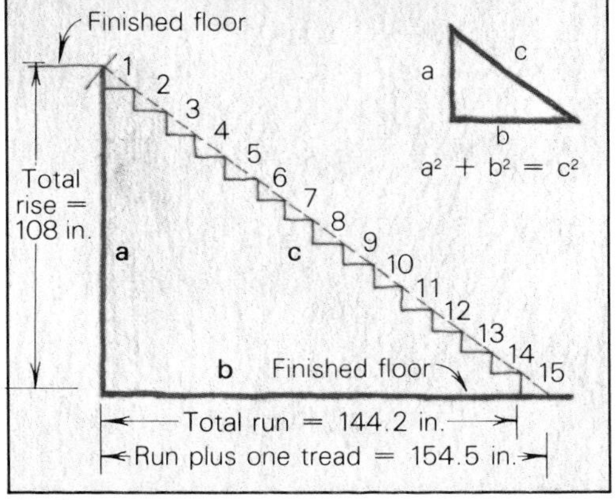

Finished floor

Total rise = 108 in.

$a^2 + b^2 = c^2$

Total run = 144.2 in.

Run plus one tread = 154.5 in.

Total rise*	Number of risers	Riser height	Tread width	Total run
8'0"	12	8"	9"	8'3"
	13	7⅜" +	9½"	9'6"
	13	7⅜" +	10"	10'0"
8'6"	13	7⅞" −	9"	9'0"
	14	7⁵⁄₁₆" −	9½"	10'3½"
	14	7⁵⁄₁₆" −	10"	10'10"
9'0"	14	7¹¹⁄₁₆" −	9"	9'0"
	15	7³⁄₁₆" +	9½"	11'1"
	15	7³⁄₁₆" +	10"	11'8"
9'6"	15	7⅝" −	9"	10'6"
	16	7⅛"	9½"	11'10½"
	16	7⅛"	10"	12'6"

*height from finished floor to finished floor

Stair geometry. The stair described in the text is shown above left. Plugging total run and total rise into the Pythagorean theorem gives the required carriage length. Based on an ideal riser and tread, the stair chart, above right, gives the number of risers and total run for a given total rise.

one above. I still use one for quick reference in the planning stage. If the run is too long, make the treads narrower, or eliminate a riser and a tread. Either way, you'll need to run a new set of calculations.

Layout—To lay out the carriages, you first need to know what length stock to buy. If you've got a calculator, it's easiest to use the Pythagorean theorem $(a^2 + b^2 = c^2)$. But you must add the width of an extra tread to your total run to get enough length for the bottom riser cut (drawing, above). In this case, a is the total rise of 108 in., and b is the total run of 144.2 in. plus a 10.3-in. tread. The hypotenuse, c, is 188.5 in. So you'll have to buy 16-footers to allow for cutting off end checks, and avoiding large knots with the layout.

Most cut-out carriages are 2x12s because you need at least 3½ in. of wood remaining below the cut-outs for strength; 4 in. is even better. Douglas fir is the best lumber for the job because of its strength. You will need a third, or center, carriage if the stair is wider than 3 ft. with 1½-in. thick treads, or wider than 2 ft. 6 in. with 1⅛-in. thick treads.

Once you have marked the edge of one of the carriages with the 188.5-in. measurement, you are ready to lay out. I step off equal spaces with dividers (photo top right) before marking the riser and tread lines with a square. Some carpenters simply step off the cut-out lines with a square, but I don't like the accumulated error you can get this way. A deviation of more than ¼ in. between the height of risers or the width of treads can be felt when walking a flight of stairs. For this reason, it's also a violation of code.

The dividers I use are extra long. You can improvise a pair by joining two sticks with a finish nail for a pivot, and a C-clamp to hold them tightly once they're set. The easiest way to find the spacing is to locate 7.2 in. (the riser height) and 10.3 in. (the tread width) on a framing square, and set the divider points to span the hypotenuse, which is 12.56 in. (a strong 12½ in.). No matter how careful you are in setting the dividers, it will take a few trial-and-error runs before you come out to 15 even spaces. Once you do, mark the points on

the edge of the carriage. These represent the top outside corner of each tread, less the profile of the nosing.

To draw the cut-out lines, use either a framing square or a pitchboard. Most carpenters use the square, but a pitchboard can't get out of adjustment. You can make one by cutting a right triangle from a plywood scrap. One side should be cut to the height of the riser, and the adjacent side to the tread width. A 1x4 guideboard should be nailed to the hypotenuse. Align it with the marks on the carriage and use it to scribe against.

If you are using a square, set it on the carriage so that the 90° intersection of the tongue and body point to the middle of the board. Along the top edge of the carriage, one leg should read the riser increment and the other, the tread increment. Use either stair-gauge fixtures (stair buttons) or a 2x4 and C-clamps to maintain the correct settings when the square is in position against the edge of the carriage. Then, holding the square precisely on the divider marks, scribe the cutlines for each 90° tread-and-riser combination (photo second from top).

Dropping the carriages—One of the most difficult things about stairs is adjusting the carriages for the different thickness of floor finish, which can throw off the height of the bottom and top risers. Any difference should be subtracted from the layout after the tread and riser lines are marked, and carefully double-checked before you do any cutting. What you marked on the carriage is the top of the treads, but since you will be nailing treads to the carriages, you need to lower the entire member enough to make up for the difference. This is called *dropping the carriages*. If they sit on a finished floor, such as a concrete basement slab, the bottom riser will need to be cut shorter by the thickness of a tread (drawing A, next page, top right). This will lower the carriage so that when the treads and upper floor finish are added, each step will be the same height. The bottom riser will have to be ripped to a narrower width. If the treads and floor finishes are of equal thickness (B), and the carriage sits on the subfloor at the bottom,

Laying out and cutting the carriages. Starting with the top photo, Syvanen uses large dividers to mark the intersections of tread and riser lines on the front edge of the carriage. From these marks, he scribes the cut-out lines using a framing square fitted with stair-gauge fixtures. With a circular saw, he begins cutting out the carriages by notching the bottom riser for the kickplate that will anchor it to the floor. Syvanen uses a handsaw to finish the cutting. Cut-out carriages are usually made of 2x10s or 2x12s, since there should be at least 3½ in. of stock between the bottom edge of the carriage and the cutout.

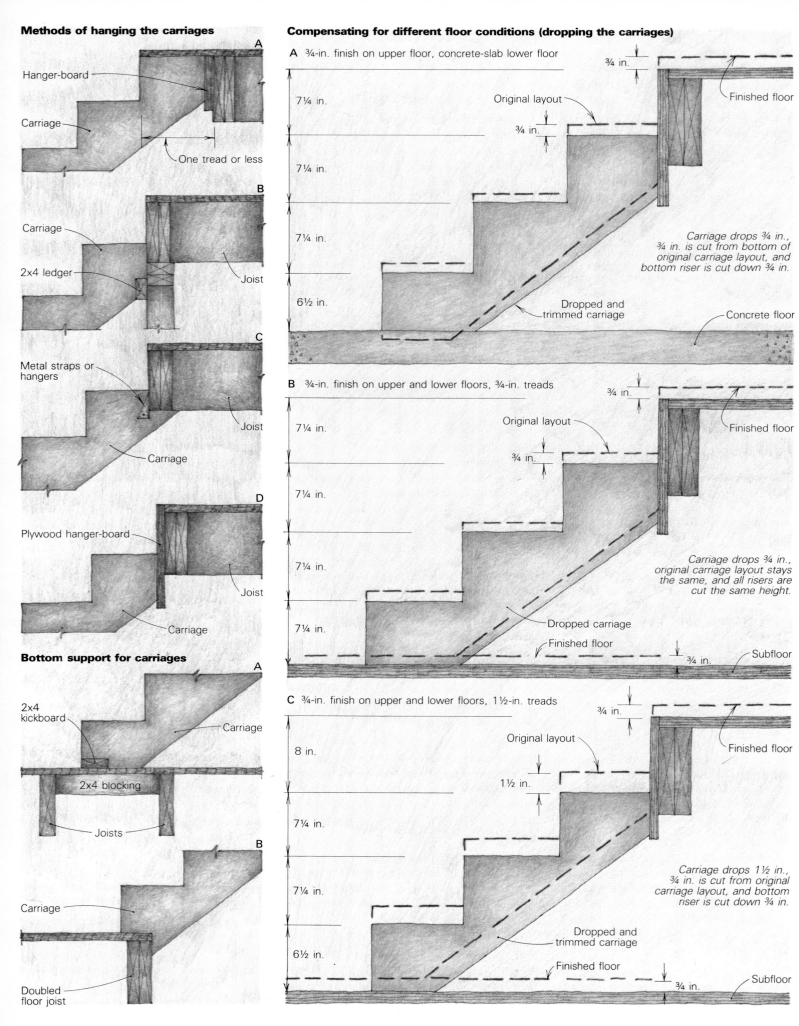

Methods of hanging the carriages

A
Hanger-board
Carriage
One tread or less

B
Carriage
2x4 ledger
Joist

C
Metal straps or hangers
Joist
Carriage

D
Plywood hanger-board
Joist
Carriage

Bottom support for carriages

A
2x4 kickboard
Carriage
2x4 blocking
Joists

B
Carriage
Doubled floor joist

Compensating for different floor conditions (dropping the carriages)

A ¾-in. finish on upper floor, concrete-slab lower floor

¾ in.
7¼ in.
7¼ in.
7¼ in.
6½ in.

Original layout
¾ in.
Finished floor

Carriage drops ¾ in., ¾ in. is cut from bottom of original carriage layout, and bottom riser is cut down ¾ in.

Dropped and trimmed carriage
Concrete floor

B ¾-in. finish on upper and lower floors, ¾-in. treads

¾ in.
7¼ in.
7¼ in.
7¼ in.
7¼ in.

Original layout
¾ in.
Finished floor

Carriage drops ¾ in., original carriage layout stays the same, and all risers are cut the same height.

Dropped carriage
Finished floor
¾ in.
Subfloor

C ¾-in. finish on upper and lower floors, 1½-in. treads

¾ in.
8 in.
7¼ in.
7¼ in.
6½ in.

Original layout
1½ in.
Finished floor

Carriage drops 1½ in., ¾ in. is cut from original carriage layout, and bottom riser is cut down ¾ in.

Dropped and trimmed carriage
Finished floor
¾ in.
Subfloor

Nearly complete, the finish stair at left is missing only its balusters, handrail, and molding under the nosing. The newel post is mortised into the first tread for stability. This stair uses both an open stringer and treads and risers that butt-join the skirtboards or stringboards that are nailed to the wall. Above, kraft paper protects the completed oak treads from construction traffic. The open stringer is mitered to the riser, and the treads overhang the stringer by the depth of the return nosing with its scotia molding beneath. The cut-out carriage that actually supports the treads is hidden behind the finish stringer and drywall blocking.

no change will have to be made for the risers to be equal.

A more confusing condition is when the treads are thicker than the finished floor (C). At this point, I usually draw a four-riser layout, at any scale, on graph paper to figure how much of a drop I need to make, and if the bottom riser needs to be narrower.

How the cut-out carriages are attached once they are raised in the stairwell also may require adjustment at the top and bottom of the carriages (drawings, opposite page, bottom left). At the bottom, I like to use a 2x4 kickboard nailed to the floor at the front edge of the riser (A). If there is a stair opening below, the carriages can be cut to fit around the upper corner of the framing (B). Stairs take a beating, and should be well secured.

At the top, the header joist usually acts as the uppermost riser, but sometimes, the floor will extend a full or partial tread width from the framing (drawing A, opposite page, top left). A 1x4 or 2x4 ledger board can be nailed to the framing (B), and if so, the carriages must be notched to fit it. Metal angles or straps can be used if the carriages aren't exposed (C). I like a hanger board because it is quick, neat and strong. I nail the carriages to a line on a piece of plywood, a riser's distance from the top. I then raise the whole business as a unit and nail it in place (D).

Give the carriage a trial fit before sawing it out. I make only the horizontal cut that rests on the floor and the vertical cut that leans against the framing at the top before trying

the carriage in the stairwell. If you are really unsure, use a 1x10 trial board. With the carriage in place, you can easily check your layout. The treads should be level from front to back, and the carriage should fit on both sides of the opening. Also make sure that the risers will all be the same height once the treads and finish floor are installed.

If everything checks out, what's left is just cutting and fitting. With this basic layout you can produce the cut-out carriages that are needed for the stairway shown above, you can dado the carriages for let-in treads, or you can just nail cleats to the layout lines. For cut-out carriages, use a circular saw as far as you can, and finish them off with a handsaw held vertically so as not to overcut the line and weaken the cut. You can nail the triangular cutouts to a 2x6 for a third stringer if the budget is tight. Use the completed carriage as a pattern to trace onto the other 2x12, and then cut the pencil line to get an exact duplicate.

Treads and risers—On a closed stairway, the cut-out carriages sit inside the finish wall stringers, which are called skirtboards, or strings. These are usually 1x10, and should be nailed hard against the wall so that the snug fit of previously installed risers and treads isn't spoiled by the skirtboards spreading when newly scribed boards are tapped into place. They should be installed parallel to the nosing line, and as high as possible without exposing any wall where the riser and tread meet. Don't nail the cut-out carriages to the

skirtboards on a closed stairway, or the mitered stringer to the outside carriage on an open stair. Instead, hold the carriages about 3 in. away from the walls, so that they are bearing only at their tops and bottoms. This keeps the treads and risers from splitting as a result of nailing too close to their ends. Skirtboards and risers can be made of pine to ease the budget, but the best treads are oak. The standard thicknesses for treads are $1\frac{1}{8}$ in. and $^{13}\!/_{16}$ in.

I like to rip all of my treads and risers to width before beginning the assembly. Keep the risers a hair narrower than what's called for. Crosscut both risers and treads to 1 in. longer than the inside dimension between the skirt boards. This allows them to fit at a low enough angle to get a good scribe and still have a little extra to cut off. If the stair is open on one side, the treads will have to be rough-cut long enough to leave a $\frac{1}{2}$-in. scribing allowance on the closed side, and some overhang on the open side, which gets a return nosing. The risers will need at least a 45° miter to mate with the open stringer. Use a radial arm saw or handsaw for this.

Stair assembly usually begins at the bottom. The first two risers are fit and nailed, and then the first tread is pushed tightly against the bottom edge of the riser for scribing. For the stair pictured above, I first had to cut the open, or mitered, stringer. It was pine, and was laid out like the carriages with the exception of the vertical cuts, which extend beyond by the thickness of the riser material to form

Stepladder stairway

A cleated or dadoed stairway at an angle of from 50° to 75° is considered a stepladder stairway, or ships's stairway. The rise on these ladders can be from 9 in. to 12 in. As the angle of the carriages increases, the rise increases. A 50° angle should have about a 9-in. rise. A 75° angle should have about a 12-in. rise.

As with all other stairs, the relationship between the height of the riser and the width of the tread is important. But in this case, their relationship is reversed. The tread width on stepladder stairs will always be less than the riser height.

If you know what riser height you'll be using, the easy formula for calculating the tread width is this: tread width = 20 − 4/3 riser height. If you use a 12-in. rise, then the tread will be 4 in. You can also rearrange this formula to solve for riser height (riser height = 15 − 3/4 tread width) if you know what tread width you want. This is useful if you are limited in how far out into a room the base of the stair can come.

A simple way to lay out a stepladder stairway is to lean a 2x6 or 2x8 carriage at about 75° in the stair opening. Lay a 2x4 on edge on the floor, and scribe a pencil line across its top edge to transfer a level line from the floor onto the carriage. Mark the vertical cut, at the top, in a similar manner.

Make these cuts and set the carriage back in place. Measure the floor-to-floor height and divide by 12 in. to get the number of risers. You'll end up with a whole number and a fraction. For instance, if your total rise is 106 in., the number of risers will be 8.83. That's close enough to call it 9 risers. Divide this back into the total rise of 106 in. to get an accurate riser dimension of 11.77 in., or about 11¾ in.

The easiest way to lay out the carriages is to make a story pole using dividers set at about 11¾ in. to step off nine equal segments within the 106 in. It might take a few tries adjusting the dividers to get it to come out just right. You can even find the riser dimension without the math by making a few divider runs up and down the story pole.

With the carriage in place, mark each tread from the story pole. A level line at each mark locates the treads. A bevel square set at the correct angle will also work. The treads can be cleated or let in with a dado. —*B.S.*

an outside miter. I'm most comfortable making these cuts with a handsaw. Keeping this angle slightly steeper than 45° allows the faces to meet in an unbroken line, without any interference at the back of the joint. The miters were predrilled, glued and nailed with 8d finish nails (photo previous page, right).

The treads are initially cut to overhang the open stringer, by the same dimension as the nosing. Then a cross-grain section is cut out so that a mitered corner is left at the outer edge. This accommodates a return nosing. I also like to put a piece of nosing at the rear of the tread where it overhangs, although you can just round off the back end of the return nosing. If the mitered stringer doesn't snug up perfectly under each tread, don't worry. The crack will be covered by the cove or scotia molding that runs under the nosing. It's more important to get good bearing for the tread on the cut-out carriage, and a little shimming or block-planing will help here.

Once beyond the open side of the stair, you will be fitting treads and risers on a closed stairway, scribing to the skirtboard on each side. With a 1-in. allowance for scribing, set the scribers at ½ in. for the first side. Set the tread or riser with the side you are going to scribe down in place on the carriage and against the skirtboard. The other end will ride high on the other skirt. If you are working on a tread, make sure it is snug to the riser along its entire length. Risers should sit firmly on the carriages to get an accurate scribe. I use a handsaw to back-bevel the cut on risers, but I keep the cut square at the front of the treads where the nosing protrudes, and then angle it the rest of the way. If necessary, use a block plane to make sure the cut fits.

Next, get the inside dimension at the back of the tread or lower edge of the riser, depending on which you are fitting. A wood ruler with a brass slide works well here. Transfer this dimension to the board you're working on, set the scriber to the remaining stock, and mark the board. Cut the boards the same way you did the first time. Cut carefully, remembering that you can always plane it off, but you can't stretch it. However, don't make the cut too strong either. Trust your measure.

Careful cutting and fitting are important with this kind of stair, and a little glue, and some nails and wedges in the right place work wonders as time goes on. Risers and treads are nailed to the carriages with 8d finish nails through predrilled holes. A sharpened de-headed 8d finish nail chucked tightly into an electric drill makes a snug hole every time.

Stairs that aren't made from rabbeted stock (like housed-stringer stairs) can be kept together if you drive three or four 6d common nails through the bottom of the riser into the back of the tread. This should be done as you fit your way up the stair. Two 1x1 blocks, 2 in. long, glued behind each step at the intersection of the upper edge of the riser and the front of the tread will cut down a lot on movement too. Any gaps between the carriage and treads should be shimmed from behind with wood wedges to eliminate squeaks. □

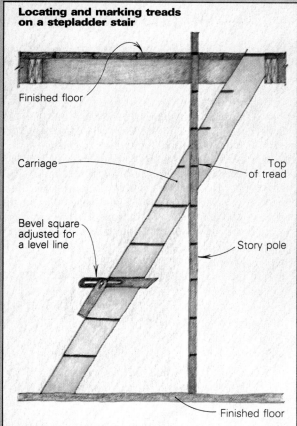

Locating and marking treads on a stepladder stair

Finished floor

Carriage

Bevel square adjusted for a level line

Top of tread

Story pole

Finished floor

Sculpted Stairway

Imagination and improvisation coax an organic shape from 42 plies of Honduras mahogany

by Jody Norskog

My brother Noel and I first met Chris Bradley and Steve Zoller in Santa Fe, N. Mex., in the summer of 1980. They were researching solar homes, looking for ways to improve a house they were designing and building, and generally checking out the Santa Fe scene. They were designing a house in Laguna Beach, Calif., and mentioned that they might have a stairway for us to build. We said sure, keep us in mind, and never expected to see them again.

Six months later we got a call from Laguna Beach; "Interested in doing that stairway?" Bradley and Zoller had been working on a design, but they weren't satisfied with the way it looked and wanted us to help.

Design concepts—Bradley and Zoller did not want a stairway framed up in the conventional fashion, and were unwilling to settle for a commercial spiral stair. They wanted a stairway that would be sculptural as well as functional—something that would be enjoyable to look at and use, but would not block the view from a large glass area.

Noel and I began debating design, materials and details. Originally we considered building the entire stairway out of steel. We later decided on wood after eliminating some other possibilities. After building models to study forms and detailing, we began exchanging ideas between Santa Fe and Laguna Beach.

The design we finally agreed on was far from conventional. We chose to support the treads and handrails with a single beam, which could be bent to form a compound curve. Imagine taking a tree trunk and bending it to the desired curve. And imagine notching the trunk and fastening a plank at each notch to make treads, and attaching a handrail on each side of the treads. This was the basic concept of the project. Our trunk would be a bent lamination.

Honduras mahogany was being used for the

Jody and Noel Norskog are partners in their firm, Norskog and Norskog. They build furniture and do architectural woodwork.

Looking as if it grew there, the Norskogs' mahogany and steel stairway winds its way gracefully to the second story. A mammoth undertaking, this stair took three months' work in their Santa Fe shop, and another six weeks at the Laguna Beach site.

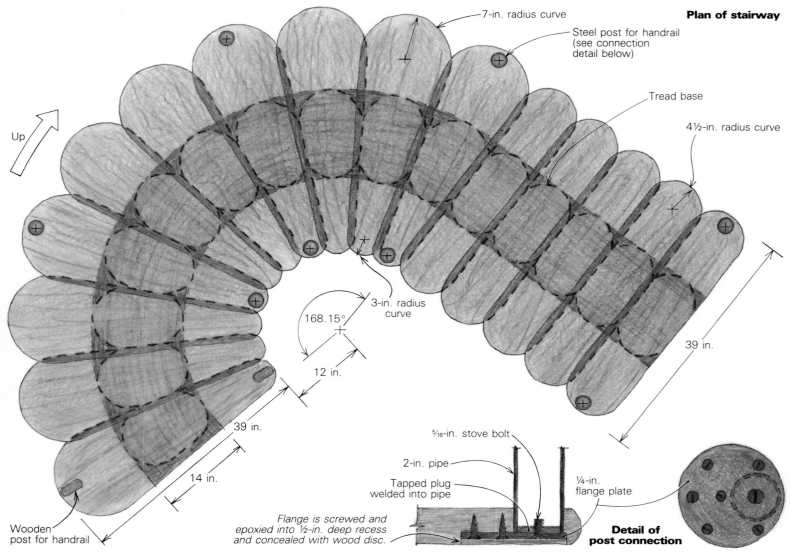

Plan of stairway

7-in. radius curve

Steel post for handrail
(see connection
detail below)

Tread base

4½-in. radius curve

Up

3-in. radius
curve

168.15°

39 in.

12 in.

39 in.

14 in.

⁵⁄₁₆-in. stove bolt

2-in. pipe

Tapped plug
welded into pipe

¼-in.
flange plate

Wooden
post for handrail

*Flange is screwed and
epoxied into ½-in. deep recess
and concealed with wood disc.*

**Detail of
post connection**

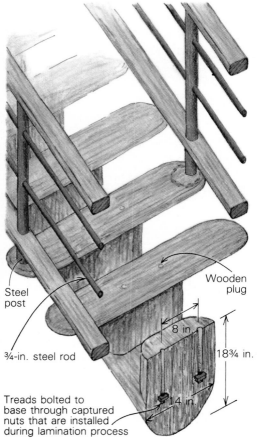

Steel
post

Wooden
plug

¾-in. steel rod

8 in.

18¾ in.

14 in.

Treads bolted to
base through captured
nuts that are installed
during lamination process

wood details and trim in the house, and it made sense to use it for the stairway as well, even though the wood had some bad qualities for the job at hand. While its crisp tissue is easy to carve and shape, it's not an easy wood to bend.

Once the design of the stairway had been reasonably defined, we worked with a structural engineer to determine what it would take to hold the thing up. After producing 30 pages of calculations, he was able to give us the information we needed—how to size the cross section of the trunk and how to design and size the hardware for connecting the trunk to the first and second floor.

The design and engineering process had stretched out over nine months before we got the go-ahead to build. Finding the wood for laminating the trunk turned out to be as hard as any other part of the project. No supplier in New Mexico could get what we needed. By doing some sample bending we determined the maximum thickness for each ply (or lamina) was ⁵⁄₁₆ in. Anything thicker was either too difficult to bend or too prone to tension failure—cracking on the convex side of the curve.

The problem was where to find pattern-grade Honduras mahogany boards ⁵⁄₁₆ in. thick by 15 in. wide by 18 ft. long. (The term *pattern grade* is peculiar to mahogany. It simply desig-

nates clear, stable stock, suitable for making foundry patterns.) Planing 4/4 stock down to ⁵⁄₁₆ in. would involve criminal waste, and re-sawing thicker, 18-ft. long boards was beyond the capability of our 14-in. bandsaw. We had to find a supplier who had lots of mahogany in the right dimensions. After much research and many telephone calls we located what we needed in a lumberyard in Long Beach, Calif. The lumberyard resawed 8/4 stock and surfaced the planks to our ⁵⁄₁₆-in. thickness. It was ironic that we bought the wood in southern California, had it shipped to Santa Fe, worked it, and then returned it to southern California.

Gluing up the trunk—The bending form for laminating the beam was an elegant piece in its own right (photo facing page). At either end there were plywood end plates with 2x4 platens for the actual working surfaces of the form, spanning between the plates. Essentially, it was like building a 2x4 stud wall to match the inside curve of the beam, and then leaning it on its side. Each 2x4 served as a clamping platen, and as a reference point for locating the position of the riser for each step. One person walked into the shop and said, "What are you guys building, an airplane?"

Few of the boards we got were wide or long enough, which meant that we had to butt sev-

Photos: Noel and Jody Norskog; Illustrations: Christopher Clapp

Work on the staircase began with milling the stock and gluing up the 42 plies for the large lamination that would be sculpted to become the stringer or trunk for the treads. The bending form, though a simple jig made from 2x4s and plywood, had to be precisely built to produce the correct curve and twist. Because plastic-resin glue takes eight hours to cure, it took 42 days to glue up the whole lamination.

eral boards together to make up each layer of laminations. We were careful to stagger the joints from one layer to the next.

After getting the clamps, wood, and form ready, we did a dry-clamping run and felt we were ready for the first lamination. We mixed up the plastic-resin glue, applied it to the boards with paint rollers, put the wood on the form and clamped two plies together.

The next day we pulled the clamps off but the glue didn't hold. We tried it again with new boards. The same thing happened. At this point we were beginning to get frantic. Bad glue? Wrong kind of glue? Should we remill the wood? Then we discovered that plastic-resin glue requires a temperature above 70°F to cure. This was January at 7,000 ft. in northern New Mexico, where it gets cold. Our passive-solar shop maintains a comfortable temperature fairly easily—but not always 70°F. With our new kerosene heater we made several more tries, all the time refining our gluing procedure. At last we were able to get a bond that we felt good about.

Each tread was to be held in place by two countersunk bolts threaded into correspond-

ing nuts embedded into the beam. The nuts were captured in the beam as it was being laminated. After the thirteenth layer, the first set of nuts was inserted. The 2x4 platens allowed us to position each nut accurately. We routed slots for the nuts and corresponding slots for bolt access. After each nut was set in the beam, the slots were taped over so they wouldn't fill with glue, and the laminating process continued. At the thirty-first layer, the second set of nuts was embedded into the beam. We used rectangular nuts made from ½-in. bar stock instead of standard hex nuts, to prevent them from turning.

Clamping time for Borden's plastic-resin glue is eight hours at 70°F. This meant that the whole lamination (42 plies in all) would grow at the rate of one ply per day. The pipe we were using for clamps could exert only enough pressure to hold down two layers of the mahogany over the curve. (We tried once to clamp three plies, but the force required was so great that the threads on the pipes began to strip.) The first ply got the glue; the second we used as a caul to even out the clamping pressure. Each caul ply became the

following day's gluing layer. Occasionally the cauls would fail in tension. There seemed to be no way of determining which board would fail except by testing it. Each day meant pulling the clamps off, sanding the entire surface, fitting boards for the next lamina, mixing and spreading glue, and clamping on the next ply. One lamination a day for 42 days.

Carving the trunk—Once the forty-second lamination was clamped down, it was time for a celebration. Even after it was notched and carved, the glued-up trunk weighed 2,000 lb., so it weighed considerably more at this point. We rented a 3,000-lb. capacity hoist for moving the beam around. Used for pulling out automobile engines, it picked up the beam, bending form and all. Once the entire thing was in the air, it was easy to disassemble the form, piece by piece.

The location of the treads was marked by the 2x4s on the form. But after all the days of clamping, the jig was pretty well abused, and we couldn't trust its accuracy. With the assistance of the hoist, we placed the beam in upright position. By drawing the plan view on

Once the lamination was complete, notches for the treads were roughed out with an electric chainsaw (top left). Then the tread-bearing surfaces were cut flat and level with a router attached to a plywood base (bottom left). With guide strips nailed to the trunk at the layout lines, it was easy to slide the router back and forth to mill the surfaces flat. To secure the stair to the first floor, a steel plate was bolted through to the framing. The bottom of the trunk was notched and relieved to receive a steel sleeve (above), which fits into the plate and is welded to it. The top of the stair is secured in similar fashion to the second floor.

the shop floor, we were able to transfer vertical riser locations up to the beam with plumb bobs and mark the horizontal cutting lines with levels.

With the rise and run determined, we used an electric chainsaw to rough out the notches for the treads to within a ½-in. of our layout lines (photo above left). To mill the horizontal surfaces of the notches flat, Noel made a long base for the router, attached guide boards to the trunk and surfaced the notches flat with a ⅞-in. straight bit, as shown in the photo below left. To give the massive trunk sculptural relief, we carved it to look like a tube, with the risers as intersecting tubes. The bulk of the carving was roughed out with the chainsaw, then cleaned up with various chisels and rasps. This exposed a pleasing lamination pattern, and led a number of people to ask how the risers were attached to the beam.

Making the treads was the only easy part of the project. Cut from 8/4 stock, they were laid out and arranged according to grain and color, and then planed down to a thickness of 1½ in. Their ends were rounded on the bandsaw and their edges routed to a half-round. The treads were then bolted to the trunk using the captured nuts that we had glued into the beam while it was being laminated.

To hold the stairway in place, our engineer called for a 10-in. square steel sleeve to be inserted into each end of the trunk. This was no problem at the top of the trunk because it was straight. But the curved bottom required a curved rectangular steel sleeve. We fabricated one (photo above) by heating four ¼-in. steel plates and bending them to the correct curvature before they were welded.

To avoid having to make separate bending forms for the handrails, and to get more accurate results, the Norskogs clamped L-shaped brackets to each tread at the location of the handrail (left), and then clamped the laminae to the brackets. Tubular-steel rails are bent to shape and then cut to length (top right). Then each rail is fitted between the posts and welded in place (above right) between the top and bottom rails of wood.

Handrail—The balustrade was designed to complement other handrails in the house—a wooden rail at top and bottom, with two intermediate ¾-in. steel-rod rails. The balusters we made from 2-in. steel pipe. Like the trunk, the wooden handrails are bent laminations. But instead of constructing a special bending form for gluing them up, we used the stairway itself as a form by making L-shaped brackets, and clamping one to each tread. Then we clamped the handrail plies to these brackets to get the right compound curvature for the rail (photo above left).

For laminating each handrail, we picked out an 8/4 board and ripped it into strips thin enough to bend to the required curve and twist. We kept the strips together in the same order in which they were cut, rolled glue on each strip, and clamped them back together. This procedure ensured that the original grain pattern would remain intact, and that the gluelines would not be highly visible.

The stairway changes from spiral to straight ten steps up. This transition made for some of the most difficult parts of the project. It causes problems for the handrail not only be-cause there is a transition from the curve to the straight, but also because there is a change in the angle of incline.

Looking at the stairway from the side view, the inside handrail has a convex transition and the outside a concave one. Instead of laminating these tricky compound shapes, we glued up stock large enough to carve solid sections as transitions between the curved and straight handrails.

Delivery and installation—This was also part of the job (a bigger part than we bargained for), so we rented a truck with an enclosed box. Loading the 2,000-lb. trunk was no problem with four men and a hoist on casters. We were able to roll the trunk out the shop door right into the waiting truck.

Unloading was a different story. The house where the stairway was to be installed is on one of the world's steepest habitable hillsides. And there's no easy access, just a long walk almost straight up. We lost sleep for nights trying to figure out a suitable way to get the trunk off the truck, up the hill and into the house. When we arrived at the building site, we mustered all the men who were working on the house, and asked them to help haul the mahogany behemoth up the hill. We didn't tell them it weighed a full two thousand pounds. There was a lot of moaning and groaning, but we got it in place at last with nothing more sophisticated than grunt consciousness and brute force.

The installation went smoothly, but it took longer than we expected, as installations sometimes do. With the help of a welder, we set our mounting plates on the first and second floors (photo facing page, top right). The beam was lifted into place with another engine hoist. After positioning the trunk, the steel sleeves were attached to the ends of the beam. With the trunk in place, the steel sleeves were welded to the mounting plates. Final sanding on the trunk started, and treads were bolted down. Once the treads were down it was time to fit, mount, and shape the handrail (photos above right).

We returned to Santa Fe after six weeks in Laguna Beach. The stair was complete except for final sanding and finishing. The mahogany got six coats of Watco oil finish. □

Concrete Spiral Staircase

A massive stair made by casting treads in a precision mold and bending thin-wall tubing

by Dennis Allen

This circular staircase was inspired by the stone stairways built in Europe during the Middle Ages. Designer Paul Tuttle wanted to create the sense of timeless solidity that massive stone steps evoke, and the two-story greenhouse in the Douglas residence overlooking Santa Barbara, Calif., gave him his chance. The room needed both a stairway and a sculptural focus, so Tuttle captured the medieval aura with the 10-ft. high, 7-ft. dia. concrete spiral stair shown in the photo at left.

Even before I had seen the design, Tuttle asked me if I would be interested in building this staircase. My first impulse was to say no. A poured-concrete spiral stairway seemed impossibly difficult. But once I saw his drawings, I got excited by the challenges its construction presented. Along with two of my associates, I agreed to build it. But I still wasn't really sure I could.

Layout—The first thing we had to do was to determine the number of steps, the rise of each step and where the beginning and end of the spiral would fall. We eventually decided on a landing and 15 steps of 7½-in. rise. Each one is a 25° segment of a circle.

Next we made a full-scale mechanical drawing of one of the steps, including sleeves for holding the balusters, a cutout for toe space, indentations for carpeting and a sleeve for the center column. This drawing (facing page, left) proved to be an indispensable reference in all stages of the project.

Making the concrete form—After we figured out what one of our concrete treads had to look like, we had to build a form that would duplicate it 15 times. The form had to be durable, yet flexible enough to be taken apart after each pour and reassembled for the next one. Carpenter John Bunyan and I thought about our options and eventually came up with the one shown on the facing page, right.

The form's main components were the base, the sides, the pivot end and the outer end. The base was a 3-ft. by 4-ft. piece of ½-in. plywood that we covered with plastic laminate (for more about working with plastic laminates, see *FHB* #9, pp. 39-41). We made the

The stair's concrete treads are threaded on a steel column and locked in place by steel balusters. Each of the 300-lb. steps was cast individually in a mold lined with plastic laminate.

Mechanical drawing of a step

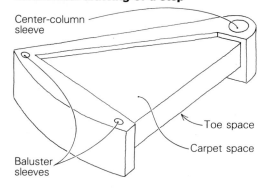

Center-column sleeve

Baluster sleeves

Toe space

Carpet space

Steelwork

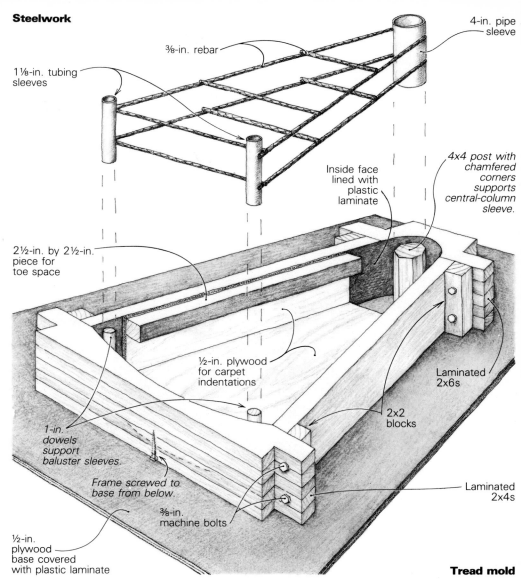

1⅛-in. tubing sleeves

⅜-in. rebar

4-in. pipe sleeve

2½-in. by 2½-in. piece for toe space

Inside face lined with plastic laminate

4x4 post with chamfered corners supports central-column sleeve.

½-in. plywood for carpet indentations

Laminated 2x6s

1-in. dowels support baluster sleeves.

Frame screwed to base from below.

2x2 blocks

⅜-in. machine bolts

Laminated 2x4s

½-in. plywood base covered with plastic laminate

Tread mold

form's sides out of fir 2x8s, with 2x2 blocks screwed and glued to their ends. We made the tight-radius pivot end from five 2x6 blocks laminated together with yellow glue. We cut out its 4-in. radius arc on the bandsaw. The outer end was made up of five 2x4s. It had a 21-in. arc of 3½-ft. radius cut out of it. We let the tails run long on the end pieces so that they could be bolted to the end blocks on the sides of the form.

Once we had the two sides and the tight-radius end piece bolted together, we lined its inside face with plastic laminate. This part of the form stayed together as a unit throughout the casting process. We lined the outer end, and then bolted it to the side pieces. We put the entire assembly on the base, and screwed the two together with 26 1½-in. screws.

Next we screwed plywood pieces to the bottom and lead edge of the form to create an indentation for carpeting. The shape of these pieces had to be carefully worked out so that the carpeting would flow from one step to the next without any offset. We planned for the bottom of the form to be the mold for the top of each step, so we could place the largest of the ½-in. let-in pieces for carpeting on the bottom of the form. Pouring the steps upside down also let us trowel the bottom of each tread—the largest and most visible expanse of concrete on each step. Finally, a 2½-in. square piece of wood 28 in. long was covered with laminate and screwed to the top edge of the form. This would form the indented toe space on the bottom lead edge of each step.

The last parts of the form were the registration pins for the steelwork—two 1⅛-in. tubing sleeves to support the rebar, and the 4-in. pipe sleeve that would fit over the central column of the stairway. We screwed two 7½-in. lengths of 1-in. wood dowel to the base near the corners at the wide end of the form (drawing, above right). At the other end, we attached a 7½-in. piece of 4x4 with chamfered corners. These pegs, carpet inserts and toe board were screwed in from the outside of the form so they could be easily released when we stripped the form from the cured concrete.

Structural steel work and balusters—We decided to cast a short length of 3-in. pipe into the slab-floor footing to act as an anchor for the 3½-in. central-column pipe. Each step would have cast within it a sleeve of 4-in. pipe. These sleeves would slip over the center

column and become an integral part of the rebar assembly in each concrete tread. The rebar grid would also include the two 1⅛-in. tubing sleeves into which the railing balusters would slide and be secured, as shown in the drawing above..

The sleeves in each step are attached to one another by a matrix of ⅜-in. rebar. Fifteen grid assemblies were required, one for each stair, and each one had to match all the others exactly in order for the balusters and central column to fit properly.

To achieve this kind of accuracy, our steel expert, Dean Upton, welded each assembly on a jig. In a ⅜-in. steel plate, Upton drilled holes at the sleeve centers and tapped them for ½-in. bolts. Then he tack-welded posts to the plate. These posts had been turned to fit the inside diameter of the sleeves, and were bored to accept the ½-in. bolts. The sleeves for each tread were cut to length and squared. Then Upton bolted them to the jig and welded the rebar in place.

The posts that support the handrail are 1-in. OD steel tubing with a .083-in. wall. The sleeves into which they fit are 1⅛-in. OD tubing with a .049 in. wall, allowing a clearance of .027 in. That seemed a bit loose to us, but

proved to be a necessary margin during the final stair assembly.

Upton silver-soldered a ¼-in. ring (from the sleeve material) to each baluster to make sure they ended up at the right height. Their tops were cut at the angle of the stairway, and as they were installed, the balusters were rotated to match the direction of the handrail. Once the treads were in place, we plugged the underside of the baluster holes with Bondo (a brand of auto-body putty).

Picking out the pipes—Material selection required some careful sleuthing. Our basic plan called for several pipe sizes, each to fit snugly over the next. The central column is 3½-in. pipe, schedule 40. The sleeves for the steps are 4-in. pipe, also schedule 40. Pipe goes by the inside diameter (ID) while tubing goes by the outside diameter (OD). With pipe, oddly enough, the OD is constant while the ID changes with the schedule or wall thickness. A 3½-in. pipe (schedule 40) is actually 4-in. OD by 3.548-in. ID, with a wall thickness of .226 in. So depending on the schedule, we had a choice of clearances between sleeve and column. With 3½-in. pipe and a 4-in. sleeve, there is a theoretical clearance of .026 in. But

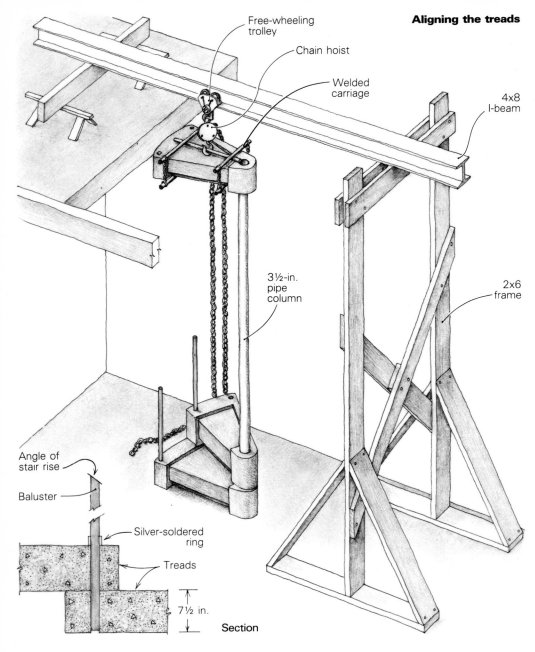

Free-wheeling trolley

Chain hoist

Welded carriage

4x8 I-beam

3½-in. pipe column

2x6 frame

Angle of stair rise

Baluster

Silver-soldered ring

Treads

7½ in.

Section

this clearance did not prevail on all of the pieces so some sleeves had to be turned down on the lathe.

Concrete technology—The concrete mix was critical for two reasons: weight and finish texture. In order to decrease the weight of each tread from more than 400 lb. using concrete with standard aggregate to about 300 lb., we used ½-in. Rocklite (The Lightweight Processing Co., 715 N. Central Ave., Suite 321, Glendale, Calif. 91203), a lightweight aggregate. But we also wanted a dense, pure white finish on each tread. This led to our using two different batches for each one. The outer inch or so is made up of 1 part white portland cement, 2½ parts 60-grit silicon sand and 2½ parts Cal-White marble sand (used mainly for swimming pools), made by Partin Limestone Products Inc. (PO Box 637, Lucerne Valley, Calif. 92356). Once we got this outer layer of white concrete in place, we filled the core of each tread with the lightweight mix.

We carefully measured all the ingredients

because slump was important—too much slump would cause the two mixes to flow together in the form. The lightweight mix for the core was 1 part cement, 2½ parts sand and 2 parts aggregate. To speed setting time, we added a little calcium chloride to each batch.

Originally we'd hoped to pour two steps per day, but found that producing one a day was quite an accomplishment. Placing the mixes in the form required two of us—one to tamp the outer mix and the other to keep the core mix from migrating to the edge of the form. We placed the concrete in layers, agitating it thoroughly after each layer to eliminate voids. Between each pour we cleaned the form, coated all dowels and wood insets with floor wax and sprayed the plastic laminate with silicone.

Surprise and delight filled us when we stripped the form from the first tread. The result was magnificent, but not what we'd expected. The plastic laminate made the surface smooth as glass, and a swarm of tiny, irregular air pockets made it look something like travertine. We were elated with this first success.

Assembly—Once the 15 treads were cast and carted to the site, our next hurdle presented itself—slipping the 300-lb. steps onto the steel column. Obviously we needed some type of device to lift the treads. It would have to be sturdy enough to carry the heavy loads, yet also adjustable so we could fine-tune the position of each tread over the center shaft.

Our solution was the homemade chain hoist shown in the drawing at left. It consisted of a 12-ft. 4x8 I-beam and a chain hoist mounted to a freewheeling trolley. We centered the beam over the column, and held it up by the stair landing on one side and a sturdy framework of 2x6s on the open end.

We fabricated a special metal carriage and harness to carry the treads as close as possible to their center of gravity. Each tread was then lifted above the 10-ft. high column, and its sleeve was centered over the shaft. Then the tread was slowly lowered into position. Once a step was in place, we would brace its outer end with a 2x4 and then one of us would tap a baluster through the aligned sleeves to secure the new tread to the one below it.

Disaster nearly befell us midway in the assembly. As we rolled the tenth tread along the I-beam, a lurch in our movements caused a sudden shift in the tread's center of gravity. Instantly the harness slipped off the carriage and the step plummeted, bouncing against several of the steps already in position and demolishing the bottom picket on its way to the floor. We were stunned. Fortunately nobody had been standing in the way of the tread when it fell. We surveyed the damage and it appeared enormous. Chunks of concrete were knocked off the treads in half a dozen places. We were so badly shaken that we packed up and went home for the day, believing the project ruined.

The next day we reassessed the damage and concluded that it wasn't as severe as it had seemed. We decided to patch the damaged edges and corners with Bondo. In some places we had to build up numerous layers of the stuff, but it worked far better than we had dared to hope. Because the Bondo was a different color from the pristine white concrete, we knew we'd have to paint the final product.

At the landing—The top step had a different shape from the others because it needed to flow into the cantilevered landing. To link the stairway to the landing, we built a triangular rebar grid to lock the top of the central steel column rigidly to the 4x12 landing girders. We welded sleeves for the balusters and the central column to this grid. The grid in turn was welded to a 4-in. by 30-in. by ⅜-in. steel plate, which was bolted to the landing framing. Then we erected a form around the grid with supports down to the floor. We were able to use several curved components from our breakdown form, but most of the pieces were new and had to be covered with plastic laminate.

We poured this last step with the same two mixes and care that we used with all the other treads, and when we took off the forms it flowed perfectly into the landing.

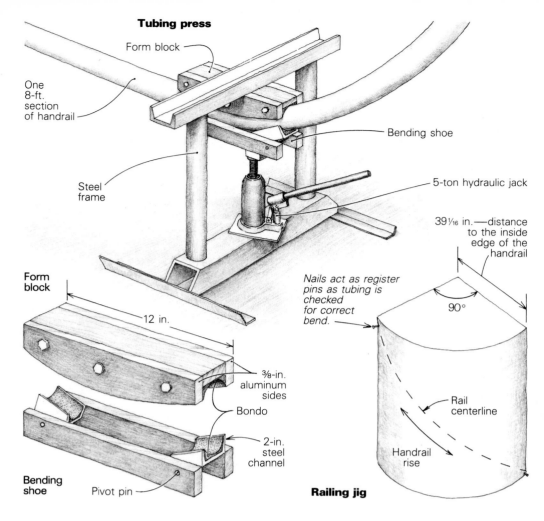

Tubing press

Form block

One 8-ft. section of handrail

Steel frame

Bending shoe

5-ton hydraulic jack

Form block

12 in.

³⁄₈-in. aluminum sides

Bondo

2-in. steel channel

Bending shoe

Pivot pin

39¹⁄₁₆ in.—distance to the inside edge of the handrail

Nails act as register pins as tubing is checked for correct bend.

90°

Rail centerline

Handrail rise

Railing jig

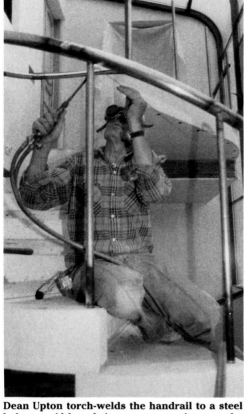

Dean Upton torch-welds the handrail to a steel baluster. Although it appears continuous, the railing is composed of short segments of steel tubing that were bent on a homemade press, and then assembled on site to create the necessary helical shape.

Handrail—Probably the most challenging part of this project was bending a 1⅞-in. dia. thin-wall tube (.063 in.) into a helical handrail. We chose this size tubing because there are stock fittings for 1½-in. pipe that fit closely enough to be used with the tubing (1.875-in. OD vs. 1.900-in. OD). At the top where the stairs meet the landing, we needed a tight return bend to blend the rising stair rail into the horizontal landing rail. We made this transition with two wide-radius elbows and a little cutting and fitting. We used another stock fitting—the half-sphere cap—to finish the bottom end of the handrail, and we used floor flanges to attach the landing rails to the wall.

The radius of the stair circle was 42 in., but the radius of the line of balusters was 40 in. The inside radius of the handrail was 40 in. less ¹⁵⁄₁₆ in. (half the diameter of the tubing) or 39¹⁄₁₆ in. Taking his cue from a tubing bender, Upton designed a press that used a hydraulic jack to generate the bending force needed to arc the straight lengths of tubing. He used an oak form block with a radius of 38⅞ in., a little tighter than the required radius to allow for some springback. As it turned out, the springback was almost nil.

The forming tool we tried out first had two spools about 12 in. apart. It bent the tubing, but it also left slight dimples at each point of contact between spool and tube. A handrail with a dimple every 6 in. was totally unacceptable (it looked like a segmented worm), so Upton made a pair of bending shoes out of

channel steel and Bondo as an alternative. They needed periodic greasing to allow the tubing to slip through as it was bent. This jig, shown in the drawing above left, worked fine. A 5-ton jack supplied the pressure.

Upton first tried to form the helix as the tube was being bent by rotating the tube a little at each bend. But it was difficult to keep track of the rotation. We could calculate how much rotation was required, but to control it was tough in a small shop. Even though it wasn't the right shape for our railing, the sculpture resulting from the first try could be mounted on a stone block and placed in front of a library.

We learned two things from this attempt. One, the press could put a wrinkle-free radius in our tubing, and two, trying to form both the radius and the helix into the full-length railing was too ambitious. Instead, Upton cut the 24-ft. tube into three 8-ft. pieces. Then he made a plywood jig that had a radius of 39¹⁄₁₆ in. (drawing, above right). This jig represented about ⅓ of a turn of the staircase, and was tall enough to allow the rise of the handrail to be marked diagonally on it. As he shaped each 8-ft. section, Upton checked its bend against the jig. This worked well, and the three pieces closely approximated the required helix plus the radius.

To make final adjustments in the helical twist, each 8-ft. section was cut into three equal pieces. After tack-welding the first section to the bottom balusters, Upton rotated

the second section slightly to create the helix. Section two was then tack-welded in place, and the third piece rotated slightly more than the second and so on until all nine parts were tack-welded in place. Each piece was aligned with its neighbor by using a short offcut of 1¾ in. tube as a dowel. Before he welded the balusters to the railing (photo above), Upton torch-welded the whole unit into one continuous piece. Then all the welds were ground down, and any little pits were filled with Bondo and sanded smooth. The finished rail is painted brick red, and appears to flow as one piece from top to bottom.

Our final job was whitewashing the treads. We wanted to preserve the texture of the concrete and to have it not look painted, so we experimented with several finishes. We finally settled on white latex paint mixed with a small amount of white portland cement. This gave the surface a little roughness to the eye, but did not destroy the glass-smooth texture to the touch. One coat completely covered the grey-green Bondo, and we were done.

The project took six weeks of concentrated effort, and it kept our attention with a series of snags and surprises. But everybody is happy with the way it turned out. The stairway cost almost $10,000—a lot for one flight of stairs, but not for a sculpture that anchors a special room. □

Dennis Allen is a general contractor living in Santa Barbara, Calif.

Outside Circular Stairway

A handsome addition without fancy joinery yields covered access to two levels

by Tom Law

In the Annapolis, Md., area, tidal creeks and rivers from the Chesapeake Bay produce some steep building lots. My clients owned a ranch-style house on just such a site. To enter the house you had to park in a space along the roadside and walk down steps onto a deck; then to reach the front door you had to walk to the far end of this deck. The owners didn't like this arrangement. The long walk to the front door was inconvenient, and the interior stairway to the basement took up valuable living-room space. An architect neighbor conceived the solution to this problem—an exterior circular stair, located closer to the road, that would provide sheltered access to both the living room and the basement. The finished stairway is shown in the photo below. I got the plans in the form of a freehand sketch. The project involved removing the interior stairs, filling in the floor, building the spiral stairway itself, and cutting a hole in the exposed concrete block wall below the deck to make a new doorway to the basement.

I started by laying out the new entry door about midway down the deck, which still allowed it to open into the living room. The right side of the door was to be the center of the new spiral-stair enclosure. Squaring over to the outside edge of the deck, I dropped a plumb bob to the ground. Using a post-hole digger, I went down about 5 ft. with a 20-in. diameter hole until I hit sandstone. I formed up an octagonal pier of the same dimension 6 in. above grade and poured concrete.

The plan called for a round central post, but I was planning simply to glue and nail the treads in place and felt that an octagonal post would make it easier to shape them to fit. The post was a 10x10 Douglas fir timber, 26 ft. long, which I had to special-order. Once I got it inside my shop I laid it across two sawhorses and chamfered the corners with a Skilsaw. This required tacking a fence on the post and making an initial pass at the greatest depth of cut, and then making an additional rip on the adjacent side of the timber to cut through. To remove the saw marks and even out irregularities I used a belt sander—first across the grain and then lengthwise, trying to get the chamfered surface straight and uniformly wide.

Setting the post was easy, in spite of its length and weight. My son Greg and I slid the post down the hillside steps right up to the pier, and with a rope tied to the top, upended it until we could reach it while standing on the deck. With Greg steadying it from the top, I bear-hugged the bottom end, lifted it onto the pier and seated it over the ½-in. dia. steel pin protruding from the concrete. With the post roped to the deck rail, slightly out of position, I cut out a half-octagonal pocket into the deck band, slid the post into place and nailed it.

On top of the post I cut a pocket for the cross-beam of double 2x6s to carry the roof of the stair enclosure, then beveled off the top of the post for a neater appearance. The roof joists were set on a ledger on the wall and cantilevered over the beam (drawing, facing page). All the sheathing for roof and walls was 2x6 tongue-and-groove fir. Letting the roof deckings run long, I found the center of the post and swung a 3-ft. radius arc. To trim the decking to this line, I used a handsaw for the plumb cuts on the joists, and a reciprocating saw for the curve. I cleaned up the sawn edges by sanding to the scribed line with my belt sander. This new section of roof was finished with flat-seam terne metal held tightly under the soffit, and no changes were made on the existing roof.

Treads and risers—The new basement door was to be directly under the new living-room door, so I marked its location and punched a hole in the basement wall to find the floor level. With a 2-ft. level on top of a straightedge I transferred this elevation to the post. Next I measured the total rise on the post, and calculated the number of risers required to get a riser height between 7 in. and 8 in.—safe, comfortable limits for a single stair rise. I decided on 12 risers. To make sure I had divided correctly and converted the hundredths into sixteenths, I set my dividers and stepped up the post, making sure the top of the last riser would be exactly at the deck level. This required several attempts. An error of only ¹⁄₁₆ in. can make a difference of ¾ in. on twelve steps. When everything was right, I marked and numbered the location of each tread on the post. That done, I went back to the shop.

To lay out a stairway of this complexity, I like to draw full-size plans. On top of two workbenches, I laid a big sheet of clean cardboard

The completed stairway addition to this ranch-style house gives access to the basement without wasting interior space. The new enclosure also offers a covered entry at the deck level much nearer to street parking than was the previous front door.

and with trammel points, I swung a 3-ft. radius arc the width of the stairway. I had a cut-off from the post so I placed it right over the center and traced the octagon onto the cardboard. From the basement landing to the living-room deck, the stairs needed to spiral 180°, so I marked off a semicircle. Using the dividers, I stepped off 11 equal segments, the correct number of treads for 12 risers. Connecting these marks to the centerpoint gave the size and shape of each tread, as shown at right. These radius lines were actually the front face of each riser, and I drew in the line of the 1½-in. nosing on the tread for clarity. Now the pattern was drawn, and all I had to do was transfer the lines on the cardboard to wood. This is where the octagonal post is better than a circular one. You can make straight or angled cuts instead of rounded ones. I did no mortising and used no fancy joinery, but some of the intersections of the treads made excellent connections, almost locking themselves into place on the facets of the octagon. I rounded the outer ends of the treads with the reciprocating saw and belt-sanded to the line where necessary.

Finally, I grooved the underside of each tread to receive the top edge of the riser. Assembly went very well. Starting at the bottom, I put a little glue on the end of the first riser and toenailed it against the post with galvanized common nails, using a drift pin to set the heads below the surface of the wood. Galvanized nails were needed not for their rust resistance but for the coarse shank that would make them resist withdrawal as the treads flexed in use. Placing the first tread over the riser with plenty of glue in the groove, I nailed it securely. After the first two treads were in position and the glue had cured, they were strong enough for me to sit on, and as I added each step, I could walk up and down those already in place. With each new tread I also checked the tread heights against the layout lines.

The next thing was to cut the basement door in the block foundation. I used a masonry blade in my Skilsaw rather than knocking out the concrete block and laying new half-block in a sawtooth pattern. Cutting masonry with an abrasive blade creates clouds of dust and particles. You need an open work area and a good dust mask. This is why I made these cuts before enclosing the stairs. When the wall was cut through and the new door and jamb set, I bridged over to the post for the landing at the bottom of the stairs to form the entry to the basement. This bridge was supported at the block wall on a ledger, and by the post and a temporary prop to the ground on the stair side.

The enclosing curtain wall was also 2x6 fir decking. At the top, each piece was nailed to a joist or to blocking; the bottom was nailed to a plate on the deck, and as I progressed downward, it was nailed to the risers and treads. The boards were held flush at the top and allowed to extend long on the bottom. Again, I applied glue to all joints for strength. The tight joints between the steps and the post allowed no play in the stair, and the vertical fir sheathing connection to the roof added further rigidity and shear strength. The bridge into the basement

Frances Boynton

Spiral stairway in plan

Tongue-and-groove vertical siding nailed to ends of treads and risers

9½ in.

3 ft.

Up

Octagonal post

The spiraling pattern of treads and risers was laid out full scale in the shop. The bottom of each tread was plowed ½ in. deep to accept the top of the riser; the riser bottom was glued and face-nailed to the back of the tread. When the steps had been glued and nailed to the octagonal post, they were rigid enough to hold the weight of the builder, even without siding to support them. The roof for the stairs is carried by the double 2x6 beam let into the top of the post.

was suspended by the vertical sheathing, which was nailed to the deck joists above; then the temporary prop was kicked out. On the deck level, I cut a doorway into the shell of the stairway on the street side, with a large casement window across from it.

When all the siding was in place with the ends long at the bottom, I marked out the helix, transferring the bottom edge of each riser to the outside of the sheathing. The resulting series of points wrapped around and up the cylinder. Using an ⅛-in. strip of wood, I connected the points to create a line. Starting at the top to get the help of gravity, I just sawed down the length with the reciprocating saw. I took a lot of care with this cut so that it

wouldn't require any further dressing. The stairway was complete.

After cutting out the living-room wall and setting a new jamb and door, the only thing left to do was to remove the old stairs, cover the framed-in opening with oak flooring and finish it like the rest.

The work was done during October and November, and the delightful weather added to my enjoyment of the project. I had a helper for two days when I was mixing concrete and setting the post, but the rest of the time I worked alone. Including repairs to the inside floor and walls, the job took me eighteen days. □

Tom Law is a builder in Davidsonville, Md.

Octagonal Spiral Stairs

A complicated stairway built and installed in separate halves

by Tom Dahlke

The house Gary Therrien built for himself is octagonal, with a truss-roof system that required no load-bearing interior walls. The floor of the large, open main level emphasizes the shape of the house: oak boards laid to form wedge-shaped sections that were edged with mahogany strips running to the house's corners. In the center of the floor is yet another octagon: the opening for the staircase that leads down to a lower-level bedroom, the entry, workshop and playroom. Therrien wanted a spiral stairway, but the curves of the standard round or oval spiral wouldn't work with the angles and straight lines of the rest of the house. He felt, and I agreed, that it would be more appropriate to build a stair whose outside edge followed the shape of the opening in the floor. The fact that this octagon didn't have equal sides was something I didn't want to think about yet.

The basic spiral staircase isn't all that hard to lay out or build, but I was dreading the thought of putting together the unsupported octagonal stringer for this one. I also realized early on that the walls around the opening on the bottom floor would make it impossible to stuff a fully assembled unit in place, I would have to build the stairway in two vertical halves and install them one at a time.

The opening was elongated—6 ft. 6 in. wide by 7 ft. 6 in. long. Each end had three 31-in. sides joined by two 43-in. sides. The stair treads had to spiral 360° over a total drop of 9 ft. 3 in. I decided on 14 treads with a rise of 7⅜ in., and began by drawing up a plan view

The opening for this stairway was an elongated octagon in the center of an octagonal house. The floor on the main level was laid in wedges to emphasize the unusual shape, and the owner wanted an octagonal spiral staircase to carry out the idea. The completed stair makes a full 360° spiral. Its unsupported stringer describes an eight-sided figure that's echoed by the rail.

160

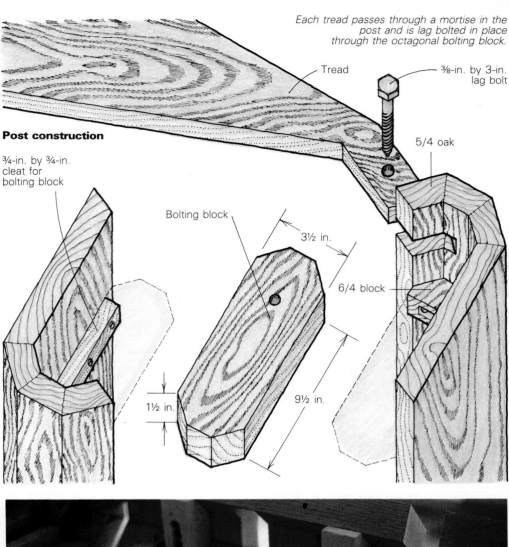

Tread

⅜-in. by 3-in. lag bolt

5/4 oak

Post construction

¾-in. by ¾-in. cleat for bolting block

Bolting block

3½ in.

6/4 block

1½ in.

9½ in.

Assembling the post. The two halves of the post were coopered up separately. Some of the blocking inside is reinforcement, right, and some will support the octagonal blocks that the treads will be bolted to. The blocks also help align the halves when they are assembled. Mortises, each one different, have already been cut for the treads. Above, the post is strapped together dry, before the treads are installed.

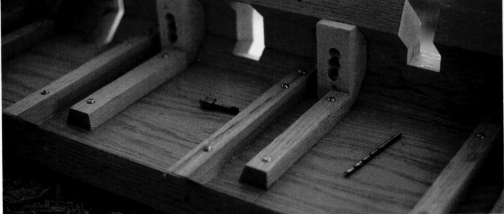

and two elevations that showed where each tread would join the post and at what angle. The treads were laid out at 22½° intervals.

I set up my table saw (which has a mortising attachment), my jointer-planer, and my bandsaw in the living room, and did all the work on site.

The hollow post—I planned to rabbet the treads into the outside stringer. Since most of the stringer would be unsupported by anything else, the treads had to be locked immovably to the central post. Because of this, and because of the space problem, I decided to cooper up a hollow post. This would let me fasten the treads very sturdily on the inside. It also made it fairly simple to assemble the stairway. I would glue up two vertical halves

of the post, and bolt the treads to 1½-in. blocks shaped to fit inside the hollow space (drawing, above). These bolting blocks would rest on cleats screwed into the post. Other bracing would serve as guides when I glued and screwed the post halves together.

I used eight boards 9¼ ft. long and 1¼ in. thick to make the hollow octagonal post. To get the correct sectional dimensions, I ripped two of them to a width of 8½ in. and six to a width of 2½ in. Next, I tilted the blade on my table saw to an angle of 22½° and beveled the two edges of each piece. Then I put them together dry with strap clamps while I marked the post for the mortises. I had to scribe each mortise on both the outside and inside of the post to determine the angles at which the treads were coming in. To do this, I marked

their locations in plan view on one end of the post, and measured down the inside and outside surfaces. I also drilled holes every 15 in. along the edges of the two halves of the post. These were for the countersunk screws that would hold the pole together after assembly. Then I took the strap clamps off, and cut the mortises on each half with the bandsaw where possible, or with handsaws and chisels. No two mortises were alike, and I had to cut them precisely so the treads would mate with the stringer. When all the mortises were cut, I strapped the halves together dry again to have a look (photo above left).

Treads and stringer—No two treads are alike, either. Their outlines depend on the angles at which they meet the post, and the

Photos facing page: Donna Coveney; Illustration: Vince Babak

The treads were installed while the two halves of the post lay across sawhorses, top. The stringer sections, which had previously been routed out, were temporarily attached to the treads, and then carefully trial-fitted, above, to achieve a smooth joint between sections.

shapes imposed on their outer ends by the rising and angling stringer. I cut the treads long enough to fit into ½-in. deep mortises in the stringer, then made sure each one fit the post. Finally, I rounded their edges with a router and finish-sanded them with 220-grit paper. With the post halves resting on sawhorses in the cellar, I fit the treads into their respective mortises and bolted them to the interior blocks with 3-in. lag bolts, as shown in the photo at left.

Treads in place, I began cutting the stringer to fit their ends, and from here on, my drawings were worthless. Each joint required a compound-angle cut. I knew that all the miters had to be 22½° (actually 67½°, but for setting the blade angle on a table saw, whose protractor reads from 0° to 45°, you have to use the complement of 67½°, which is 22½°) for the stairway to make 135° turns, but the vertical angles varied because the rise of the segments varied. (The segments of the octagon were not equal, and this was the only way to be sure that the front and back edges of the treads would remain within the width of the stringer.) I'm sure there is a way to figure this stringer out on paper but I could only get part of it, and wound up trial-fitting instead.

I used three 12-ft. oak 2x12s for the stringers, with the best one in the middle where it is most visible. I rough-cut the segments to length one after the other down a board so the grain would be continuous on both sides of a joint. Then, with the post halves bristling with treads still on the sawhorses, I used my drawings and some simple calculations to lay out the mortises on the stringers that would house the wide ends of the treads. I routed out the mortises and set the rough-cut stringers in position on the treads. I predrilled each segment so it could be bolted to the treads with 3½-in. lag bolts countersunk and plugged with 1-in. oak plugs.

From here on, I could concentrate on the joints between stringer segments. I started at the bottom and worked up, cutting and recutting each segment until all the joints were tight (photo bottom left). I predrilled them for six 2-in. screws, to be countersunk and plugged with ½-in. oak plugs. Then I beveled the stringers' edges and used my table-saw mortising jig to drill dowel holes for the rail balusters. Finally, I sanded the sections with 220-grit paper.

Installation—On the day of the raising we had ten people on site, and we needed every one of them to muscle the two halves of the stairway into place. The stringer segments that could be installed later were left off, but the segments at the top had to be installed before the stair could be positioned. At the third and fourth steps from the top, the stringer passes within ¼ in. of a wall, and after we'd maneuvered the first half in where it belonged, we lag-bolted through the stringer segment into the studs. Two 1-in. dia. oak pins through the bottom octagonal bolting block hold the post in place on the floor.

With one half secured, we slid the other

into place to check the fit one last time. It mated perfectly, so we glued, clamped and screwed the halves together.

After the remaining stringer segments were glued, screwed and bolted in place, I plugged all holes with oak plugs and sanded them smooth. The bottom section was drilled to sit on a 1-in. dia. oak peg set in the flooring.

The top of the post is held in place by a landing support cut from the stringer stock. A one-of-a-kind joint was cut so the 2x12 oak would join the post flush. To continue the effect of the spiral, we built the landing in two tread-size sections to give the illusion of two more steps.

Bannister and balustrade—The owner wanted the bannister to look like a continuous, floating ribbon, beginning over the post, going around the opening, then down and around the stair. I bought five rough-cut 14-ft. full 2x6 mahogany boards, then resawed them down to 1¾ in. by 5 in. I took the lengths and angles directly from the stringer, and sawed each segment so the grain would carry around the turns. I beveled the top edge of each section to get three surfaces of equal width, and routed a ¼-in. bevel on the bottom edges. All of the segments were predrilled to accept the dowels for the balusters. The joints were glued and screwed where they were easily accessible, and splined where it was harder to get at them. The connection between the horizontal rail at the top of the stairway and the first angled rail was made with dowels for the sake of simplicity.

The balusters are 1¼ in. square, and they are about 6 in. o.c., but because the sections of rail are different lengths and I wanted the balusters to be evenly spaced along each section, I had to vary the spacing a bit. I drilled ½-in. dia. holes 2 in. deep in both ends of each baluster to accept ½-in. dowels. Normally, I would mortise a baluster in, but the stairway's very shape helped make the rail so sturdy that the doweling, which is much simpler, was strong enough.

The balusters couldn't be cut to a single length to fit between the bannister and the stringer. I had to custom-cut each one to length, in the process making sure that I cut the correct angle on each one's top and bottom. To get the clean ⅛-in. bevels on all four edges, I hand-planed each baluster.

We had trouble deciding how to resolve the top end of the rail. The post ends at floor level with an oak and walnut burl cap sitting on top. At first we thought the rail would look good if it appeared to emerge out of the post. Well, once done, it didn't, so I brought out a chainsaw and did some remodeling. The result was to have a short 6-in. piece dip down at the same angle as the piece opposite where it starts to go down the stair. A much better solution. At the base of the stair, the rail simply ends with the last tread. □

Tom Dahlke is currently building timber-frame houses in Canton, Conn. Photos by the author, except where noted.

Ticksticking

by Sam Clark

Fitting a large panel—such as a section of plywood subfloor or countertop—into an irregular space isn't easy. It can be done by making a paper pattern, by laying out some sort of grid on which to plot points, or by cutting the piece oversize and then laboriously trimming away the excess. Often a better method is ticksticking, a nautical carpentry technique I learned from fellow builder Henry Stone, who discovered it in an old yachting magazine.

Ticksticking can be used to reproduce any flat shape quickly and accurately. It's good theater, too. You make some apparently nonsensical hieroglyphs on a scrap of plywood and a stick, and some equally arcane scratches on the stock to be cut. Then you saw out a shape without the intervention of ruler, bevel, level or mathematics. Your audience—which surely will have gathered by now, and which will have been making unkind comments at your expense—falls silent. Then there's applause, as the piece goes in the first time with no trimming, all 20 facets perfect.

The applause is entirely undeserved, however, because the method couldn't be simpler. Suppose you want to cut a countertop to fit against a wall that takes several jogs to form a niche at a window opening. No moldings will conceal the joints; the fit must be exact. Take a scrap of any thin sheet material; ¼-in. plywood is ideal. It's best if this scrap is at least one-third as big as the area to be measured. Secure it to the cabinet in the plane the counter will occupy. The scrap can be positioned anywhere on this plane, at any angle, but it is convenient to align one or more edges with the eventual location of the counter. In this case, the front edge of the scrap overhangs the cabinet 1 in. because the counter will eventually do the same.

Make a tickstick—just a thin stick or a piece of lath about 4 ft. long, with a point at one end. Lay the point on one of the critical lo-

Sam Clark is the author of The Motion-Minded Kitchen, *published by Houghton Mifflin (1983).*

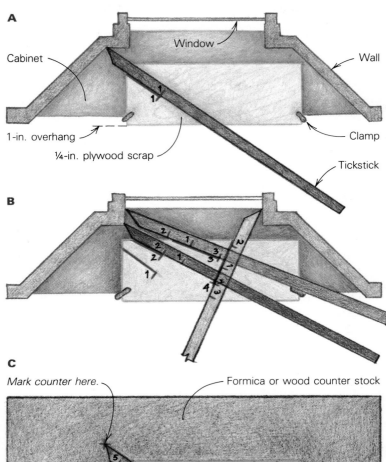

A
Cabinet
Window
Wall
1-in. overhang
¼-in. plywood scrap
Clamp
Tickstick

B

C
Mark counter here.
Formica or wood counter stock
Align flush.
Clamp
Plywood scrap

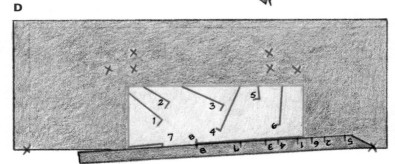

D

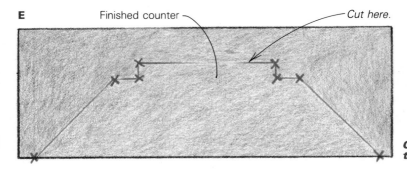

E
Finished counter
Cut here.

cations, say the left rear corner. Let the body of the tickstick fall anywhere on the scrap; it doesn't matter where. Hold the stick firmly, and with a sharp pencil draw a line on the scrap along the left edge of the stick.

Mark the stick and scrap at first point.

Without moving the stick, make a hash mark across the line you just drew and on the stick **(A)**, at the same point along the line. It doesn't matter where along the line you choose, as long as the two hash marks meet. Label both hash marks #1.

Mark the other critical points.

Now reposition the tickstick, say with the point at the left corner of the window bay. Again mark along the left edge, make two more hash marks, and number them #2. In like fashion, mark and number all critical points along the perimeter **(B)**. For a curve, approximate by fixing many points along the arc.

You'll end up with a stick with numbered hash marks, and a scrap with lines crossed by numbered hash marks. Now remove the scrap from the cabinet, and clamp it on the countertop to be cut **(C)**. This can be done near the cabinet, or in your shop. To mark the tickstick, the scrap was centered in the space to be fitted with its front edge projecting 1 in. for the countertop overhang. To transfer these points onto countertop stock that is cut to rough length, just center the plywood right and left, with its front edge flush with the front edge of the stock. Now put the tickstick to the right of line #1, with hash mark #1 on the line touching hash mark #1 on the stick. Mark the counter right under the point of the stick. Do the same at line #2 and all the other lines **(D)**. Connect all these points, along with the one that marks the counter's length on its front edge **(E)**, and cut along the resulting line. Install the piece. Turn to the audience. Bow. □

Align the scrap and stick at #1; mark counter.

Align and mark other counter points.

Connect the dots.

Building a Helical Stair

Laying out a spiral stringer with a little help from the trig tables

by Rick Barlow

A house I recently finished near Telluride, Colo., needed a spiral stairway to wrap halfway around its 4-ft. dia. stone chimney. I decided that a helical stringer supporting the treads by fabricated steel brackets would satisfy structural requirements and complement the house's contemporary design.

In theory, making a laminated wood helix isn't that difficult. Basically all you do is wrap successive layers of wood around a cylinder so that it spirals upward at a constant angle. But to glue up a helical bent lamination like the one I wanted for my stair stringer would require a cylindrical bending form larger than the chimney itself. I could have framed up such a form from curved plywood plates and 2x studding, but that would have eaten up a lot of time and materials.

I chose instead to make a bending form from five pieces of ¾-in. plywood and a strip laminated from three layers of ¼-in. plywood (drawing, below). The trick was to space the on-edge pieces of ¾-in. plywood at equal intervals along the proposed rise of the stairway, and to locate on each of these supports—or bulkheads—a 10-in. wide plane that would lie precisely ¾ in. below the surface of my theoretical half cylinder. To these five planes I could attach three layers of ¼-in. plywood. Glued together, this rigid plywood curve would be the platen, or base, of the bending form. To it I'd clamp the mahogany plies that would make the helical stringer.

In section, I wanted the stringer to be 5 in. wide by 6 in. deep, with its corners radiused with a ¾-in. roundover bit. Plies ¼ in. thick would easily make the required bend, and I used 20 of them 6 in. wide.

The centerline of the helical stringer would lie 42½ in. out from the centerline of the chimney cylinder. This meant that a typical 36-in. wide tread would clear the stonework by about ½ in. Subtracting half of the stringer thickness (2½ in.), gave me a 40-in. radius for the concave side of the stringer. This dimension then was the radius of the outside (convex side) of the bending form.

To create the helix, the first and fifth supports of the bending form needed to support the stringer against vertical edges. The second and fourth needed 45° edges, and the stringer would run over a horizontal edge at the middle support.

I nailed the five plywood bulkheads to the floor with 2x4 blocks and braced them diagonally to handle the forces they'd have to withstand during gluing and clamping. Next, I bent the three layers of ¼-in. plywood over these supports. The first layer was screwed and glued to the bulkheads, and the next two glued on top. I made sure that during this process the whole assembly stayed properly aligned throughout its length.

I built the form so that its centerline would be 5 in. off the floor. This way the stringer could run long at each end and be trimmed to fit the floor and rim joist later. I covered the plywood helix with a strip of visqueen so the mahogany wouldn't get glued to the form.

Laying out the stringer—The laminations for the stringer started out as 16-ft. long mahogany boards. I had them milled to a thickness of ¼ in. and ripped into 6-in. widths. To determine the length of these pieces, a little trigonometry was required.

First, I drew the stringer in elevation as if it were a straight piece of wood (drawing, facing page). The height of the triangle thus formed was the rise of the stringer. This is the total rise (distance from the first floor to the second) minus one riser height, as in a standard staircase. The total rise (110 in.) breaks down into 14 rises of 7.86 in. (about 7⅞ in.). Therefore the total rise minus one rise (7⅞ in.) for the top step (landing) results in a stringer rise of 102⅛ in.

The base of the triangle represents the distance traveled by the stringer in its run around the chimney. In this stair, the stringer winds through a 180° arc around a 42½-in. radius at the center of the lamination. Its run, therefore, is $\pi \times 42\frac{1}{2}$ in., or 133½ in.

The length of the stringer itself is the hypotenuse of this right triangle. The Pythagorean theorem $(a^2 + b^2 = c^2)$ tells us it is 168 in. long, just over 14 ft.

Since the outer plies travel larger radii than those at the center, they required slightly longer pieces of wood. I ordered the 16-footers to be sure of sufficient length.

The drawing of the flattened helix also helped me determine various angles. By finding the cotangent of the triangle ($\cot A$ = adjacent/opposite = 133.5/102.125 = 1.3072), I got the pitch angle of the stringer, 37° 25′.

The complement of this angle is 52° 35′. This is the angle that the stringer makes when intersected by a plumb line, an angle I needed

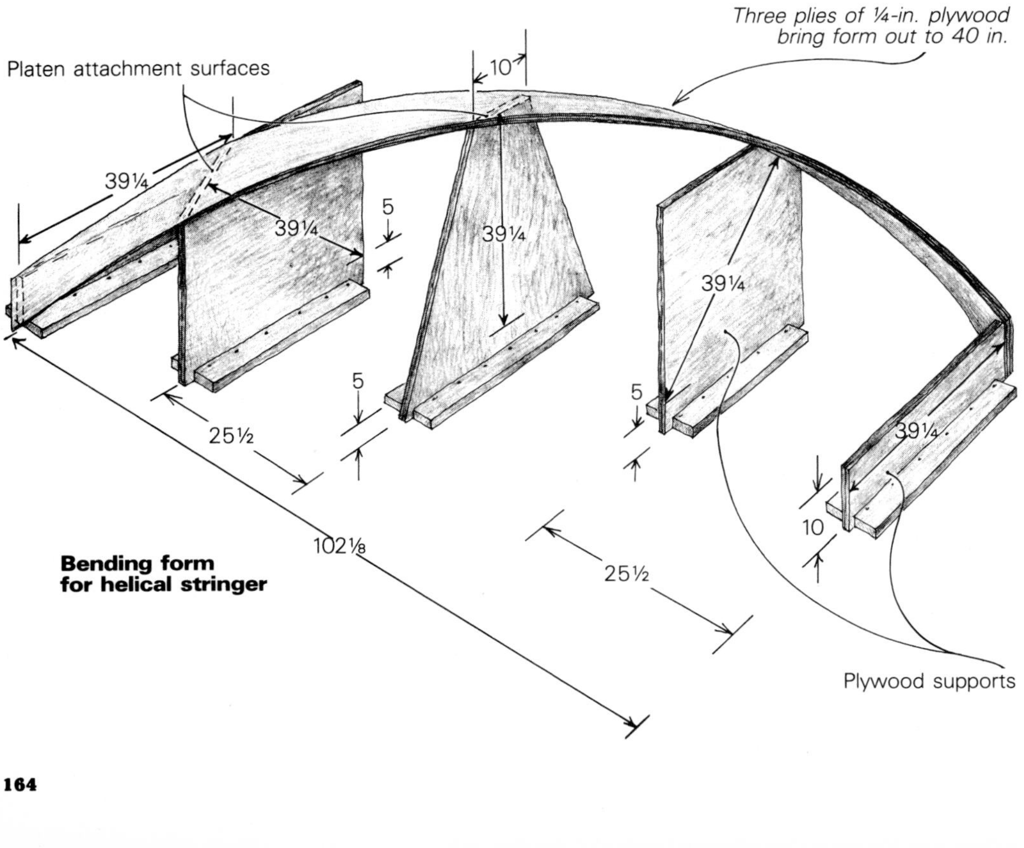

Three plies of ¼-in. plywood bring form out to 40 in.
Platen attachment surfaces
10
39¼
39¼
5
39¼
39¼
5
5
39¼
25½
10
102⅛
25½
Plywood supports

Bending form for helical stringer

Photos: Rick Barlow; Illustrations: Christopher Clapp

to know later when I counterbored the underside of the handrail to receive its posts.

I laid out the treads on the drawing of the straightened helix, and divided the 102⅛-in. rise of the stringer into 13 equal rises of about 7⅞ in. each. I figured the distance along the stringer from one tread to the next by dividing 168 (the stringer's length) by 13. This distance is 12.96 in.

Gluing up—Now, with the mathematics out of the way, I started gluing up the stringer. Since I wanted the glue to stick to the steel tread-support plates as well as to the wood, I used a resorcinol-formaldehyde glue. It's a hassle to use, and almost impossible to remove from tools and hands when it sets up, but it's strong stuff.

After spreading the glue with a small roller, I laid three or four 16-ft. long mahogany plies on the upright center support, clamped them there and then bent each side down to the ends of the plywood helix. You need lots of clamps. I have eight bar clamps, four handscrews, four smaller bar clamps and ten C-clamps. I borrowed about this many again from a friend and still had what I think is the minimum number of clamps for the length of this stringer, one clamp about every 3½ in.

Resorcinol glue has to cure at 70°F or above. It will set faster at temperatures higher than this, but it won't bond properly at temperatures below 70°F. Since I was doing this laminating in a house under construction in the wintertime at 9,000 ft. in the Colorado Rockies, I needed to camp out at the job site for seven days and nights to make sure the correct temperature was maintained while the laminating was going on. I glued up three or four layers at a time, letting them dry overnight before removing the clamps.

To rout the slots that would receive the steel tread-support plates, I laminated the stringer without glue between the third and fourth laminations and between the 17th and 18th laminations. This let me separate the first three and the last three plies from the center core of the stringer, rout out slots for the plates, insert and secure the plates, and then glue these inner and outer laminations onto the center one. The entire stringer was shaped and sanded before the plates were inserted so the steel would not get in the way.

I made two guide templates for my router to rout out the slots. One was for the thirteen slots on the convex curve of the stringer, and the other was for the thirteen slots on the concave curve of the stringer. Each jig was made to rout a 3-in. wide slot ¼ in. deep. To locate and orient each slot, I hoisted the center core lamination of the stringer into the position it would ultimately take, trimmed off the ends to fit, and temporarily secured it there

Barlow's helical stairway winds its way around a 4-ft. diameter stone chimney. The single stringer was laminated from 20 plies of ¼-in. mahogany. The treads, made of plywood and solid cherry end boards, are supported by brackets fabricated from ¼-in. steel plate.

Plan

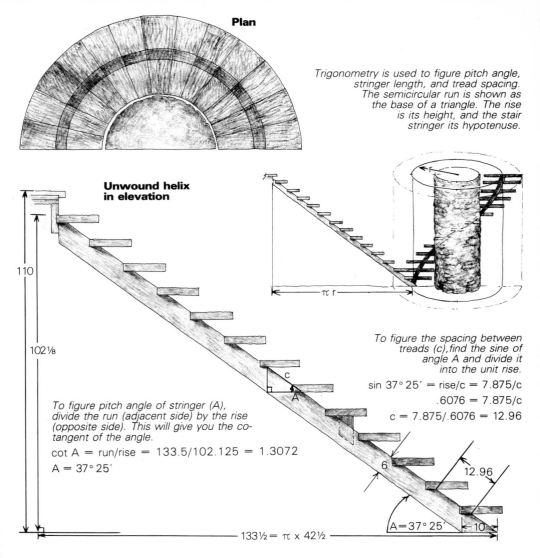

Trigonometry is used to figure pitch angle, stringer length, and tread spacing. The semicircular run is shown as the base of a triangle. The rise is its height, and the stair stringer its hypotenuse.

Unwound helix in elevation

110

102⅛

To figure pitch angle of stringer (A), divide the run (adjacent side) by the rise (opposite side). This will give you the cotangent of the angle.

$$\cot A = \text{run/rise} = 133.5/102.125 = 1.3072$$
$$A = 37°\,25'$$

133½ = π × 42½

To figure the spacing between treads (c), find the sine of angle A and divide it into the unit rise.

$$\sin 37°\,25' = \text{rise}/c = 7.875/c$$
$$.6076 = 7.875/c$$
$$c = 7.875/.6076 = 12.96$$

c

A

6

12.96

A=37° 25′

10

Before its treads are attached to the steel-plate brackets, the stairway is installed. The temporary midway support will be replaced by a steel pipe that will be grouted into a masonry planter below.

laminating two ½-in. thick pieces onto a ¾-in. thick core, yielding a total thickness of 1¾ in.—the same as the cherry. The ¾-in. middle ply of plywood was cut 1 in. shorter on each end, to form a ¾-in. by 1-in. groove. I cut tongues on the cherry end boards and screwed and glued them to the center pieces.

I bandsawed the ends of the treads to conform to the semicircular plan view, and rounded over all the edges with a ½-in. piloted router bit. Finally, I fastened each tread to its mounting plate with four lag bolts.

Bending the railing—The form I built to laminate the handrail was like the one I used for the stringer, only with a larger radius. The handrail would be just under 16 ft. long, so I could use the same boards. I used eight laminations ¼ in. thick and 3 in. wide, making the handrail 2 in. by 3 in. in section. To accept the handrail posts, I counterbored two ½-in. dia. holes in each tread at the centerline radius of the handrail, and spaced them evenly 7¾ in. apart. I then counterbored the underside of the handrail. These holes had to be drilled at an angle of 52° 35′ (the complement of the pitch angle). To find the proper spacing for

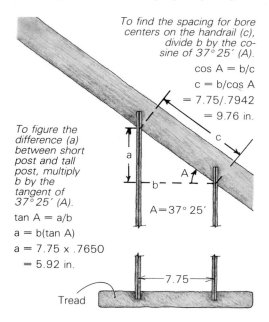

To find the spacing for bore centers on the handrail (c), divide b by the cosine of 37° 25′ (A).

$$\cos A = b/c$$
$$c = b/\cos A$$
$$= 7.75/.7942$$
$$= 9.76 \text{ in.}$$

To figure the difference (a) between short post and tall post, multiply b by the tangent of 37° 25′ (A).

$$\tan A = a/b$$
$$a = b(\tan A)$$
$$a = 7.75 \times .7650$$
$$= 5.92 \text{ in.}$$

A = 37° 25′

the bore centers here, I divided 7¾ (the horizontal spacing) by the cosine of the pitch angle, as explained in the drawing. Then ½-in. dia. steel rods, painted flat black, were inserted in the treads. Since each tread gets two rods, one must be longer than the other. By multiplying the tangent of the pitch angle by 7¾, I found the regular difference in length between the long and short rods.

With the help of friends, I worried the rods into their holes one at a time, starting at the bottom rod and working up to the top. By using a rubber mallet and some clamps, I brought the handrail down onto the 26 rods.

All that remained was to carpet the plywood center pieces and metal tread plates. Since the treads are visible from below, this was a difficult job, but it turned out fine.   □

Rick Barlow is a contractor in Colorado.

with clamps and a 2x4 leg at its midpoint. Then using a level, I drew plumb lines on each side of the 26 slots. This way I was sure they would be plumb in the finished staircase.

These slots were 3 in. wide, and their sides followed the curve of the stringer, so the legs of the plates also had to be slightly curved. I hit the middle of each leg with a sledge while it rested on two other plates along their edges. One or two whacks did the trick.

I drilled holes in these plates for four 8d nails to hold them onto the stringer. These and the resorcinol-formaldehyde glue made for a solid bond between metal and wood.

With the tread-support brackets installed, I put the stringer back onto the form and glued the inner and outer laminations onto the core. They were already shaped and sanded, so I had to align the pieces precisely, and use pine blocks to keep the soft mahogany from being crushed by the clamp jaws. I was also careful to spread the glue so not much would ooze out of the joints. What squeeze-out there was, I wiped off before it dried. This assembly was left to dry overnight (photo above).

Installing the stringer—After a little more finish sanding, I cut the outer plies to fit the floor and rim joist. A 180° helix like this one

needs support at its midpoint. I drilled a 2-in. hole underneath the stringer and inserted a galvanized pipe that sits discreetly in some rockwork below.

When the stringer assembly was firmly attached in place, the horizontal tread plates were welded to the vertical plates protruding from the stringer. Using a universal level, I held the horizontal plates level in both planes, while a welder attached the plates from underneath. After cleaning up the welds with a file and wire brush and before attaching the treads, I oiled the stringer and painted the plates and the support pipe flat black.

Making the treads—Working out the dimensions of the tapered treads required more figuring. The radius to the outside of the treads was 60½ in. Multiplying by π gives the length of a 180° arc, or 190 in. Dividing this by 13 treads gives 14.62, so the outside arc of each tread is 14⅝ in. Similarly, I found the length of the inside arc of each tread to be 5¹⁵⁄₁₆ in. Nosing added an inch, so each tread is 36 in. long with a 15⅝-in. arc at its outer end and a 6¹⁵⁄₁₆-in. arc at its narrow end.

I planned to carpet the centers of the treads, and leave solid cherry end boards on each end. I made a center of AC plywood by

Building Louvered Shutters

Jigs and careful planning make quick, accurate work of a potentially tedious job

by Rob Hunt

A few years back, I was asked to build movable louvered shutters for a Victorian house that was built in 1888. The house's new owners were restoring the building, and they wanted new shutters made to match the old ones that had deteriorated over the years. Thanks to this first commission, we've been building shutters for a number of clients. Out of necessity we've found ways to make the work go quickly and accurately.

Many early homes had shutters. Hinged to open and close over windows, and with movable louvers, they served to protect the windows from bad weather and to diffuse the incoming light. Changing the position of the louvers changes the flow of light and air through the room, giving you a range of lighting conditions to choose from. Shutters can do a lot to enhance windows, and they give you more control in adjusting the amount of

daylight you want in a room, and more control over the ventilation day and night.

The frame of a shutter consists of two vertical members, called stiles, and at least two horizontal members, called rails. The shutters shown here have two central rails in addition to the top and bottom rails, dividing each shutter into three sections. The louvers are beveled slats with round tenons that fit into holes in the stiles. Each set of louvers pivots as a unit thanks to a vertical rod coupled to the louvers by an interlocking pair of staples.

Cutting the louvers to uniform size, calculating the amount of overlap (the spacing between each louver) and mounting the louvers accurately within the frame are critical steps in the production process, and we've been able to increase both the accuracy and speed of the job by using layout sticks and several jigs that I'll describe as we go along.

Layout—The first step in laying out louvered shutters is to make precise measurements of the opening where the shutters will go. Measure from the top of the window or door jamb to the sill on each side, then measure across the jamb at top and bottom and at several places in the middle. Taking extra measurements for height and width is especially important in older houses, since their jambs are seldom square.

If the jamb is out of square by ⅛ in. or less, I build the shutters to fit the smaller measurement. If the skew is worse than this, I build them ⅛ in. smaller than the largest measurement and then trim them to fit the opening after they are completed.

When you measure the height of the opening, remember that the sill is beveled. Run your tape to the point where the outside bottom edge of the shutter will hit the sill. Then

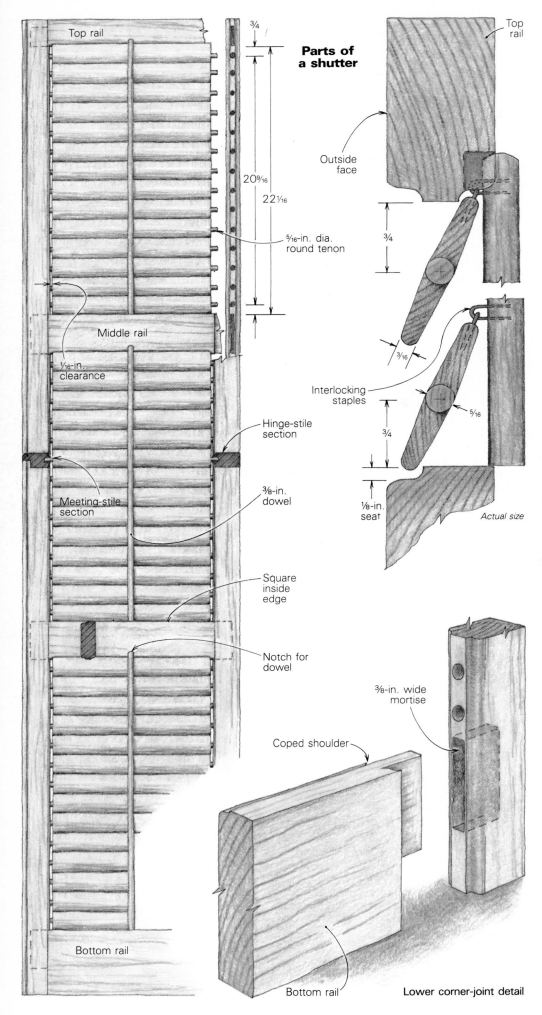

Parts of a shutter

Top rail

20⁹⁄₁₆

22¹⁄₁₆

³⁄₄

⁵⁄₁₆-in. dia. round tenon

Middle rail

¹⁄₁₆-in. clearance

Hinge-stile section

Meeting-stile section

³⁄₈-in. dowel

Square inside edge

Notch for dowel

Bottom rail

Top rail

Outside face

³⁄₄

³⁄₁₆

Interlocking staples

³⁄₄

⁵⁄₁₆

¹⁄₈-in. seat

Actual size

Coped shoulder

³⁄₈-in. wide mortise

Bottom rail

Lower corner-joint detail

when the shutters are built, you can bevel the bottom edge of the rails.

For projects like this, which require a number of identical, precisely cut pieces, we find layout sticks very helpful. They serve as full-size templates and contain all the required dimensions and joinery details. The horizontal stick we made for these shutters is basically a full-size sectional drawing. It shows the width of the opening, the width of the stiles, the length of the rails, the louver dimensions, the bead on the stiles and the coped mortise-and-tenon joint that joins stiles and rails. All these measurements are critical, since each pair of shutters has to swing closed along a ¼-in. rabbet. And finally, the louvers need to operate smoothly, without binding.

The vertical layout stick for these shutters is marked off to show the length of the stiles, the location of the rails, and the stile holes that accept the tenoned ends of the louvers.

The spacing of the louvers is determined by louver size and the degree of overlap desired, the overall height of the shutter and the width and number of rails per shutter. The shutters shown here are 79¾ in. high and have four rails. The top rail is 2 in. wide; the two center rails are 3 in. wide; the bottom rail is 5½ in. wide. Measuring between the top and bottom rails gives us 72¼ in.; so to get three equal shutter bays and allow for the 3-in. wide central rails, each bay must be 22 ¹⁄₁₆+ in. high (the plus means a heavy ¹⁄₁₆ in.).

The next thing to figure out is the number of louvers you need, and the spacing between them (actually the spacing between bore centers for the holes in the stiles that receive the round tenons), so that the holes for the tenoned ends of the louvers can be marked on the stiles. Louvers on most Victorian shutters are 1¾ in. wide, and this is the width we used (Greek Revival style shutters have 2¼-in. wide louvers).

The top and bottom louvers in each section must be located first, since their positions determine how much the louvers can close. Closing the louvers onto the bead is best for shedding water, but if you want to shut out the light, you need more pivoting clearance at top and bottom, which will allow the louvers to seat against each other. We centered the holes for the top and bottom louvers ¾ in. from the top and bottom rails, sacrificing complete closure for a ⅛-in. seat against the bead on the rail (top detail drawing).

Once the holes for the top and bottom louvers have been located, we can figure the spacing for the rest of them. We want the louvers to overlap about ¼ in. when closed to block the light and shed water, so the tenon holes in the stiles should be approximately 1½ in. apart, give or take a very small amount. The next thing that we do is to measure the distance between the top and bottom hole in each section (20⁹⁄₁₆+ in.) This we divide by 1½ in., yielding (thanks to my calculator) 13.713 spaces per section. You can't have fractions of a louver, so I divide 20⁹⁄₁₆ in. by 14, and get 1.469, or 1¹⁵⁄₃₂ in. between louvers. This is the spacing that I check by locating

Illustrations: Christopher Clapp

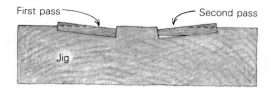

First pass   Second pass

Jig

Shaping louvers. Slats of white pine or cypress are beveled by running them through a jig (drawing above and photo at right) that's clamped to the planer bed. Grooves in the jig are angled at 6° so that the planer removes a narrow wedge of wood to make the bevel. Flipping the slat and running it through the other groove completes the beveling job. A second jig, far right, acts as both fence and stop as the squared tenons of each louver are rounded with a plug cutter. The shoulders of the tenons have already been cut on the table saw.

the centers for the dowel holes on my vertical layout stick.

Another way to space your louvers is to adjust your rails so that 1½ in. will divide into the bay height equally. But for most of the shutters we've built, I've been matching existing shutters, so rail sizes and locations have been fixed.

Cutting louvers—With layout work done, it's time to make the parts. We start with the louvers. The completed louver needs to have two bevels on each side, and the long edges have to be rounded over. Each louver also has to be tenoned at each end to fit into the holes in the stiles.

To produce the slats for the louvers, we first take rough 8/4 stock—either white pine or cypress—and mill it down to a net thickness of 1¾ in. Then we rip strips ⅜ in. thick by 1¾ in. wide and surface ½₂ in. off of each side.

To bevel the louvers, we use a special jig, sometimes called a slave board, that we clamp to our planer bed. It's made from a piece of 1¾-in. thick oak that's slightly longer than the planer bed and about 7 in. wide. Two shallow, 1¾-in. wide grooves in the top face of the slave board are what make the jig work. The groove bottoms are angled 6° off the horizontal, and this slightly angled running surface for the wood strips lets the planer waste a narrow triangular section from the strip to create the bevel (photo top left). Each strip of wood is fed through one groove, then flipped and fed through the other, giving you all four bevels with just two passes through the planer, as shown in the drawing above.

To round the edges, we run the beveled strips through a ³⁄₁₆-in. bead cutter on the shaper, using feather boards to hold them in place. Now all that remains is to cut the slats to length and cut round tenons on their ends. The shoulders of the louvers should clear the edges of the stiles by at least ½₂ in. or else they will bind when they are painted. We use a clearance of ¹⁄₁₆ in. to stay on the safe side.

We cut the louvers to a length that includes the ⅜-in. long tenon on both ends. Then we cut to the shoulder lines of the tenons on the table saw, holding the louver on edge against the miter gauge and setting the blade height to leave a ⅜-in. square tenon on both louver

With a hollow-chisel mortiser, the stiles are mortised to receive rail tenons. Holes have already been drilled to receive the louver tenons.

ends. The tenons get rounded with a ⁵⁄₁₆-in. plug cutter chucked in a horizontal drill press (Shopsmith). As shown in the photo top right, we use a fence to guide the louver into the plug cutter and a stop to keep from cutting into the shoulder.

Stiles and rails—We start with rough 6/4 stock, joint one edge and one face, and then surface it to a net thickness of 1¼ in. (frames for some shutters may be as thin as 1⅛ in.). The stiles for these shutters are 2 in. wide, and they're mortised to house tenons on the rails. We cut the mortises with a hollow-chisel mortiser (large photo above) and the tenons on a small, single-end tenoner.

The holes in the stiles for the louvers get drilled at this time too. Rather than mark and

drill the stiles directly, we've found it more foolproof to make a drilling jig. It's the same length and width as the stile, and the hole centers are transferred onto it from the vertical layout stick and then drilled out on a drill press. We attach the template to the stile with C-clamps, then start drilling, making sure the depth of the stile holes is ⅛ in. deeper than the length of the tenons.

The lower edge of each rail on the inside face of the shutter is left square, while the upper edge needs to have a short notch in its center to accept the vertical dowel. The notch and the square edge allow the shutter to open and close securely.

The remaining inner edges of the frame get an ovolo bead. This means that the shoulders on the rails have to be coped where they meet

the molded edges of the stiles. Coping is tough to do by hand, so we use a shaper fitted out with a three-wing coping cutter (a Rockwell 09-128 male sash) on a stub spindle. The stub spindle allows the rail's tenon to pass over it, so we can cope one shoulder at a time.

The last pieces to make are the dowels that will be attached vertically to each bank of louvers. We make these on a shaper with a ½-in. beading bit.

Assembly—Before putting everything together, we need to make the jigs that will hold the louvers in uniform position. The drilling template for the stiles can be converted to a louver jig by cutting ¼-in. grooves across the center of the holes. We make another jig just like this one and use the pair to hold the louvers. In the photo above left, both jigs are lying on the table while I test-fit stiles to rails.

Assembly isn't really that tricky when you use these jigs, though you have to work faster than the glue that's used on the mortise-and-tenon joints (we use West Systems Epoxy, made by Gougeon Co., 706 Martin St., Bay City, Mich. 48706). We first glue and assemble one side of the shutter completely, pressing all stile-to-rail joints home. Then we lay this sub-assembly on top of the jigs, apply glue to the exposed rail tenons, and engage them in their mortises. The trick here is to leave just enough clearance for the dowels at the ends of the louvers, as shown in the photo below left. Here's where fast work is important. Have your louvers ready, get them all engaged in their holes and set in the jig, then close the joints between stile and rails.

You can cut the dowel to length after stapling it to the louvers, or you can cut before stapling. The length of the dowel is the distance between the rails plus the length of the groove in the rail at the top of the section. Round the top of each dowel so that it will fit into the groove when the louvers close.

The dowel receives staples at intervals equal to the distance between louvers; and the uppermost staple on each dowel should be located far enough down the dowel so that the dowel fits into its groove in the upper rail when the louvers are closed. We use an Arrow T25 stapler with ⁹⁄₁₆-in. staples. It's a stapler that is used for putting up small wire, so the staples don't sink all the way in.

Next we attach each dowel to its section of louvers by shooting staples into the louvers through the staples already in the rod (photo below right). We learned, after some mishaps, that grinding the bevel off the staple points made them shoot straight in, with a minimum of splitting.

We usually leave painting the shutters to someone else, but it's important to seal the wood with a wood preservative before the finish coats are applied. Spray application is far better than brush-on because of the shutters' many movable parts. □

Rob Hunt is a partner in Water St. Millworks and a cabinetmaking instructor at Austin (Tex.) Community College.

Assembling the shutters. **Above, test-fitting the mortise-and-tenon joints between stiles and rails before final assembly. The grooved jigs on the table will hold the louvers as the shutter is assembled, as shown at left. Rails are joined tightly to the right stile, but loosely to the left one, providing clearance so the tenoned louvers can be inserted in the stile holes. The final step, below, is stapling the vertical rod to each louver section. Double-stapling keeps louvers aligned with each other, so they can be opened and closed as a unit.**

Photos: Jane Hunt

Foundations and Masonry

A Pier and Grade-Beam Foundation
Advice from a contractor on one way to build on a steep slope

by Michael Spexarth

As good home sites get harder to find, many builders are looking at hillside lots that used to be considered unbuildable. A builder who can cope with the special foundations required on steep sites can often buy a lot for a modest sum, and build a house that enjoys an attractive view of the scenery below.

Here in the San Francisco Bay area, chances are good that you'll find a pier and grade-beam foundation under just about every new hillside home. This type of foundation links a poured concrete perimeter footing to the ground with a matrix of grade beams and concrete pilings, some up to 20 ft. deep. The resulting grid grips the hillside like the roots of a giant tree (see pp. 191-193). Slopes in excess of 45° can be built on with this kind of foundation. And in areas where there are landslides, expansive soils or earthquakes, a pier and grade-beam foundation may be required by local building codes.

As a rule pier and grade-beam foundations need more reinforcing steel than conventional foundations, and require special concrete mixes. On the other hand, they usually call for less formwork than perimeter foundations do. What's more, the footings for grade-beam foundations don't have to be level.

Soil survey—A soil engineer's report is usually required by local codes, and even if it isn't, it's smart to get one. Because the tendency on steep sites is to overbuild, the survey's recommendations may reduce the number and size of piers, which designers or engineers with less knowledge of geology will spec to err on the side of caution. This kind of needless overbuilding wastes money.

To make the survey, the soil engineer will make test drillings or trenches at various spots around the site with a drill rig that typically bores a 6-in. dia. hole. Or he will bring in a backhoe. At various depths, core samples are taken with a cylinder that is lowered into the test hole. Although they will vary from site to site, these holes are usually 20 ft. deep or less because that's the limit of most drilling rigs used to bore the finished holes. The engi-

neer will take the soil samples back to the lab and analyze them to determine their bearing and expansion characteristics.

These tests, taken together, give a picture of the soil types and conditions at various depths, and of the natural water courses. Armed with this data, the soil engineer makes recommendations about the number and the diameter of the piers needed, and their depth. He specifies the steel-reinforcing requirements and the minimum distance between the piers. In northern California, the typical soil survey costs between $1,000 and $1,500.

Site preparation—The first order of business is to get rid of unwanted plants, loose topsoil, logs and other debris that isn't considered a permanent part of the landscape. Once the lot is cleared, mark stumps, rocks and areas of soft, wet soil with stakes and flagging. The safety of your drill operator could depend on his knowing what's where.

Upslope lots are usually more difficult to deal with than downslope lots. Here's the rule of thumb around here. A foundation on a downslope lot will probably cost twice as much as a foundation on a level site; a foundation on an upslope lot will be three times as expensive. In addition to moving the weight of the entire house uphill, the builder has to deal with the problem of hauling off tons of dirt.

Most soils are compacted in nature, and when drilled into or excavated, they can expand to two or three times their undisturbed volume. A 16-in. dia. by 18-ft. deep pier hole will contain about 1 cu. yd. of compacted soil, which will become as many as 3 cu. yd. of loose soil. So if 15 pier holes are drilled, you might have 45 cu. yd. of loose soil covering the site, or a volume 20 ft. by 30 ft. by 2 ft. deep. This soil can change the contour lines in the site plan, the elevation of the house, the loading on the foundation, and the drainage of the hill. So you have to move it out of there.

In the Bay area, some building departments allow loose soil to be evenly distributed up to 3 ft. deep if the slope is less than 18°. Such soil-dispersal zones must be away from natural drainage channels. On steeper slopes, the loose soil has to be removed or buttressed with a retaining wall to keep it from creeping downhill. Other building departments ask the soil engineer to recommend soil removal or dispersal, then the county inspectors check the project to see if they agree. Each jurisdic-

tion has its own policies governing drilling spoils. You should find out what they are to avoid trouble later.

Most builders stockpile the soil during construction, and then use it for landscaping once the building is finished. If the soil is unusable, or there's just too much of it, it has to be hauled off. We usually rent a dump truck and a tractor with a front-end loader to move the stuff, and take it to the nearest fill site.

Layout—The standard batter board used to locate the corners of perimeter foundations isn't the best solution for layout on slopes. Batter boards don't work well because most pier holes are drilled by a tractor-mounted auger, and as the operator maneuvers the rig around the site, chances are good that the batter boards and some of the survey stakes will be crunched into little splinters. In addition, 12-in. survey pins marking the exact location of piers can shift around in loose topsoils. A D6 Cat weighing around 10 tons may push loose topsoil downhill 1 ft. to 3 ft. as it lumbers around the site. So although the pins may appear to be in the correct relationship to one another, they might have moved downhill with the topsoil. This can cause a lot of problems when you start building your forms.

In order to ensure the correct siting of the house, and to guarantee exact layout points for drilling each pier, I arrange to get my surveyor on the site as the holes are being drilled. The surveyor centers the transit over a survey hub at a fixed elevation away from the drilling area, and then uses a chain (surveyor's tape measure) to locate the corners of the foundation and piers as calculated degree angles are turned. This ensures that no matter what soil movement has affected the preliminary stake positions, each pier hole is located precisely, right at the time it's drilled. The extra cost involved in having the surveyor on site is usually negligible compared to the additional labor and material necessary to do the whole job over again if your pier holes are drilled in the wrong places. If you can't get your surveyor on the site during drilling, the best alternative is to place reference points outside the drilling area to double-check stake location as the pier holes are being bored.

Drill rigs—There are two basic types of drill rigs: tractor or crawler mounted, and truck-mounted. Within these two categories you'll

Dirt flies as a tractor-mounted auger pulls soil from a 20-ft. deep pier hole. Rigs like this can work on slopes as steep as 45° or a little more. The ever-present helper with a shovel positions the auger, plumbs the drill shaft and directs dirt away from the hole.

Rebar for grade-beam and pier cages is assembled on site with jigs nailed to the top of sawhorses. The slots in the jigs position the #4 steel as #3 rebar stirrups are tied in place. Finished grade-beam cages await installation atop the woodpile. A rebar offcut used as a stake holds the woodpile in place.

find rigs that are as different as the people who operate them. The truck-mounted rigs have an auger attached to a boom that can reach several drill targets from one setup point. Such trucks do less damage to a site than a tractor, but they can't negotiate slopes greater than about 30°.

Tractor-mounted augers can drill in terrain that is inaccessible to trucks, and they can work on slopes up to 45° or slightly more. Sometimes an operator will anchor his tractor's winch cable to a stout tree or to a deadman driven into the ground uphill. Some operators carry a power pole on their cat. They auger a hole at the top of the slope for the pole and hook the cable to it. This lets them hang on the side of a steep slope as the holes are drilled. It's a chilling sight.

Drill operators generally charge by the hour, by the footage (the cumulative depths of the holes they drill) or by the job. They base their fees on the capabilities of their equipment, on the difficulty of the job and on the kind of soil they're working in. The rig that's pictured in this article cost $130 per hour, and it took 12 hours to drill 16 pier holes. And with a crawler-type rig, there is an additional delivery charge (often called travel time) tacked onto the hourly rate—$150 for this job. In rocky locations, you can expect a surcharge that will cover the cost of broken equipment, such as auger teeth.

Although these rigs can drill holes up to 20 ft. deep, soil conditions can change every few feet. One hole can be 6 ft. deep and hit rock, and 8 ft. away the next hole might be 18 ft. deep before the same layer of rock is engaged. The variables are surprising, so un-

less you're good at poker (or bridge), do this type of bidding liberally.

During drilling, soil lifted out by the auger is deflected away from the hole by a helper using a shovel (photo, p. 172). The helper will also need a hose to water down the auger bit. Water lubricates the cutting action, reducing the time needed to bore the holes, and lengthens the life of the equipment.

The loose dirt is left at the bottom of the hole because it's too far down to clean it out. Because pier and grade-beam foundations work by friction rather than direct bearing, this loose soil isn't usually a problem. But several years of drought can cause the soil around the piers to shrink and reduce the friction that holds the house on the hill. Enough shrinkage, and the house begins to settle.

You can limit the possibility of soil shrinkage by pouring a few gallons of water and a third to a half-bag of portland cement into each hole after the final depth has been reached and the auger has been removed. Then have the operator re-insert the auger for a minute's worth of mixing. The auger works like a giant milkshake machine to blend cement, water and tailings. When it's lifted out again, the mix should be thin enough to ooze off the drill bit. Once it hardens, this slurry forms a pad that will take direct bearing, and so lessen the chances for foundation settlement during dry years.

Once the holes are drilled, we cover them with plywood to keep out loose topsoil and debris. If there's even a remote chance that small children might wander onto the site, we stake the plywood to the ground and cover it with loose topsoil to make the area uninter-

esting. We also notify the neighbors that there are deep holes covered with plywood on a lot nearby, and we post the area.

Concrete and steel—A pier and grade-beam foundation has a lot of steel in it, and an equally large volume of concrete—100 cu. yd. is not uncommon. These foundations are a lot like icebergs, with their bulk concealed below the surface. With this much steelwork, the builder who sets up an efficient way to fabricate the beam and pier cages can save money.

On a recent job we assembled #4 rebar cages on sawhorses using #3 rebar stirrups and spirals. We easily bent the #3 steel with a rebar bender, and wired the straight pieces in place with standard foundation tie wire (photo above). Some foundations call for heavier steel, and anything over #4 (½-in. dia.) is difficult to bend. If your specs show a lot of #5s, #6s, #7s, consider having the stirrups and spirals fabricated at a metal shop and then delivered to your site for cage assembly.

The foundation can be monolithic, or it can happen in two pours. This latter way produces a cold joint between piers and grade beams. For this job we poured the piers first, and let the cages run long for ties to the grade beams. We suspended each pier cage in its hole with a rebar offcut placed under a stirrup, as shown in the top left photo on the facing page. After the piers were poured, we bent the ends over and tied them into the beam cages (photo facing page, top right).

Forms—When we build our forms depends on the configuration of the rebar. If the grade-beam steel is a series of interconnected cages, we tie them together and then build the forms around them. This allows for adjustment if any piers are out of alignment. The outside of the forms defines the shape of the house, and registers the placement of the mudsills.

If unconnected individual bars are used for the grade-beam steel, it's easier to build the forms first and then place the rebar. Since the forms are built according to the contours of the site, nothing has to be level. We pre-drill the mudsills, insert the foundation bolts and then place them in the wet concrete (photo bottom right). Leveling the house occurs when the foundation walls, sometimes called pony walls, are built.

Pumped concrete—The standard pumping mix in this area contains ⅜-in. pea gravel in a six-sack mix. This aggregate size allows you to use the lower-cost grout pumps, which have a 2 or 3-in. dia. hose. These pumps are pulled behind a pickup, and can pump the concrete 150 ft. up a 45° slope. The six-sack mix has one bag more portland cement than the usual mix. The richer mix is needed to achieve a higher compression rating than you get with the smaller-size aggregate.

You should have water available for lubricating the pump hose, washing loose soil out of the forms and, if pouring the beams separately from the piers, for removing any dirt that may have accumulated on top of them. For cold-joint connections, an epoxy bonder applied to the top of the piers just before the grade beams are poured will help ensure a good bond between the two pours.

Pumpers usually know about how much water they want in the mix, and they will check the mud in the transit mixer for water content. A stiffer mix can clog a dirty hose, and it makes the pump work harder. Since adding more water will lower the psi rating of the concrete, try to hire a pumper who keeps his equipment in good condition. Tell the batch-plant dispatcher you are using a pumper on a steep slope. Adding a water-reducing agent will keep the mix soupy enough to pump and still achieve the required psi compression rating when it cures. Around here this admixture costs only about $2 per cu. yd., and it adds about 1½ in. of slump to a load.

But I still prefer a stiffer mix because it won't run downhill as much as a wet batch. We make an initial pass around the forms, filling them half full. With a stiff mix, the concrete is able to carry some of its own weight by the time we pour in the rest.

If you happen to hear that glorious sound of bursting forms and splitting stakes, halt the pumping immediately. If you wait for 30 minutes to an hour, the mix should set up enough to allow you to continue, despite a weak spot in the formwork. In the meantime, fill other areas of the foundation. Even in mild weather, stripping forms the next day won't keep your engineer happy, but it allows you to get your form lumber back in usable shape. □

Michael Spexarth is a general contractor in El Cerrito, Calif.

A pier cage hangs on a rebar beam (above left) as concrete is pumped into the hole. The protruding bars will be bent over to tie into the grade-beam steel. Once the concrete piers have set up, the pier-cage steel is bent over to tie into the grade-beam cages, as shown at top right. Because there are so many splices involved in the steelwork, Spexarth first installs the cages, then constructs the forms around them. Redwood mudsills are cut to length, and placed near the appropriate footing.

As the wet concrete for the grade beam is screeded flush with the top of the form, the sills are embedded in the wet concrete and checked for side-to-side level. The J-bolt protruding from the sill will be worked down into the concrete, and the nut will be tightened after the beam cures.

Rubble-Trench Foundations

A simple, effective foundation system for residential structures

by Elias Velonis

Although it was first used extensively by Frank Lloyd Wright early in the 20th century, the rubble or gravel-trench foundation has largely been ignored by builders since Wright's time—perhaps because it represents a different way of thinking about what it takes to support a house. The conventional poured-concrete or block perimeter wall attempts to solve a building's load-bearing requirements in monolithic fashion by creating a solid, supposedly immovable and leakproof barrier extending from a footing poured below frost line to 8 in. or more above grade. But since freezing water expands 9% by volume with a force of 150 tons per sq. in., monolithic foundations are unlikely to survive in frost country unless they include a footing-level perimeter drain backfilled with washed stone, which carries away water that might collect and freeze under or against the foundation wall.

The two functions of load-bearing and drainage are solved separately with a solid foundation, but the rubble-trench system unites these two functions in a single solution: the house is built on top of a drainage trench of compacted stone that is capped with a poured-concrete grade beam. The grade beam is above the frost line, but the rubble trench extends below it, and the building's weight is carried to the earth by the stones that fill the trench (drawing, facing page, center). The small airspaces around each stone allow groundwater to find its way easily to the perforated drainage pipe at the bottom of the trench. Atop the grade beam, a short stemwall of concrete block, poured concrete or pressure-treated wood is built to support the floor framing. Or you can pour a slab. More about this later.

While this foundation system has been time-tested in many of Wright's houses, acceptance by building officials and the codes they follow is still not assured. In *The Natural House* (Horizon Press, New York, 1954), Wright speaks of what he calls the dry wall footing. "All those footings at Taliesin have been perfectly static. Ever since I discovered the dry wall footing—about 1902—I have been building houses that way.... Occasionally there has been trouble getting the system authorized by building commissions."

The disapproval of a building inspector usually arises from a lack of familiarity with the technique, since the Uniform Building Code states clearly that any system is acceptable as long as it can "support safely the loads imposed." When I first approached our local building inspector with plans for a rubble-trench foundation, he studied them quietly for a moment, ahemmed in good New England fashion, and said, "Yep, that looks as if it oughta work." And so it will, except in what Wright calls "treacherous soils," which I would judge to be any soils with a bearing capacity of less than 1 ton per sq. ft.

Determining the bearing capacity of a soil without engineering analysis is a matter of common sense and experience. If the earth in the trench is dry, seems to be well drained, feels solid when you jump on it, and is a mixture of gravel, rock, sand or clayey sands, it will very likely carry all the weight your house can bear on it. If, on the other hand, your heels sink several inches into soft clay, loose sand or fine silt when you jump into the trench, you'd better consult a soils engineer.

Construction—Assuming you've got stable soil, bulldoze the area of the house level, clearing all topsoil away and saving it for fin-

---

'All those footings at Taliesin have been perfectly static. Ever since I discovered the dry wall footing—about 1902—I have been building houses that way.'   —Frank Lloyd Wright

---

ish grading. If you have a sloping site, you will have to cut a level shelf in the hill, graded away from the house on all sides. This will ensure a good path for surface runoff. Lay out your foundation in the conventional manner (see "Site Layout," *FHB* #11), but make sure the batter boards are set up far enough outside the lines of the building that the backhoe will have room to maneuver. Sprinkle a line of lime 4 in. inside the strings that define the building's outer edge. This white line represents the center of the masonry wall that will rise up from the on-grade footing, or grade beam, and it provides the backhoe operator with a centerline to follow with his bucket. Ask the excavator if he has a narrow bucket for the backhoe—16 in. to 20 in. is perfect for

most soils. A wider trench gives you more bearing in softer soils, but it also takes more stone to fill it.

Have the backhoe operator cut the trench with straight sides, as deep as the frost line at the high point and sloping down to one or more outlet trenches along the perimeter (drawing, facing page, bottom). These should run away from the building and out to daylight at a slope of at least 1 in. in 8 ft. If you have a level site, I recommend running trench drains to a drywell, if your water table isn't too high. A drywell is a hole filled with a combination of small (1½-in.) stone and coarser rubble. You can base the depth and diameter of your drywell on the drainage qualities of your soil and the surface runoff you expect. Compute this from average-rainfall data and figures from the site's percolation test.

Clean up all your trenches by hand, making sure that their bottoms are flat and that they slope toward the drain line. Disturbed soil at the bottom of the trench may settle unevenly, so tamp the bottom firm with a pneumatic tamper or the heels of many boots.

Next, pour in a few inches of washed stone, and lay 4-in. dia. perforated PVC drainage pipe on top of it in the foundation and outlet trench. Make sure that the pipe follows the slope without dips that could restrict the flow of water. A ½-in. block taped to the end of a 4-ft. level makes the job of sloping the rigid pipe quite a bit easier. When the bubble reads level, you've got a 1-in. in 8-ft. slope.

I place the perforated pipe with its holes down (that is, at 4 o'clock and 8 o'clock), as I would in laying out a leach field, because as the trench fills with water, I think this orientation gets rid of it quicker. On the other hand, a case could be made for putting the holes up. It would take longer for them to silt up, but this shouldn't be a problem in good soils.

Now begin filling the trench with washed stone, taking care not to disturb the pipe as you cover it. I use 1½-in. stone because it's easy to find and easy to shovel, but larger washed stone is okay, too, as is the occasional clean fieldstone. (This is where the technique gets the name rubble trench.) Tamp the stone every vertical foot or so to make sure it is compact. To this end, I have even driven a loaded dump truck along the filled trench to make sure it was well settled, although this seemed to have little effect.

The outlet trench need not be filled with

Washed stone is dumped straight from the truck into the foundation and outlet trenches.

stone except for a foot in all directions around the pipe. Cover this stone with hay, burlap or tar paper as a filter, and backfill it with the original soil. If the pipe is running to daylight, be sure to leave its end exposed on a bed of stone. You want it to drain freely, so don't cover it with soil. Cap the end of the pipe with wire mesh to keep out rodents.

The grade beam—After the drains are installed and the trenches filled with stone, you're ready to build the forms for the grade beam. For one-story wood-frame structures, a 16-in. wide by 8-in. deep grade beam with three runs of ½-in. rebar is more than adequate. For a two-story structure, increase the depth of the beam to 10 in. or 12 in., and add two more lengths of rebar in its upper third.

Restring the lines from the batter boards and place the form boards on edge beneath them. I use 2x8s or 2x10s and brace them with stakes every 3 ft. Level the top edges of the form boards all around, nailing in 1x2 spreaders every 4 ft. to 6 ft. across the top to hold the forms in place. To reinforce the corners, use metal strapping or plumber's tape (perforated steel strapping), nailing it around outside corners. Place three runs of ½-in. rebar spaced evenly along the bottom of the beam, wiring them securely at the joints, and stagger these joints around the perimeter. Wire short pieces of rebar across these runs every 6 ft. Bend the lengths of rebar around all corners rather than splicing them there. Lift the rebar about 2 in. off the bottom of the beam with small stones. For a two-story building, prepare two more perimeter runs of rebar, and put them alongside the forms, ready to be dropped in the top third of the beam during the pour.

As the concrete is poured into the form, vi-

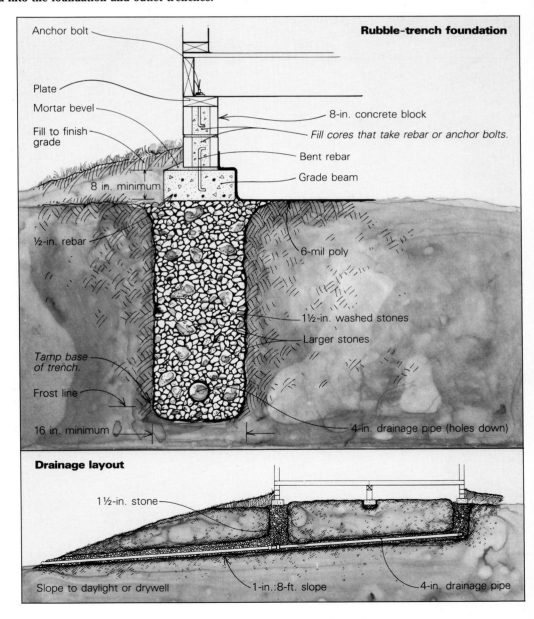

Rubble-trench foundation

Anchor bolt
Plate
Mortar bevel
Fill to finish grade
8 in. minimum
8-in. concrete block
Fill cores that take rebar or anchor bolts.
Bent rebar
Grade beam
½-in. rebar
6-mil poly
1½-in. washed stones
Larger stones
Tamp base of trench.
Frost line
16 in. minimum
4-in. drainage pipe (holes down)

Drainage layout

1½-in. stone
Slope to daylight or drywell
1-in.:8-ft. slope
4-in. drainage pipe

Illustrations: Jackie Rogers

Forming and pouring the grade beam. At left, forms for the grade beam are being set up on top of the stone-filled trenches. The 2x10 form boards are held together by steel plumber's strapping and by 1x2 wood stretchers nailed across their tops. Below left, rebar has been placed between the forms, and tied off to spreaders. The transit mixer is discharging its load of concrete, which is being spread and leveled by the crew of Heartwood students.

brate it well with a short piece of 2x4 to get rid of air pockets. Screed along the tops of the forms to get a level surface. If you intend to build a masonry wall, rough up the top of the grade beam with a broom before the concrete cures to ensure a good bond between it and the mortar. In areas where high winds are a problem, set 1-ft. lengths of bent rebar vertically in the top of the grade beam. They will anchor the stemwall. For a pressure-treated stemwall, place anchor bolts for the sill plate every 6 ft., and 1 in. from the end of each plate member.

If your design calls for a slab floor, you can pour the grade beam and slab at the same time, using a turned-down or "Alaskan" slab. For this you need only build outside forms, as shown in the second drawing below. For

houses with wooden floors, however, the height of the stemwall will determine the height of the crawl space, which should be at least 26 in. Since this may make the level of the finished floor higher than you want, raise the grade around the perimeter by 1 ft. or so, which will help slope it away from the house for surface runoff. Leave adequate ventilation ports on the stemwall, and when the building is roofed, cover the earth in the crawl space with 6-mil poly. Place anchor bolts for the plate in the usual manner, and parge the outside of the wall that will be above grade for a clean appearance. You might also add a bevel of mortar between the stemwall and the grade beam, to shed any water that might want to find its way between them.

Insulation alternatives—If you've designed an energy-efficient building and want to extend the insulation down below grade on the exterior, modify the foundation as shown in the first drawing at left. In an area where the frost line is 4 ft. deep, dig the trench only 2 ft. to 2½ ft. below initial grade, as you will be raising the finished grade around the building by 1½ ft. to 2 ft. Fill the trench with stone to 8 in. below initial grade, and then pour the grade beam to the surface. Lay up four courses of block (surface-bonded block walls work very well in this application) and build the floor.

When you're ready to insulate the exterior of the building, glue 2-in. or 3-in. thick panels of rigid foam insulation to the exterior of the stemwall down to the grade beam, then lay more foam panels on a sloping bed (at least 3 in. in 1 ft.) of stone, 2 ft. to 4 ft. around the entire perimeter. This insulation apron preserves a large bubble of relatively warm earth beneath the house, tempering the crawl space with its warmth. Cover this apron with 6-mil poly to protect the foam from water, and backfill up to finish grade. The exposed foam above grade and below the siding should be covered with asbestos board or the equivalent, or parged (see pp. 179-181) with a surface-bonding compound troweled on ⅛ in. thick over a wet coat of Styrofoam adhesive. The adhesive helps eliminate expansion cracks in the surface bonding between panels, but if cracks do appear, they can be caulked.

The only reservation I've heard from other builders about the rubble-trench foundation is that it might settle unevenly. In non-uniform soils this might be a problem, although the reinforcing in the grade beam is ample to span a good deal of uneven settling. We've built four rubble-trench foundations over the last five years, and none has shown the slightest sign of settling, cracking or frost damage. The great advantage of the system, of course, besides speed and relatively low cost, is that instead of building a massive wall underground, you just pour stones into a trench and are free to carry on building above ground. □

Elias Velonis is the founder and co-director of Heartwood Owner-Builder School in Washington, Mass. Photos by the Heartwood staff.

Superinsulated rubble trench

Fill to finish grade

6-mil poly

2-in. or 3-in. foam

1½-in. stone

An Alaskan slab

2x forms Mesh

4-in. slab

½-in. rebar

6-mil poly vapor barrier

1x4 stakes and bracing to hold forms

¾ in. to 1½ in. of stone

The finished grade beam, with its forms stripped.

Insulating and Parging Foundations

Covering concrete walls with rigid foam insulation and troweling on stucco requires experience with the materials

by Bob Syvanen

If you've got the idea that a builder's skill is an unchanging body of knowledge passed down through the generations, think for a minute about insulating a foundation from the exterior. Even in cold climates, what you used to see between the bottom of the siding and the grade was the bare concrete foundation wall. But these days, with estimates of heat lost in a house through the foundation running as high as 30%, what looks like concrete is more likely parging, or stucco, applied over rigid foam insulation.

Insulating the outside of foundations has been a problem for a lot of builders, including me, because many of the materials and methods are new. Although rigid foam-board insulation doesn't look like much of a problem, it isn't as simple as it first appears. Polystyrene is the insulating material most often used. It comes in 2-ft. wide panels and handles like plywood, but it's a lot lighter. You can cut it with anything from a knife to a table saw. But polystyrene foam is produced in two forms: expanded and extruded. Expanded polystyrene (EPS), also known as beadboard, is more susceptible to soaking up moisture than its extruded cousin, say the researchers on one side of this controversy. This could lead to a considerable loss in R-value. Although I used expanded polystyrene on the job shown here, I think the extruded version is probably the better bet despite its higher cost.

There are more than 100 makers of EPS; but extruded polystyrene is made in the U. S. by only three companies: Dow Chemical (Midland, Mich. 48640), whose blue-tinted Styrofoam is often called blueboard; Minnesota Diversified Products (1901 13th St. N. E., New Brighton, Minn. 55112), whose yellow product is trademarked Certifoam; and U. S. Gypsum (101 S. Wacker Dr., Chicago, Ill. 60606), the makers of pink Foamular.

Whichever brand you use, the process of applying it is the same, and so are the problems. For instance, asphalt-based products dissolve most foams, so my usual method of waterproofing a foundation is suddenly out

the window. And to complicate things even more, polystyrene needs to be protected above grade from impact as well as from deterioration by ultraviolet light from the sun.

I wanted a protective coating that was easy to install, good looking and long lasting. There are many commercial systems—fiberglass panels, super stucco mixes, and even a rigid insulation with a factory-applied coating that can be attached to concrete forms before pouring—but I wanted to use more traditional materials (see *FHB* #18, p. 8, for alternatives).

I first used asbestos board cemented on the foam, but it is fragile, hard to repair and impossible to glue, and required a lot of fitting time at corners, doors and windows. I also tried a latex-cement product applied directly on the foam. Unfortunately, it didn't age well. In fact, I have repaired not only the job I did with it, but several others in my area.

I finally settled on covering the foam with cement-stucco, called parging where I live. When stucco is used for exterior wall finish on a house, it is usually done in three coats like plaster. I was determined to come up with a single application process. Although parging and surface-bonding mixes can be applied directly to the insulation, I don't trust the bond, and want a thicker parging for durability. This means using some kind of lath.

On the first parging job, I used small-mesh chicken wire. I stretched it over ⅜-in. wood lath at 12 in. o. c. both horizontally and vertically to hold it off the surface of the insulation. Chicken wire wasn't the answer. Although my mason got the chicken wire to support the cement out of sheer stubbornness, the diamond-shaped pattern showed through a little, and there were some shrinkage cracks.

I refined the system by using metal lath,

and by reducing the thickness of the wood lath to ¼ in. This worked much better for the mason, but the lath strips were still tedious to install. Next I eliminated the wood lath and applied the metal lath directly to the foam. What I ended up with is a protective coating that is long lasting, attractive and relatively easy to install. Although it is a little expensive, after seeing some of the jobs using cheaper materials other local builders and I have done, I think it's worth the cost.

Since the insulation and lath-work usually fall to the carpenter or contractor who is on the site every day, the only sub I use is my mason, who is much faster and neater than I am with a trowel. I am used to paying anywhere from $2 to $4 a square foot for parging, although conditions vary enough that both the mason and I get the best deal when I use him on a time-and-materials basis. Since a bag of masonry cement covers 20 to 30 sq. ft. of wall, most of the expense is in the labor.

Installing rigid foam insulation—A partially earth-sheltered, passive-solar house I just completed gave me a good chance to try my new system. The plans called for its concrete walls to be insulated with two layers of 2-in. foam. One wall is 7 ft. 10 in. high, and the other three are 2 ft. high. The parging was to cover the first 2 ft. below the mudsill on all of them. Some folks also use insulation laid horizontally below grade (see *FHB* #8, p. 6), but I simply ran my panels down to the footings.

The first thing I needed was a good adhesive, since there shouldn't be any give in the plane of the insulation panels if the parging is going to last. But the high wall is also below grade and part of the living space, so it had to be well waterproofed. Since asphalt-based products can't be used with foam, I looked

Installing foam insulation. First, a waterproofing agent that also serves as a mastic is spread directly on the concrete. Temporary braces hold the foam panels in place while the mastic dries. Two-by-four nailers are used between the two 2-in. layers. The horizontal nailer is 27 in. down from the sill—the width of the metal lath that will be applied next.

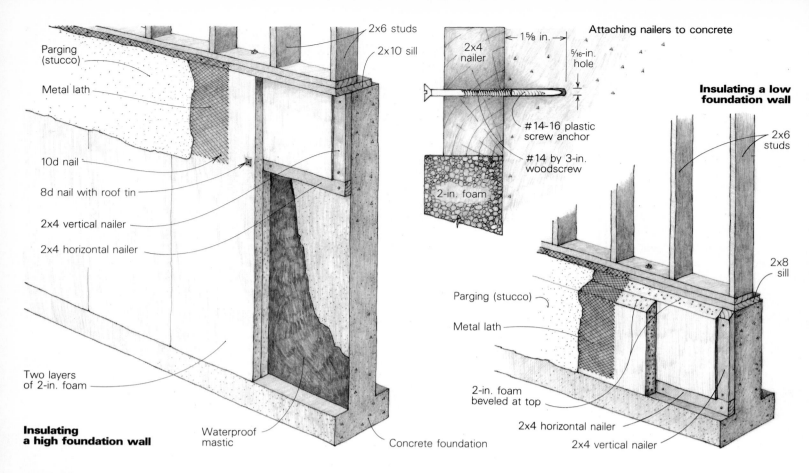

Parging (stucco)

Metal lath

10d nail

8d nail with roof tin

2x4 vertical nailer

2x4 horizontal nailer

Two layers of 2-in. foam

Insulating a high foundation wall

Waterproof mastic

Concrete foundation

2x6 studs

2x10 sill

2x4 nailer

← 1⅝ in. →

⁵⁄₁₆-in. hole

Attaching nailers to concrete

#14-16 plastic screw anchor

#14 by 3-in. woodscrew

2-in. foam

Insulating a low foundation wall

2x6 studs

2x8 sill

Parging (stucco)

Metal lath

2-in. foam beveled at top

2x4 horizontal nailer

2x4 vertical nailer

around for something else. What I found was a mastic, Karnak 920 (Karnak Chemical Corp. 330 Central Ave., Clark, N.J. 07066), which is marketed as both an adhesive and a waterproofing agent. Theoretically, you trowel the mastic waterproofing on the concrete and then press the foam panels in place. But the walls had enough irregularity that the panels contacted the mastic in only a few places, and they fell off about as fast as I put them on. I then found out that the foam has to be applied before the mastic skins over. This is enough time to apply just one or two panels and brace them with sticks, 2x4s, stones or buckets (photo previous page). When the foam was applied in this way, the adhesive held.

In this case, I installed 2x4 pressure-treated wood nailers with the first layer of foam panels in order to get nailing for the second layer. On the 7-ft. 10-in. wall, I began by placing a horizontal nailer 27 in. down from the sill. It was used to attach both the second layer of insulation and the lath, which comes 27 in. wide. Next, I attached vertical nailers above the horizontal at 24 in. o. c. because the panels are 2 ft. wide. I also filled in with nailers at windows and corners to catch the edges of the panels. A quick, easy way to fasten these 2x4s to concrete is to hold the nailer in place and drill through the wood into the concrete. Using a hammer-drill makes this almost fun.

For 2x4s, I use a #14-16 screw anchor, 1½ in. long, with a #14 by 3-in. flat-head woodscrew. Use a piece of tape on a ⁵⁄₁₆-in. masonry bit, at 3 in. from the tip, to limit the hole depth. Most hammer-drills have an attached depth guide. If the hole is too shallow, the tip of the woodscrew won't hit the con-

crete before snugging up the 2x4. If the hole is too deep, the screw won't grab the anchor.

After drilling, insert a plastic screw anchor into the hole in the nailer and turn a woodscrew a few turns into it. Then hammer the screw-and-anchor combination through the nailer into the concrete. Last, screw the woodscrew home (drawing, above center).

On the 2-ft. wall, I cut the first layer of foam to fit between the footing and the horizontal nailer and installed it before the nailers. This way, I could wedge the nailers between the foam and the footing while I fastened them.

On the high wall (drawing, above left), I cantilevered the mudsill over the concrete by the depth of the foam so that I could nail the metal lath directly to its top edge. This meant that the second vertical layer of foam tucked up underneath it, flush with its outside edge. This layer is held in place with 8d nails wherever there are nailers. The 8d nail reaches through to the nailer, and when given an extra tap, the foam compresses and snugs up the panel nicely. I use a roofing tin on each of these nails to increase its bearing surface. This is a stamped 2-in. by 2-in. flat metal plate with a hole in the center, and is typically used to hold down roofing felt on windy days. You can buy them from a roofing-supply yard or make your own by cutting out sheet-metal squares. I have also seen pins and plastic shields manufactured for this purpose.

The 2-ft. wall was insulated in a similar manner, but here the sill is flush with the outside face of the foundation wall (drawing above right). Since the foam projects past the sill, I beveled the top edge of the foam at a 45° angle. The vertical 2x4 nailers were also bev-

eled at the top before I fastened them to the concrete. The corner nailers are beveled from each direction (photo facing page, top left). To make a neat bevel cut in the foam, I snapped a chalkline the length of the foundation on the face of the panels and sawed along it with a bread knife. This bevel design worked well here because the finished grade was to come at the bottom of the bevel.

Installing metal lath—The galvanized metal lath I use measures 27 in. by 96 in. It is sold in single sheets or in bundles of ten. There are two things to keep in mind when you're working with metal lath. First, the diamond-mesh pattern is formed on an angle between the front and back of a sheet. This means that the dividing wire that is roughly horizontal forms a small lip or cup at the bottom of each hole. Make sure these cups are facing up to catch and hold the parging. It will work both ways, but things go better if the cups are up. The other thing to remember is that metal lath is sharp. I don't think I have ever worked with the stuff without cutting myself. The cuts are not bad, just annoying. Wearing gloves helps, but I find that more annoying than the cuts.

On the high wall, I nailed the mesh to the top edge of the sill. The bottom of the mesh nails through the foam into the horizontal nailer 27 in. below. I used leftover 3d shingle nails at the top, and 10d commons on the bottom and along the edges wherever I had a nailer. The 2-ft. walls were fastened similarly, but because of the beveled top, I had to bend the lath before I nailed it in place.

For corners, expanded corner bead—the plasterer's version of a metal sheetrock cor-

Photos: Bob Syvanen; Illustration: Frances Ashforth

ner—is the best way to go because it forms a neat, stiff straight line. But I didn't have any on the job, so I pre-bent the lath at 90° before installing it (photo top right). Bending sheet metal, particularly metal lath, on the job site isn't hard if you think of how it's done in the shop and duplicate the procedure. The shop uses a brake, which is a cast-iron table and a bar that folds the sheet metal over the edge of the table. On site, I sandwich the sheet between two 2x boards and "break," or fold, the piece that sticks out over the bottom 2x using a scrap block about 2 ft. long. Nailing the sheet metal to the top of the bottom 2x keeps it from creeping out as the bend progresses. This system is particularly good for metal lath because you don't have to handle the material constantly as you bend it.

Parging—Parging is not impossible for a novice, but a good finish takes experience. The first job I did turned out okay, but there was lots of room for improvement, so I went to school by watching mason John Hilley.

The parging he uses is a one-coat stucco with a steel-trowel finish. Other finishes might work better, but I am satisfied with this one. The mix he uses is 16 shovels of sand per bag of masonry cement. He doesn't have any trouble using up a batch that size before it begins to set. Masonry cement is a mix of portland cement, hydrated lime and additives that combine with water and sand to form mortar or stucco. For a parging mix, use Type M for higher compressive strength and greater resistance to water.

Large expanses of stucco are usually worked with darbies and floats. For foundations, though, a standard mason's trowel is easier. The mud is picked up on the bottom surface of the trowel and immediately applied to the lath. The free hand assists by pushing against the top face of the trowel, forcing the mud into the mesh. It is a quick process—pick up, apply, press. With each pressing motion, the excess cement gets pulled along with the sliding trowel (photo center right).

At the same time that the parging is applied, it should be roughly surfaced, to establish an even thickness. As with brick jointing and slab work, compressing the material is what finishing is all about. This requires a bit of pressure, but it should be with good control. Use two hands on the trowel, one on the handle and the other on the flat of the blade, and keep your arms straight.

The finishing is done when the shine leaves the surface of the parging. The trowel is dipped into water, shaken once to get the excess off, then pressed against the surface of the stucco using both hands (photo bottom right). Try to get a smooth finish in just a few strokes so you don't overwork the cement.

A cloudy, cool day is best for parging because the mix can be worked longer before it sets. If the parged wall is in direct sun, mist the surface with a pump-up garden sprayer filled with water to keep the surface of the stucco from drying out too fast, which will cause shrinking and surface cracks. □

Preparation. This foundation corner (top left) is ready for lath. Cutting the double bevel on the top of the corner nailer to match the bevel on the two layers of foam requires much less work later when the stucco is finish-troweled. An 8d nail and its roof tin, which acts like a large nailhead, are just visible at the bottom of the photo. This same corner is ready for parging once it is wrapped in metal lath (top right), which is pre-bent on a brake. The lath is nailed to the sill at the top, and through the second layer of polystyrene into the horizontal 2x4 nailer at the bottom.

Parging. Mason John Hilley forces the stucco mud into the lath (above), using two hands and the weight of his body. Just one coat of parging is used, but it is troweled twice. The first time is a rough troweling. When the shine disappears, the surface is smoothed with the same trowel, dipped frequently in water. This finish process is also done with one hand on the face of the trowel for direct pressure (right), and with arms held straight for good control.

Dry-Stack Block

Precision-ground concrete blocks make it easy to build a wall

by Rob Thallon

Designers have been trying for years to develop a mortarless concrete-block system that could be used by unskilled builders. The concrete blocks in use today look quite uniform, but their dimensions actually vary so much that mortar is necessary not just to hold them together, but also to make up for their irregular sizes. Mixing and applying the mortar to the joints in a block wall require skill and time (see *FHB* #15, pp. 44-47), and the process accounts for 20% to 30% of the material and labor in a masonry project. Manufacturers have recently developed mortarless, interlocking block for industrial and commercial buildings. I use it in house construction. It's called dry-stack block, and it can be laid up as easily as the plastic toy blocks in a Lego set.

Dry-stack blocks look very much like ordinary concrete blocks, but they are consistently a full 16 in. long and 8 in. high (regular blocks are an inexact ⅜ in. less in each direction to allow for the mortar joint). During the manufacturing process, the dry-stack blocks I use are sent through a machine that grinds the top and bottom surfaces to a tolerance of 0.005 in. These parallel, exact and smoothly ground surfaces are what allow the block to be laid up so regularly without mortar.

Most dry-stack blocks have interlocking tongues and grooves at their ends to help align and secure them during placement. Besides standard blocks, there are also bond blocks for bond beams (these have knockouts to accept horizontal rebar), and half blocks. Special corner blocks are manufactured without tongues for finished outside corners (drawing, facing page, center). Where the block remains exposed, its edges are usually chamfered to create a hand-tooled corner that's less likely to chip. It is also possible to have the face of the block ground and sealed to create a smooth, marble-like appearance.

There are three essential differences between the ordinary mortar-laid block and the dry-stack. First, the dry-stack method uses mortar only at the joint between the footing and the first course of block. This mortar joint at the base lets you set the first course absolutely level. Second, ordinary block is usually grouted (filled with concrete when the wall is complete) in only the cells containing reinforcing steel (rebar), while dry-stack blocks are usually grouted in every cell. This locks the blocks in place, and also fills the bond beams completely without having to pour them individually (drawing, facing page, top).

Third, you have to be careful with ordinary block walls to be sure that fallen mortar (as distinct from grout) doesn't hang up in the rebar or clog the bond-beam channels. This usually means that you have to build the wall in 4-ft. vertical increments so that the grout completely fills the appropriate cells. A dry-stack wall, however, can be grouted all at one time because there is no mortar to clog the steel or to plug up the cavities. Grouting tall dry-stack walls all at once can save a lot of time, especially if you use a concrete pumper.

When the dry-stack system was first introduced in the Eugene, Ore., area, about half the projects were questioned by the building department. The building official wanted to see calculations proving that the dry-stack system is as strong as a regular block-and-mortar wall. This is reasonably easy to demonstrate by showing that the compressive strength of the block is greater than that of mortar.

Residential applications—I had seen dry-stack block used successfully on several houses before I had the opportunity to try the system myself. I had designed a house for a steep site, with a complex foundation and several retaining walls. It looked as though using dry-stack blocks would allow a significant saving on labor. In addition, my client wanted a warm-colored block, and not having to use mortar meant we wouldn't have to mix colored mortar to match the block.

Before ordering the block, I asked the supplier about various coloring agents, but everything they showed me gave the blocks a bland uniform color—they looked phony. As an alternative, the manufacturer (Willamette-Greystone Inc., P.O. Box 7816, Eugene, Ore. 97403) suggested using scoria, a brownish-red volcanic aggregate found in Oregon's Cascade Mountains. This seemed to be just what I wanted, so I ordered a special run of blocks.

When the blocks finally arrived at the building site, I was surprised and disappointed. Instead of the rich, red-brown color I had expected, the blocks were pink. Evidently a slurry of scoria dust and cement had come to the surface as the blocks were extruded and vibrated during the manufacturing process. We eventually remedied the problem by sandblasting the finished wall.

With the footings poured and blocks on hand, we began building the walls. Our crew consisted of an experienced block mason and two laborers. I worked part time. The mason and I were anxious to see just how easily the dry-stack block could be laid up—he from a professional's point of view, and I from the perspective of a novice. On this job, we used almost 3,000 blocks and finished the foundation walls and three large retaining walls in about two weeks. The mason estimated that it would have taken four weeks using regular block and mortar.

First course—Getting the first course level is the most important part of the whole process. If you don't get it right, you'll be fighting your mistakes for the rest of the job. So the first rule is to have good footings, flat and within ¼ in. of level.

Mark the corners of your building on the footing, just as you would for an ordinary block-and-mortar wall, and check for square. It's a good idea to lay out at least one wall on the footing without mortar to test the blocks for length. We found that our blocks varied enough in length to accumulate a ½-in. error in a 20-ft. run if we didn't pay attention. By laying out the blocks dry, we could see how big a gap we had to leave between blocks to make things come out even.

After setting the corner blocks in mortar, we stretched out the mason's line and got down to laying the first course. We found that the work proceeded more easily than we expected, because the vertical joints don't require any special attention. This is a boon for the inexperienced mason. All you need to do is to lay two tracks of mortar along the footing, set the block on the mortar and level in both directions (drawing, right). The smooth surface of the blocks makes leveling easy. As

Dry-stack and conventional block walls compared

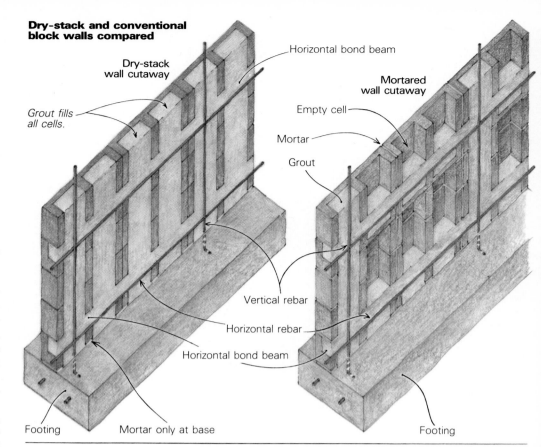

Dry-stack wall cutaway

Grout fills all cells.

Horizontal bond beam

Mortared wall cutaway

Empty cell

Mortar

Grout

Vertical rebar

Horizontal rebar

Horizontal bond beam

Footing

Mortar only at base

Footing

Four types of dry-stack block

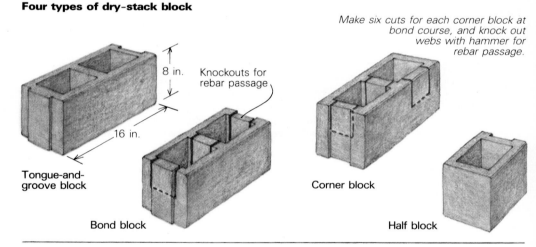

8 in.

16 in.

Tongue-and-groove block

Knockouts for rebar passage

Bond block

Make six cuts for each corner block at bond course, and knock out webs with hammer for rebar passage.

Corner block

Half block

Laying the first course

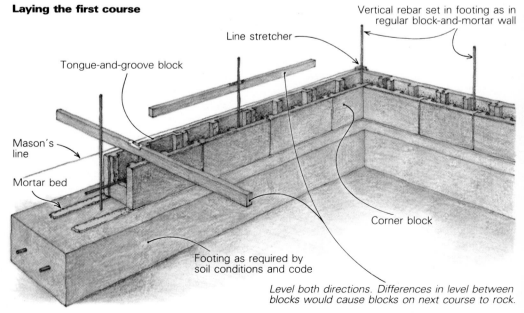

Vertical rebar set in footing as in regular block-and-mortar wall

Line stretcher

Tongue-and-groove block

Mason's line

Mortar bed

Corner block

Footing as required by soil conditions and code

Level both directions. Differences in level between blocks would cause blocks on next course to rock.

Photos: Rob Thallon; Illustrations: Christopher Clapp

	Dry-stack block walls (actual cost)	Standard block-and-mortar walls (estimated cost)
Materials		
Mortar	$ 30	$ 300
Block	2,847*	2,477
Grout	(32 yd.) 1,481	(20 yd.) 920
Steel (2500 ln. ft.)	461	461
Subtotal	**$4,819**	**$ 4,158**
Labor		
Laying block	$2,700	$ 5,400**
Grout-pump truck	165	(2 lifts) 330
Grout labor	200	300
Subtotal	**$3,065**	**$ 6,030**
Total	**$7,884**	**$10,188**

*2,414 8-in. regular, 345 8-in. half, 175 12-in. regular.   **Based on the mason's estimate of cost-per-square foot at about $2, a conservative figure. The 1979 Western Edition Building Cost File quotes a figure of $3 per square foot.

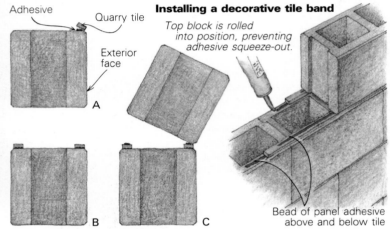

Installing a decorative tile band

Adhesive Quarry tile

Exterior face

Top block is rolled into position, preventing adhesive squeeze-out.

A

B   C

Bead of panel adhesive above and below tile

A quarry-tile inlay makes a thin red stripe around the house. The blocks have been sandblasted and sealed, revealing their volcanic aggregate.

we worked, we checked for length every 4 ft. or so and either tightened or loosened the joints slightly to come out even at the corner.

As block walls grow—Once the first course was laid, we built up the corners, carefully plumbed, as a guide for subsequent courses. The weight of the top blocks kept the lower blocks from moving and allowed us to stretch a mason's line tight from corner to corner as a guide. We kept the line about a string's width from the wall, so that accumulating error from successive blocks wouldn't force the string slightly out of line as they touched it. We found that we could set the blocks into place so rapidly that moving the string line became a significant part of the work.

In fact, the work sometimes went so fast that, in our enthusiasm, we made mistakes. The beauty of a mortarless system is the ease with which such mistakes can be corrected. At one point we dismantled a large portion of a 6-ft. tall fireplace footing and ash dump that had gone awry, and put it back together again in less than an hour.

We had some high walls on this project. One, which we built with 12-in. block, was more than 12 ft. tall, and we had several 8-in.

block walls over 8 ft. tall. Walls this high can get pretty wobbly before they are grouted, so we spread a double bead of panel adhesive between every fourth course for stability.

Bond beams—The only major cutting we had to do was notching the corner blocks at bond courses to let in the rebar. We used a mason's saw for this, and for the other minor cutting chores required for vents, bolts and the like. On small jobs, a circular saw with a carborundum blade would work fine.

The rebar required for a dry-stack wall is the same as that required for a block wall with mortar. The minimum requirements are listed in local building codes. For retaining walls up to 4 ft. high or 12 ft. long, we used one #4 bar at 32 in. o.c. vertically and one #4 bar at 24 in. o.c. horizontally (every third course with 8-in. blocks). Beyond these limits we had the wall engineered. Masonry suppliers have brochures listing rebar requirements compiled by the National Concrete Masonry Association.

To form an opening for a window or door, we supported a 2x8 formboard at the appropriate level with temporary posts, then laid up a steel-reinforced and grouted bond beam to serve as a lintel.

A tile feature strip—Just before the last course of block was laid, we installed a narrow tile feature strip. I wanted a thin band of color built into the wall itself to complement the horizontal water-table band at the top of the wall (photo above) so I had a local tile shop split 4x8 red quarry tiles into four lengthwise sections. I sandwiched these between two courses of block. The two pieces with finished edges were used on the exposed side of the wall, and the other two pieces on the hidden side.

We dusted off the top surface of the block and then laid a bead of panel adhesive about 10 ft. long near the inside edge of the tile. I aligned the tiles carefully along the chamfered edge of the block (drawing, above) and rolled them back onto the adhesive, forcing the excess toward the center of the block. This process was then repeated less carefully on the inside of the block and finally on top of the tiles with the final block.

The panel adhesive turned out to be an indispensable part of building the walls. It held all the tiles in position, bonded the top two courses together, and made the wall very rigid. This added rigidity was especially important during the grout pour.

Dry-stack block suppliers

If you're ready to build with dry-stack blocks, you need a local supplier because the blocks are just too heavy to ship economically more than about 150 miles. But finding a local supplier can be frustrating. We called all the block companies listed by the National Concrete Masonry Association as sources for interlocking block and ground-surface block, and asked each company if it could supply the mortarless units. It seems that relatively few builders ask for dry-stacks, so there aren't many made. As a result, the blocks can be hard to find.

The blocks we did find vary in price, design and dimensional tolerances. As a general rule, the closer the tolerances, the more expensive the block. In addition to the basic 8-in. high by 16-in. long by 7⅝-in. wide wall block, each supplier had a full line of sash, corner and bond blocks.

The McIBS Co. is the most active force in the mortarless-block industry. This company has perfected a special liner that can be used in conventional block-molding machines. The blocks made with these liners are double tongue-and-groove on their ends, tops and bottoms, and they maintain dimensions within 0.003 in. This close tolerance, coupled with the tongue-and-groove arrangement on all the hidden surfaces of the block, allows a wall made with them to be sealed with products designed for conventional block walls.

Of the mortarless blocks we found, McIBS are the most widely available, and the most expensive—from 30% to 80% more than conventional blocks. They are currently available in California, Colorado, Illinois, Indiana, Missouri, Nevada, Texas and Wisconsin, with several more states soon to join the list. Write to McIBS Inc., 130 S. Bemiston, St. Louis, Mo. 63105 for more information. For $1, they'll send you a brochure detailing their products and how they are used.

In Texas, builders can find interlocking blocks with tolerances held to ⅛ in. at two places: Valley Builder's Supply, Inc., P.O. Drawer Z, Pharr, Tex. 78577; and the Barrett Co., Rt. 3, Box 211 BI, San Antonio, Tex. 78218. Both suppliers make blocks with tongue-and-groove joints cast into their ends. They cost only about 3% more than conventional blocks. Both companies recommend surface-bonding the finished wall (see *FHB #12*, pp. 34-37). Similar blocks are available in Minnesota at the Charles Friedheim Co., 3601 Park Center Blvd., Minneapolis, Minn. 55416, and in Iowa at the Marquart Block Co. 110 Dunham Place, Waterloo, Iowa 50704.

Oklahoma builders can find dry-stack blocks at the Harter Concrete Products Co., 1628 W. Main St., Oklahoma City, Okla. 73106. Harter grinds its blocks to ¹⁄₃₂-in. tolerance, and then cuts a ¾-in. deep slot in the top and bottom of each block to accept a plastic spline. The splines help to align the blocks vertically. The cost is currently about $1.30 per unit for the basic block, and surface-bonding the finished wall is also recommended.

Yet another type of dry-stack block is made by the Buehner Block Co., 2800 S.W. Temple, Salt Lake City, Utah 84115. The block is 8 in. by 16 in., but only 5⅝ in. wide. The blocks hold to ⅛-in. tolerance, and use an interlocking system of plastic rings for alignment during placement.

Dry-stack blocks end up in a wide variety of projects—from houses to roadside sound barriers, from racketball courts to Holiday Inns. They might even become the universal building component that Frank Lloyd Wright and his son Lloyd envisioned in the 1920s (see *FHB #14*, p. 71). The blocks are simple to use, and they offer the thermal storage capacity that is an essential element of passive-solar design. —*Charles Miller*

of the asphalt-base sealers recommended by masonry suppliers would also work.

We sprayed the exposed walls with a two-part application of clear acrylic sealer. First, we applied a coat of relatively inexpensive Stone Glamour, then we sprayed on a finish coat of Mex-Seal for the reflective surface we wanted. These sealers bring out the blocks' color much as an oil enhances wood, and protect the block from the deteriorating effects of water penetration. For longest life, an exposed block wall should be resealed every five years or so, depending on the severity of the thaw-and-freeze cycles in your climate.

We didn't do anything to seal the exposed cracks between the blocks, even though we worried about the problems they might cause. I was afraid that capillary action would pull water through the cracks and cause moisture problems inside the house, and that the moisture in the cracks might freeze and fracture the blocks.

We resolved the first problem by sealing the inside of the walls with Thoroseal wherever they enclosed living space. We decided to ignore the second potential problem because the climate here isn't very severe. In the three years since we finished the walls, there has been no cracking.

In a climate where the combination of moisture and freezing is liable to cause problems, I would seal the exposed joints with clear silicone caulk. The caulk could be spread between blocks as the wall is laid up, or applied to the grooves between the blocks' chamfered edges after the wall is assembled. Either one of these procedures would increase construction time and expense, but the job would still be quicker, cheaper and easier than laying up a wall with mortar.

Grouting—Before we scheduled delivery, we calculated the amount of grout we needed with the following formulas:

for 6-in. block:
  number of full blocks/110 = cu. yd. grout;

for 8-in. block:
  number of full blocks /90 = cu. yd. grout;

for 12-in. block:
  number of full blocks/50 = cu. yd. grout.

On our job, for example, the calculation was: (2,414 + 345/2 [half blocks])/90 + 175/50 = 32.2 cu. yd. grout.

Because we needed so much grout (about 60 tons), we decided to hire a concrete pumper to get the grout from the trucks to the walls. We completed in about two hours a job that would have taken at least two days if we had done it by hand.

The only problem we encountered was on one of the tall walls. When we filled the cells, the weight of the mud blew out the side of one of the lower blocks, which was probably already cracked. Grout spurted out all over the place. We were able to repair the block and then refill the wall, a little at a time. If you have a wall 8 ft. high or taller, I recommend

grouting it up to about 5 ft., filling the shorter walls, and then returning to top off the tall walls after the first pour has had time to set up a bit.

Cleaning—You will inevitably slop some grout over the sides of the block. It's easy to rough-clean the surface by scraping it within 24 hours after the grout is poured. If you have the chance, clean the walls immediately with a light water spray and a soft brush. This can save a lot of hard work later on. To remove the grout stains completely, use a masonry cleaner like muriatic acid.

On this particular job, we wanted to remove both the stains and the pink slurry that formed the surface of our dry-stack blocks. We decided to sandblast only after trying several chemical cleaners without success. The blasting produced the desired results and cost only $320 for the whole job.

Waterproofing—You waterproof dry-stack block the same way as regular block. Below grade, we used Thoroseal, a water-base sealer, which we brushed on in two coats. Some prefer to apply one coat with a trowel, but this requires more skill. I'm sure that any

Cost comparison—When we finished the project, all of us who had worked on it were impressed with the dry-stack block. The mason was sure we had cut our labor time significantly by using the dry-stacks (chart, facing page), and he thought that they would result in a 50% labor saving on an average project. The laborers liked it because they got to lay some block themselves, which broke the drudgery of their usual lot—lugging heavy objects around the site all day.

What's wrong with this system?—Availability, that's what. Dry-stack blocks are so heavy that long-distance shipping is prohibitively expensive. Consequently, the blocks have to be manufactured close to their point of use. Although makers of standard concrete block are liberally scattered around the country, relatively few have the grinding or molding equipment necessary for making dry-stacks (see the sidebar above). And unfortunately, there isn't a comprehensive list of manufacturers that make the blocks. So if you're interested in the dry-stack system, get out the Yellow Pages and do some dialing. □

Rob Thallon is a partner in the architectural firm of Thallon and Edrington in Eugene, Ore.

Building a Fireplace

One mason's approach to framing, layout and bricklaying technique

by Bob Syvanen

I have been involved in building as a designer and carpenter for over 30 years, but building a fireplace has always been a mystery to me. I recently had the chance to clear up the mystery by observing, photographing and talking to my mason friend, John Hilley, as he built three fireplaces. I now understand more clearly than before what I should do as a carpenter and designer to prepare a job for the mason. I also know I can build a fireplace.

The job actually begins at ground level, with a footing (drawing, facing page). A block chimney base carries the hearth slab, upon which the firebox and its smoke chamber are built. The chimney goes up from there.

The importance of framing—As a carpenter, I've had to reframe for the mason too many times. This is usually because the architect or designer didn't realize how much space a fireplace and its chimney can take up, and how this can affect the framing around and above it. We'll be talking about a fireplace built against a wall, which is a pretty simple arrangement, but planning is still important.

Most parts of the country have building codes that specify certain framing details. In Massachusetts, where I live, code requires that all framing members around the fireplace and chimney be doubled, with 2 in. of airspace between the framing and the outside face of the masonry enclosing the flue.

The modified Rumford fireplaces that Hilley usually builds are my favorites because they don't smoke, they heat the room about as well as a fireplace can, and they look good. The firebox is 36 in. wide by 36 in. high, and the two front walls, or pilasters (returns) are 12 in. wide, for a total masonry width of 60 in. From the fourth course above the hearth, the rear wall of the firebox curves gently toward the throat. It's harder to lay up than a straight wall, but I think it looks a lot better. The back hearth is 20 in. deep and about 18 in. wide at the back—not in line with Count Rumford's proportions (*FHB* #3, pp. 40-43), but the minimum allowed by the Massachusetts code.

To figure the full masonry depth, you have to add to the 20-in. back hearth 4 in. for the back-wall thickness, 4 in. for the concrete-block smoke-chamber bearing wall, and 4 in. for the concrete-block substructure wall, for a total of 32 in. Thirty-six inches is better, because it gives extra space for rubble fill between the back wall and the block. Using

these dimensions, the chimney base is 36 in. by 60 in. Add a front hearth depth of 24 in. (16 in. is minimum), and clearance of 2 in. on each side and rear, and you get a total floor opening that's 64 in. wide by 62 in. deep. In situations like this one, where the fireplace is on a flat wall and the chimney runs straight up, with no angles, the framing is simple—double the framing around the openings and leave 2 in. of clearance around the masonry.

To locate the flue opening in the floor above the fireplace, find the center of your layout and drop a plumb line. This determines the side-to-side placement of the flue. Its depth is determined by the depth of the firebox. The flue will sit directly over the smoke shelf, and is supported in part by the block and brick laid up behind the firebox's rear wall. The framing for the chimney depends on the flue size. An 8x12 flue requires a minimum 18x22 chimney (a 1-in. airspace all around, inside 4 in. of masonry). Once the ceiling opening is framed, you can establish the roof opening by dropping a plumb bob from the roof to the corners of the ceiling-joist opening.

Wood shrinkage is something you should take into account when you're framing around the hearth. I think the hearth looks and works best if it's flush with the finished floor. Since it is cantilevered out from the masonry core (see below), and isn't supported by the floor framing, shrinking joists and beams can leave it standing high and dry. I've seen fireplaces built in new houses where the 2x10 floor joists rested on 6x10 beams. The total shrinkage here could leave the hearth an inch above the finished floor. A better framing system is to hang the joists on the beams and thereby reduce the shrinkage 50%.

From footing to hearth—The fireplace really begins at the footing, which is usually a 12-in. thick concrete slab 12 in. larger all around than the chimney base, and resting on undisturbed soil. The footing for this fireplace, therefore, is 48 in. by 72 in. Between it and the concrete hearth slab is a base, usually of 8-in. concrete block if it is in the basement or crawl space. To make sure the hearth comes out at the level you want it, the height of this base has to be calculated to allow for the 4-in. thick reinforced-concrete hearth slab, the bed of mortar on top of it, and the finished hearth material—in this case, brick.

Before pouring the hearth slab, the opening

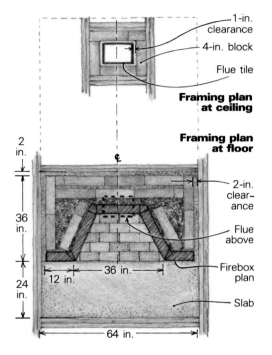

Framing plan at ceiling

1-in. clearance
4-in. block
Flue tile

Framing plan at floor

2-in. clearance
Flue above
Firebox plan
Slab

2 in.
36 in.
24 in.
12 in.
36 in.
64 in.

in the top of the concrete-block base is covered with a piece of ½-in. plywood that is supported by the inside edges of the blocks, leaving most of the course exposed for the slab to bear on. Cover the holes in the block with building paper or plastic, and build the formwork, secured to the floor joists, to support the cantilever at the front of the hearth. Then pour your 4-in. slab over a 12-in. grid of ⅜-in. rebar located 1 in. from the top.

Once the hearth slab has cured, it's time to lay up the structural masonry core that will support the chimney. Only the firebox, pilaster and lintel bricks will be visible on the finished chimney, so Hilley used 4-in. concrete block for the core. The blocks should be laid at least 4 in. from the face of the firebox brick and far enough in from the line of the front wall to allow for the pilaster bricks. Hilley sets a brick tie in each course to tie the pilasters in with the block.

Before beginning the brickwork, Hilley nails vertical guide boards (drawing, p. 188) to the face of the studs that frame the walls on each side of the fireplace opening, from floor to 12 in. above the lintel height. These boards are the thickness of the finished wall, and they locate the face of the fireplace. He marks off the brick courses up to three courses above the lintel on each guide board, starting from the

Illustrations: Christopher Clapp

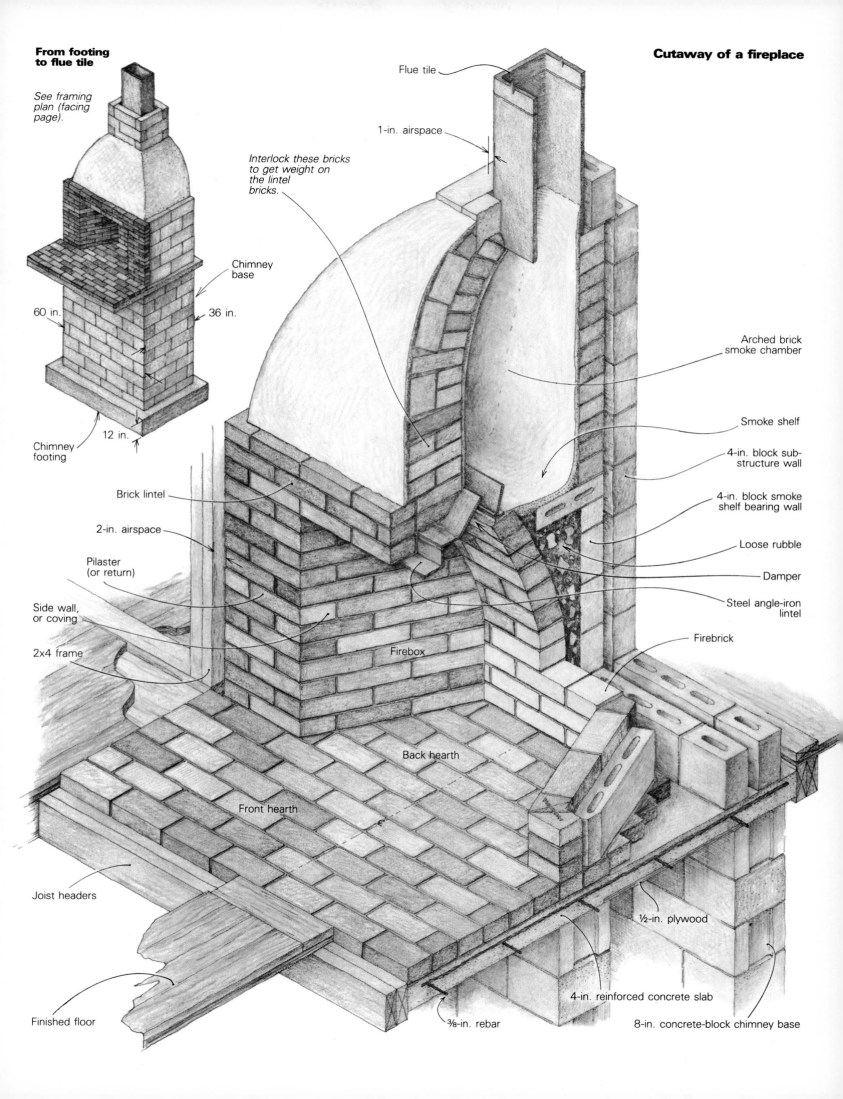

From footing to flue tile

See framing plan (facing page).

Chimney base

60 in.

36 in.

12 in.

Chimney footing

Interlock these bricks to get weight on the lintel bricks.

Cutaway of a fireplace

Flue tile

1-in. airspace

Arched brick smoke chamber

Smoke shelf

4-in. block sub-structure wall

4-in. block smoke shelf bearing wall

Loose rubble

Damper

Steel angle-iron lintel

Firebrick

Brick lintel

2-in. airspace

Pilaster (or return)

Side wall, or coving

2x4 frame

Firebox

Back hearth

Front hearth

Joist headers

½-in. plywood

4-in. reinforced concrete slab

8-in. concrete-block chimney base

⅜-in. rebar

Finished floor

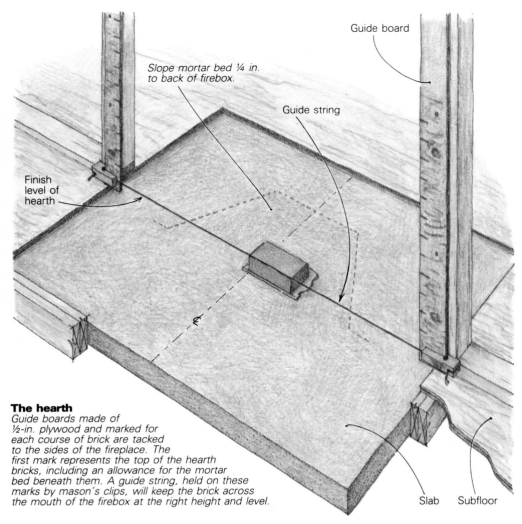

Slope mortar bed ¼ in. to back of firebox.

Guide board

Guide string

Finish level of hearth

Slab   Subfloor

The hearth
*Guide boards made of
½-in. plywood and marked for
each course of brick are tacked
to the sides of the fireplace. The
first mark represents the top of the hearth
bricks, including an allowance for the mortar
bed beneath them. A guide string, held on these
marks by mason's clips, will keep the brick across
the mouth of the firebox at the right height and level.*

Firebox layout

Mitered corner

├─9 in.─┤─9 in.─┤

20 in.

├──── 18 in. ────┤──── 18 in. ────┤

Square corner

¢

├── ½ in. ──┤

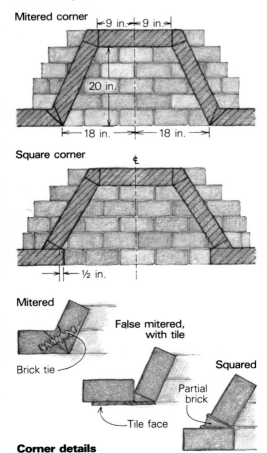

Mitered

False mitered,
with tile

Brick tie

Squared

Partial
brick

Tile face

Corner details

The firebox is being laid to the penciled layout, starting with five courses of the back wall.
Notice the curve starting at the fifth course
of the back wall. The cut-brick piece for the front
mitered corner will be alternated from front
wall to sidewall on each course to maintain a
strong bond. The V-shaped gap at the rear will
be filled with rubble. Brick ties every couple of
courses hold the joints together. The brick ties
in the concrete block will secure the brick front
wall (or return). The small torpedo level will
be used to level the back wall.

hearth, which on this job is 1 in. above the
subfloor. Once the guide boards are marked,
Hilley uses a guide string on mason's blocks
to control the height and alignment of the
brick courses as he lays them up.

Hilley picks sound, hard used brick for the
firebox and hearth. The hearth is laid to the
guide string in a good bed of mortar (drawing,
top left). The firebox walls will be laid on this
brickwork, so it extends beyond their eventual positions. Hilley likes to slope the hearth
toward the back wall about ¼ in. to keep water from running into the room if any rain
finds its way down the chimney. As with all
brickwork, small joints look best, so pick your
bricks for uniform thickness (see *FHB #14*,
pp. 32-35).

Laying out and building the firebox—With
the hearth laid, Hilley finds the centerline of
the opening, and marks off 2¼ bricks on each
side for a 36-in. opening. Standard bricks are
8 in. long by 3¾ in. wide by 2¾ in. deep, but
these measurements can vary, especially with
used brick. Hilley uses bricks instead of a tape
or ruler for an accurate layout, because 4½
used bricks (two times 2¼), laid end to end,
don't always total exactly 36 in. The line of
the back wall is 20 in. from the front line, and
its length is figured by counting a little more
than a brick on each side of the center line.
Hilley pencils these lines on the brick hearth.

The lines for the diagonal sides of the firebox are drawn between the ends of the front
and back lines. Where the side line meets the
front line at the juncture of pilaster and firebox wall, you can draw either a mitered corner, or a square corner (drawing, bottom left).
I like the look of the mitered corner, and I
think the time it takes to cut the bricks is
worth it. Cutting brick with a masonry blade
in a skillsaw is easy when the brick is held
securely between two cleats nailed to a plank.
Both pieces of the cut brick are used, so cutting halfway through from each side is a better way to go.

One way to achieve a mitered look without
cutting is to start a full brick at the front corner and butt the front return brick to the back
corner of the starting brick. The triangular
gap in front can be filled with mortar and covered with a tile facing, finish parging, stone,
or the like, as shown in the drawing at left.

When Hilley is doing a square-cornered fireplace, he brings the side walls to a point ½ in.
back of the edge of the return. This gives a
neat line, which is very important with used
brick because its width can vary from 3½ in.
to 4 in.

Firebrick isn't required when you're building a firebox like this one, but Hilley uses it
because heat-stressed common brick sometimes fractures violently. Most people don't
like the look of firebrick in a Colonial fireplace, so he uses it only for the first six or
eight courses—just high enough to cover the
hot spot of a fire. You can see this blackened
hot spot on the back wall of any fireplace.
After a few fires, the firebricks soot up and
blend in with the used brick in the rest of the

fireplace. Hilley doesn't use refractory cement with the firebrick, but he does keep his mortar joints under ¼ in. thick.

Hilley begins by sprinkling sand or spreading a piece of building paper on the brick hearth. This simplifies cleanup later. Then he lays up four courses of the back wall plumb, level, and parallel to the front—a small brick wall about 20 in. wide by about 11 in. high. The fifth course is a tad longer. It's also tilted or rolled in slightly by troweling on more mortar at the rear of the joint than at the front. This is the beginning of the curved back wall (photo facing page).

Next, five courses of the mitered side and front wall are laid up using the angle-cut brick at the front corners and by cutting and butting the rear brick to the back wall. The way to do this at the back wall is to score each end brick in the back wall with the tip of the trowel as you hold the brick in the rolled position. The coving is plumb, so the trowel should come off the bricks of the coving below and follow through in a plumb line, as shown in the drawing below. The scratch is very visible, and cutting is done with a brick chisel or the sharp end of a mason's hammer.

The two pieces of angle-cut brick at each front corner should fit together tightly where they show, and the V-shaped gap behind should be filled with mortar and a piece of brick. Hilley also likes to use a brick tie across this corner every couple of courses. This corner can get out of plumb easily, so a constant check with a level is a must. If a running bond

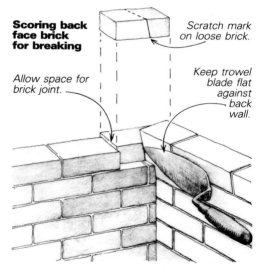

Scoring back face brick for breaking

Scratch mark on loose brick.

Allow space for brick joint.

Keep trowel blade flat against back wall.

is to show on the lintel course over the opening, you will have to watch the bond on your pilasters so that it will flow right into the bond on the lintel course.

Continue by rolling a few courses of the back wall, then building up the side walls. The roll will produce a gentle curve up to the damper, and it will make the back wall wider at lintel height than it is at the base. Each back-wall course is a little longer than the one below it, which is why the end bricks have to be marked in place for cutting. When a back-wall course needs to be a tad longer than two bricks, Hilley stretches it by setting a half-brick, or less, over the middle of the back

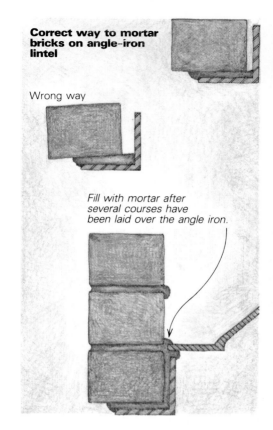

Correct way to mortar bricks on angle-iron lintel

Wrong way

Fill with mortar after several courses have been laid over the angle iron.

course below. The stretch, in other words, is accomplished in the middle of the course, not at its ends.

It is important while you're laying up the firebox to keep the side walls plumb. (In a square-cornered fireplace, the front and back walls are laid up first, a few courses at a time. The side walls are filled in.) You also must keep the back wall parallel with the hearth bricks. To do this, eyeball down the face of the back wall as it is laid, or measure from front to back on each side.

At the top of the firebox, the width of the opening from the outside face of the lintel brick to the rear face of the back-wall brick should be around 16 in. Hilley's formula for the amount of roll to give each back-wall course is simply experience. This is how most masons work. I'm always amazed at the way they seem to come out exactly where they want to be with exactly the right-sized opening, with no measuring at all. A novice might want to make a cardboard template to use as a guide, or spring a thin strip of wood against the first few courses to see how the curve projects up to lintel height.

Standard firebrick is thicker than used brick, so the back-wall courses will be higher than the side-wall courses. But the height should even out by the time you reach the lintel because the upper back-wall courses are tipped or rolled forward. As the back wall approaches lintel height, you can see how its courses relate to those of the front and side walls. By varying the joints, the wall heights can be adjusted to match.

When the firebox is at lintel height, Hilley fills in the space between the concrete-block wall and the back face of the firebox almost to the top with loose rubble. The rubble acts as a

The lintel. Side walls, back wall, and angle-iron lintel are at the same height to support the damper. The first course of bricks over the lintel overhangs the flange of the angle iron, and these bricks have to be laid up carefully so they won't roll forward. Pieces of building paper tucked at the ends of the angle iron serve as expansion joints.

heat sink, and more important, keeps the firebox positioned while allowing for expansion. A little mortar thrown in now and then will keep some of the rubble in place if a burned-out brick ever has to be replaced.

The lintel—A very important step in fireplace building is the proper installation of the angle-iron lintel. In this 36-in. fireplace, Hilley used 3-in. by 3-in. angle iron, which he installed with its ends bearing 1 in. or so on the pilaster bricks with a minimum of mortar—just enough underneath to stabilize it. The lintel's ends must be free to expand, and to ensure this Hilley tucks rolled-up scraps of building paper at each end. They act as spacers, keeping mortar and brick away from the angle-iron ends, and allow it to move.

The bricks in the first course above the lintel overhang the steel, and they have to be laid carefully (photo above) so that they won't roll forward. To help keep them from rolling, Hilley doesn't trowel any mortar behind them until a few courses have been laid, as shown in the drawing above. This eventual filling in, though, is important. Hilley feels that it prevents distortion of the angle iron from excess heat.

The damper—The damper should be sized to cover the firebox opening. The opening should be about as wide in front as the damper's flange, and from 2 in. to 5 in. narrower at the rear, depending on the damper's shape. The front flange rests on the top edge of the angle iron, and the side and back flanges rest on the firebox brick. The damper should be set in a thick bed of mortar on the brick and angle-iron edge, after three lintel courses are laid up, as shown in the photos at

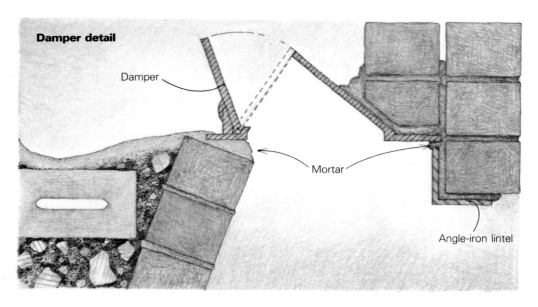

Damper detail

Damper

Mortar

Angle-iron lintel

The damper is mortared in place after three lintel courses are laid up. The space between the back wall of the firebox and the concrete-block core is ready for loose rubble fill, as shown in the drawing above.

The smoke shelf behind the damper is a 1-in. mortar cap over 4-in. concrete blocks on top of the loose rubble fill behind the firebox. The damper side is higher than the rear so any rainwater will drain away from the opening.

Laying up the smoke chamber is not fussy work. Hilley uses soft brick and concrete block, and then he parges the smoke shelf and chamber walls with mortar.

center left. As with the angle-iron lintel, it is important to keep masonry away from the ends of the metal to allow for expansion.

Smoke chamber—The smoke chamber is the open area behind the damper, where cold air coming down the chimney bounces off the smoke shelf at the bottom and is deflected upward, along with smoke rising from the firebox. As a base for the smoke shelf, Hilley lays a flat course of 4-in. concrete block on top of the rubble and concrete-block back wall. He sometimes lays a few concrete blocks, dry, directly on top of the loose rubble behind the rear wall. Then about 1 in. of mortar is smoothed out to make the smoke shelf's surface. Rainwater will puddle up here, so pitch the shelf away from the firebox and trowel it well. (Accumulated water will eventually evaporate or be absorbed into the masonry.)

The smoke chamber (drawing, p. 187) is formed by rolling the bricks of each course inward until the opening at the top is the size of the chimney flue tile. Hilley rolls the bricks a few courses at a time, alternating the corner bricks to maintain a bond.

Where the rolled brick courses meet at a corner, Hilley breaks off a piece of the lead corner for a better fit. He uses soft, spalling used bricks for this work. They are easy to shape, and it's not fussy work. In fact, Hilley had me hold up a sagging wall while he finished an adjacent supporting corner. A wall will collapse if laid up too much at one time.

Hilley says rolling the bricks to meet an 8x10 flue should give you a smoke chamber 24 in. to 36 in. high. Don't reduce from damper size to flue size too fast, and keep the smoke chamber symmetrical. Hilley once built a fireplace with the flue on the right side of the smoke chamber. This created unbalanced air pressures in the chamber and caused little puffs of smoke on the right side of the firebox.

The inside face of the smoke chamber is parged with mortar. (Be sure you leave enough clearance for the damper to open.) A piece of building paper or an empty cement bag laid on the damper before parging will keep things clean. You don't want your damper lid locked in solid with mortar droppings. The smoke-chamber walls must be 8 in. thick, so Hilley builds out their lower part with interlocking brickwork, and the upper part with flat-laid 4-in. concrete block. Then he parges the whole business with a layer of mortar (photo bottom left).

The rolled brick and outer block shell of the smoke chamber transfer the flue and chimney weight to the lintel, keeping the lintel bricks in compression. The first flue tile sits on top of the smoke chamber, fully supported by the brick, and the chimney is built around it. Brickwork against a flue will crack as the hot flue expands, so there must be at least a 1-in. airspace between the tile and the chimney shell. If the chimney is concealed, the masonry can be concrete block. □

Consulting editor Bob Syvanen is a carpenter in Brewster, Mass. Photos by the author.

Foundations on Hillside Sites

An engineer tells about pier and grade-beam foundations

by Ronald J. Barr

Most houses built on conventional sites sit atop a spread-footing foundation. It has a T-shaped cross section (small drawing, below center), and supports the house by transferring its own weight and the loading from above directly to the ground below. Spread footings can be used on sloping sites by stepping them up the hill (small drawing, below right), but this usually requires complicated formwork and expensive excavation. For slopes greater than 25°, the structurally superior pier and grade-beam foundation (large drawing, below) may be less costly. Apart from making steep sites buildable, pier and grade-beam foundations make it possible to build on level sites where soils are so expansive that they could crack spread footings like breadsticks as the earth heaves and subsides.

How they work—The pier and grade-beam foundation supports a structure in one of two ways. First, the piers can bear directly on rock or soil that has been found to be competent. This means it's stable enough to act as a bearing surface for the base of the pier. Second, the piers rely on friction for support. This is what sets pier and grade-beam foundations apart from other systems. The sides of the concrete piers develop tremendous friction against the irregular walls of their holes. This resistance is enough to hold up a building. A structural engineer can look at the soil report for a given site, study the friction-bearing characteristics of the earth and then specify the number, length, diameter and spacing of the piers that will be necessary to carry the proposed structure.

The rigs used to drill pier holes for residential foundations have to be able to bore holes 20 ft. deep or more. Foundations have to extend through unstable surface soil, such as uncompacted fill, topsoil with low bearing values and layers of slide-prone earth. Anchored in stable subterranean strata of earth or rock, deep piers don't depend on unstable surface soil for support.

The slope of the lot and the quality of the surface soil influence the size and placement

A pier and grade-beam foundation. Deep piers extend through layers of loose topsoil to lock into the stable soil below, which can support the weight of a house. The grade beams follow the irregular contour of the site, and transfer the building's loads to the piers.

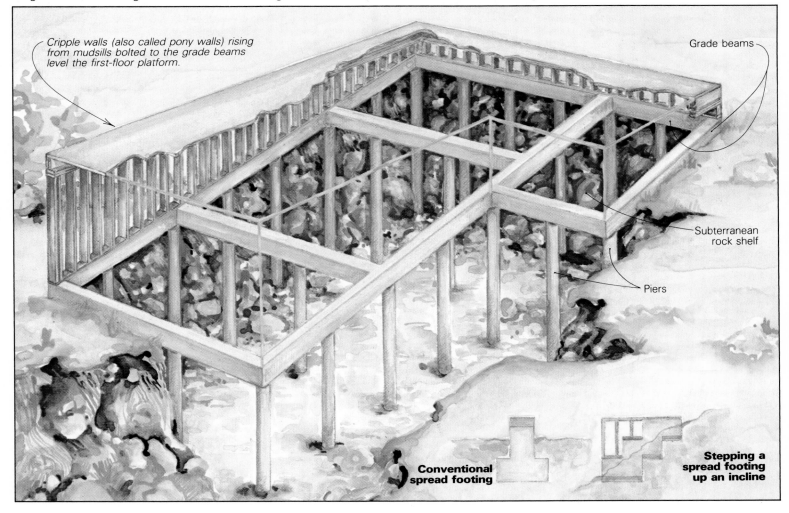

Cripple walls (also called pony walls) rising from mudsills bolted to the grade beams level the first-floor platform.

Grade beams

Subterranean rock shelf

Piers

Conventional spread footing

Stepping a spread footing up an incline

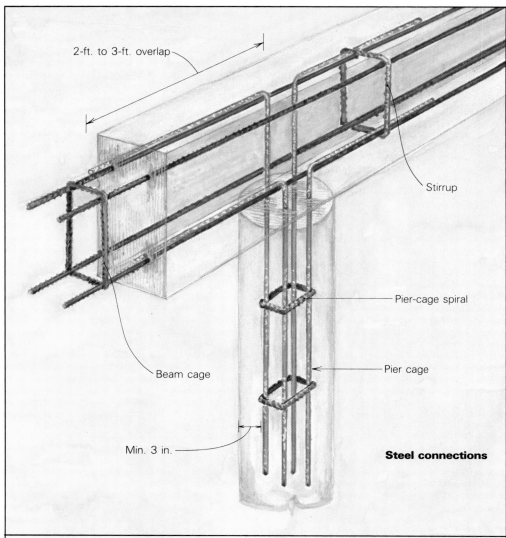

2-ft. to 3-ft. overlap

Stirrup

Pier-cage spiral

Beam cage

Pier cage

Min. 3 in.

Steel connections

of reinforcing steel in each pier (drawing, left). The steeper the lot and the more suspect the soil's stability, the bigger the steel. This is because the pier doesn't just transfer the compression loads from the house to the ground. It also resists the lateral loads induced by winds, earthquakes or by the movement of surface soils, which result in a kind of cantilever beam-action on the piers, as depicted in the drawing below left. They have to be strong enough to resist these forces—pier cages made from 1-in. rebar are not uncommon in extreme cases.

Grade beams—As the name implies, a grade beam is a concrete beam that conforms to the contour of the ground at grade level. Grade beams require at least two lengths of rebar, one at the top and one at the bottom. Foundation plans often call for two lengths at the top and two at the bottom, tied into cages with rectangular supports called stirrups. The bars inside the beams work like the chords in a truss, spreading the tension and compression forces induced by the live and dead loads of the structure above.

The sectional size of the grade beam depends on the load placed on it, and on the spacing of the piers. Pier spacing is governed by the allowable end-bearing value of the piers, or by the allowable skin friction generated by the pier walls and the soil under the site. Pier depth ranges from 5 ft. to 20 ft.; pier spacing, from 5 ft. to 12 ft. o.c. Pier diameters are normally 10 in. to 12 in., and can go up to 30 in. Beam sections start at about 6 in. by 12 in. and go up.

Because of the number of variables involved in designing a pier and grade-beam foundation, the engineering is best left to a professional. It's not happenstance that there are lots of lawsuits over foundation failures—many of them on slopes.

What the engineer needs—An engineer designing a pier and grade-beam foundation has to know what lies beneath the surface of the site. A soils engineer's report is usually required for this information (see p. 173). Sometimes the city or county building department will have records of soil characteristics in your area. The engineer also needs a copy of your house plans to calculate the loading on the foundation. If you can, tell the engineer what pier size your excavator can drill. If you're planning on making the steel cages yourself, ask your engineer about rebar diameter requirements. Although he may specify #5 or #6 rebar, which is impossible to bend in the field, it could be that more #4s (½-in. rebar) would do the job just as well.

Make sure you find out the spacing and size

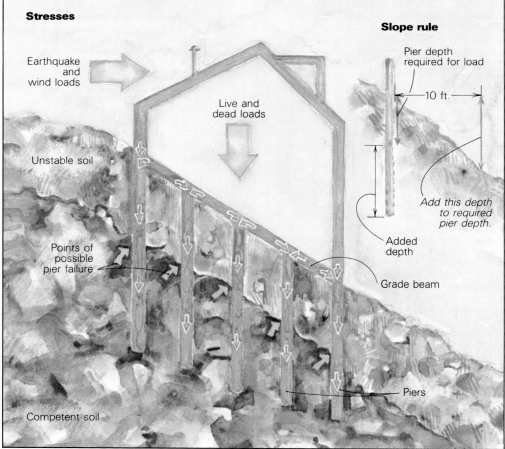

Stresses

Slope rule

Earthquake and wind loads

Live and dead loads

Pier depth required for load

10 ft.

Add this depth to required pier depth.

Added depth

Unstable soil

Points of possible pier failure

Grade beam

Competent soil

Piers

Engineering considerations. **The steel embedded in each pier must resist the lateral loads from wind or earthquakes, which are transferred to the piers at the junction of the unstable and competent soils. If piers are embedded in a slope, the amount of slope rise in a 10-ft. run must be added to the pier depth required for load.**

Pier and grade-beam in expansive soils. A well-engineered system has a 2-in. gap between the grade beam and the soil, to allow for expansion. The pier should be straight, not bell-shaped at the top, so loads are transferred down to the stable soil.

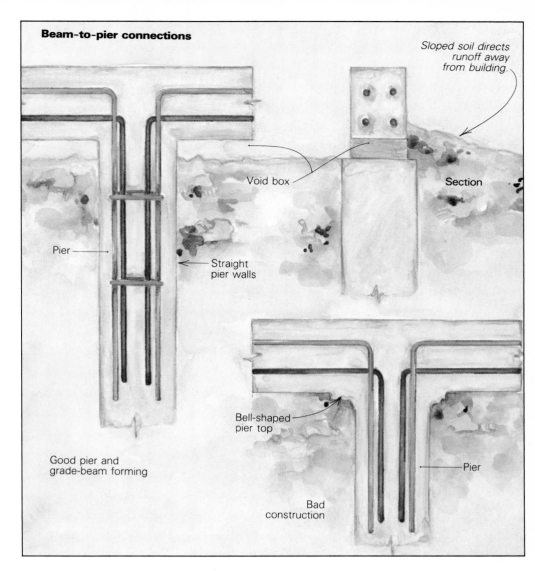

Beam-to-pier connections

Sloped soil directs runoff away from building.

Void box

Pier

Straight pier walls

Section

Good pier and grade-beam forming

Bell-shaped pier top

Pier

Bad construction

of the anchor bolts your foundation will need for fastening sills to the grade beams. For some downhill slopes, the cripple walls (drawing, p. 191) will need additional blocking, and anchor bolts placed 2 ft. o.c.

Piers on a slope must be longer to compensate for the lack of soil on the downhill side (inset drawing, facing page, bottom). Figure the additional length as follows: Tie one end of a 10-ft. long string to the stake marking the pier location, and pull it taut downhill and hold it level. The distance from the end of the string to the ground directly below has to be added to the pier hole to compensate for its location on a slope.

Void boxes—Ask your engineer whether you can cast (pour) the grade beams directly on the soil, or whether they should have void boxes under them. A void box (drawing, right) is a gap, usually about 2 in., between a grade beam and expansive soils. Some builders put a 2-in. thick piece of Styrofoam at the bottom of the grade-beam forms before the pour. Once the forms have been stripped, they pour a solvent along the base of the grade beam to dissolve the Styrofoam. Another method is to fold up cardboard boxes until you've got a stack 2 in. thick, and then lay them at the bottom of the forms. You don't have to remove them—they'll rot away.

When I worked as a building inspector, I saw what can happen to a house when void boxes aren't included in grade-beam pours over expansive soils. A couple of home owners asked me to find out why their foundation had settled in the center of the house. There was a large fireplace there, and they thought it was the culprit. But it wasn't. The fireplace showed no signs of movement. However, outside the perimeter grade-beam foundation, recent rains had caused the clay to expand, raising the foundation (and the exterior walls) as much as 2 in. in places. The center of the house had stayed put, and everything else had risen around it. None of the doors or windows worked, and every wall in the house was cracked. Voids under the grade beams would have prevented all this.

Problems—Once the foundation design has been determined and your driller is on site, it's almost inevitable that something unexpected will happen. For instance, your excavator may not be able to drill to the depth specified on the plans. Sometimes you can get away with this; other times you have to provide for another pier nearby. If you're drilling friction-bearing piers, you may have to change their diameter and spacing, or change their end-bearing condition. Check with your engineer for the correct course of action.

Drill alignment sometimes slips during drilling, and the auger can break into an adja-

cent pier hole. If this happens, check with your structural engineer. The remedy will depend on the load-bearing requirements at that particular part of the foundation.

Watch the soil as it's being removed from the pier holes. Its appearance should change in the upper half of the hole, as the drill leaves the topsoil layer and passes into the more stable soil. If you notice an abrupt change in the color or texture of the soil at the bottom of the hole, get in touch with your soils engineer right away, because a change in soil type may mean you'll need to drill deeper piers to make sure you're into solid earth. Shallow piers are the cause of most pier and grade-beam failures, so it's best to err on the side of caution and drill those holes deep.

A little while ago I drew up some plans for remedial foundation work for a house that was suffering from short piers. Actually, some of the piers were long enough to extend into solid material, and this uphill portion of the house hadn't moved. But the downhill half rested on shallow piers that weren't engaged with the competent soil layer. The awesome power of the surface layers moving downhill had pulled the house's grade beams apart and stretched their reinforcing rods like taffy. Above these cracks in the beams, the house was slowly being torn in half. Just getting a

drill rig to the site required dismantling part of the house, and the necessary foundation repairs ended up costing $60,000.

After the drilling—Grade beams should be centered over the piers. If one is out of alignment by more than a few inches, the eccentric load placed on the pier could eventually crack it in half. Rebar splices should overlap by 3 ft. And make sure that the tops of the pier holes haven't been widened to create a bell shape (drawing, above). Such a bell can be a bearing surface for expansive soils, and can cause uplift problems during wet weather.

A pier and grade-beam foundation usually doesn't need a special drainage system, but it's important to direct runoff away from the building. Banking the soil a few inches up the exterior side of the grade beam is generally all that's needed, and won't appreciably intrude into the void under the beam (drawing, top). The important thing is to make sure that running water doesn't wash away the soil around the piers. This erosion might not be critical if the piers are end-bearing, but if they are friction piers, the washing action could severely limit their load-carrying capabilities. □

Ronald J. Barr is a civil engineer. He works in Walnut Creek, Calif.

The Fireplace Chimney
Flashing and capping are the tricky parts of the job

by Bob Syvanen

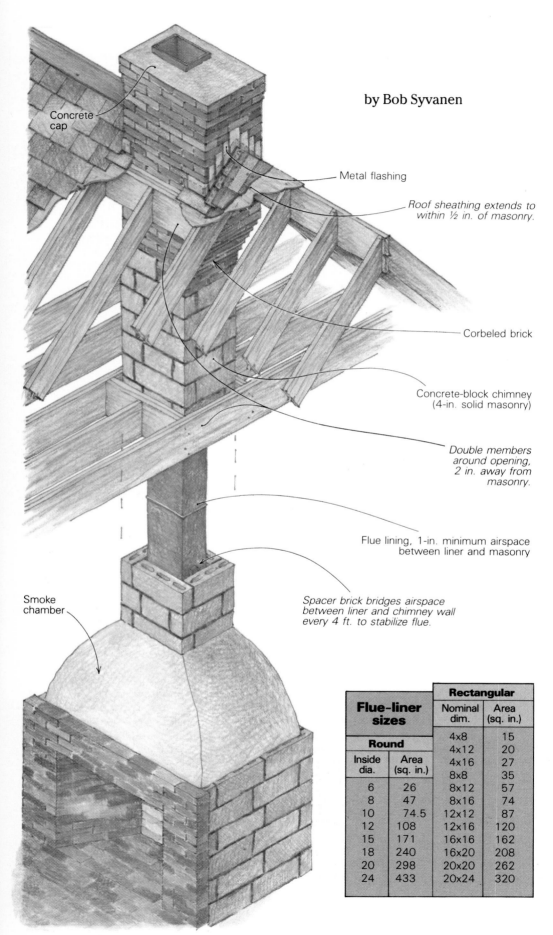

Concrete cap

Metal flashing

Roof sheathing extends to within ½ in. of masonry.

Corbeled brick

Concrete-block chimney (4-in. solid masonry)

Double members around opening, 2 in. away from masonry.

Flue lining, 1-in. minimum airspace between liner and masonry

Spacer brick bridges airspace between liner and chimney wall every 4 ft. to stabilize flue.

Smoke chamber

Flue-liner sizes		Rectangular	
		Nominal dim.	Area (sq. in.)
Round		4x8	15
		4x12	20
Inside dia.	Area (sq. in.)	4x16	27
		8x8	35
6	26	8x12	57
8	47	8x16	74
10	74.5	12x12	87
12	108	12x16	120
15	171	16x16	162
18	240	16x20	208
20	298	20x20	262
24	433	20x24	320

Throughout most of history, masonry chimneys were just hollow, vertical conduits of brick or stone. These single-wall chimneys have many problems: they conduct heat to the building's structure; they aren't insulated from the cold outside air, and so allow severe creosote buildup as the cooling gases condense; and they suffer from expansion and contraction, which lead to leakage of water (at the juncture of chimney and roof) and smoke (through cracks in the masonry).

In present-day chimneys, ceramic flue tile carries the smoke. It can withstand very high temperatures without breaking down, and also presents a much smoother and more uniform passageway, which means easier cleaning and thus less chance of chimney fires. Brick or concrete block, laid up around the tile but not in contact with it, serve as a protective and insulative layer.

Ceramic flue tiles can be either circular or rectangular in section, and are available in a number of sizes (see the chart below left). Brick, block and mortar are the only other materials you need to build a chimney (to find out how to build a fireplace, see pp. 186-190). John Hilley, a mason I've worked with for the last eight or nine years, uses type S mortar throughout the entire chimney, but some codes require refractory cement to be used for all flue-tile joints.

Requirements and guidelines—Before you start building a chimney, you've got to consider sizing, location and building-code requirements. I'll be talking about masonry chimneys that are built above fireplaces, but most of the construction guidelines also apply to masonry chimneys that are meant to serve a woodstove, a furnace or a boiler.

The Uniform Building Code ($40.75 from I.C.B.O., 5360 S. Workman Mill Rd., Whittier, Calif. 90601) contains basic requirements for chimney construction, and most states and municipalities have similar rules tailored to meet particular regional needs. I have a 1964 U.B.C. that reads the same as the 1984 Massachusetts State Building Code. It calls for a fireclay flue lining with carefully bedded, close-fitting joints that are left smooth on the inside. There should be at least a 1-in. airspace between the liners and the minimum

Consulting editor Bob Syvanen lives in Brewster, Mass. Photos by the author.

Illustrations: Christopher Clapp

Just below the roofline, the chimney wall shown above changes from 4-in. thick concrete block to brick. Block is much faster to lay up than brick, but it's usually used where appearance isn't important. The brick wall has been corbeled out against a temporary plywood form, enlarging the chimney above the roof and also centering it on the ridge. From this stage on, a good working platform on the roof is essential, right.

4-in. thick, solid-masonry chimney wall that surrounds them. Combustion gases can heat the liner to above 1,200°F, and the space between liner and chimney wall allows the liner to expand freely. Only at the top of the chimney cap is the gap between liner and wall bridged completely, and here expansion can be a problem. We'll talk about this later.

Chimney height is pretty much dictated by code. According to my Massachusetts Building Code, "All chimneys shall extend at least 3 ft. above the highest point where they pass through the roof of a building, and at least 2 ft. higher than any portion of a building within 10 ft." This rule applies at elevations below 2,000 ft. If your house is higher above sea level than this, see your local building inspector. The rule of thumb is to increase both height and flue size about 5% for each additional 1,000 ft. of elevation.

Exterior chimneys must be tied into the joists of every floor that's more than 6 ft. above grade. This is usually done with metal strapping cast into the mortar between masonry courses. No matter where your chimney is located, it should be built to be freestanding. The chimney can help support part of the wood structure, but the structure shouldn't help support the chimney. And there should be at least 4 in. of masonry between any combustible material and the flue liner.

The U.B.C. also has standards for flue sizing. For a chimney over a fireplace, the interior section of a rectangular flue tile should be no less than $\frac{1}{10}$ the area of the fireplace opening. If you're using round flue tile, $\frac{1}{12}$ the area of the fireplace opening will do because round flues perform slightly better.

Rectangular flue tile is dimensioned on a 4-in. module—8 in. by 8 in., 8 in. by 12 in., 8 in. by 16 in., 12 in. by 12 in., and so on. Standard tile length is 2 ft., though you can get different lengths from some masonry suppliers. When a mason talks about a 10-tile chimney, he usually means that it's at least 20 ft. high.

Using the flue-sizing formula isn't a guarantee that your chimney will draw properly. The chimney height and location, the local wind conditions, the firebox type, and how tight the house is are all factors that influence performance. John Hilley has found that on Cape Cod (at sea level), a 10-tile chimney atop a shallow firebox will work fine with flue sizing as low as 7% of the fireplace opening. Generally, short chimneys won't draw as well as taller ones. If you've got any doubts about what size flue tile to use, it's a good idea to ask an experienced mason or consult with your local building inspector.

Chimney construction—The fireplace chimney starts with the first flue tile on top of the smoke chamber of the fireplace (see the drawing on the facing page). As described on pp. 186-190 of this volume, the smoke chamber is formed by rolling the bricks of each course above the damper to form a strong, even, arched vault with an opening that matches the cross section of the chimney's flue tile. Since the smoke chamber will carry the weight of the chimney above it, it's got to be soundly constructed. If the rolled bricks of the chamber form an even, gradual arch, there should be no problems.

To begin the chimney, seat the first flue tile on top of the smoke-chamber opening in a good bed of mortar. Building paper or an empty cement bag on the smoke shelf will catch mortar droppings. You can remove the paper by reaching through the damper when you've finished the chimney.

Lay the chimney wall up around the flue tile, maintaining a 1-in. minimum airspace between tile and chimney wall. It's best to lay one flue tile at a time, building the chimney wall up to a level just below the top of each liner before mortaring the next one in place. Use a level to keep the liners and the chimney walls plumb. The mortar joint between tiles should bulge on the outside, but use your trowel to smooth it flush on the inside.

At every other liner, Hilley bridges the 1-in. airspace with two bricks or brick fragments that butt against the flue tile. These should extend from opposite sides of the chimney to opposite ends of the liner. This stabilizes the flue stack without limiting its ability to expand and contract with temperature changes.

If the chimney wall will be hidden behind a stud wall or in an attic, you can use 4-in. thick concrete block, which can be laid up much faster than brick. You can then switch to brick just below the roofline. Below the roof, chimney size is usually kept to a minimum to save space. If a larger chimney is desired above the roof, the chimney walls can be corbeled in the attic space (photo above left). The corbel angle shouldn't be more than 30° from the vertical (an overhang of 1 in. per course). A temporary plywood form set in place at the desired angle works well as a corbeling guide.

Coming through the roof—Roof sheathing should extend to within $\frac{1}{2}$ in. of the chimney, with structural members around it doubled and no closer than 2 in.

The next step in this part of chimney construction is to set up a good working platform on the roof (photo above right). Staging for

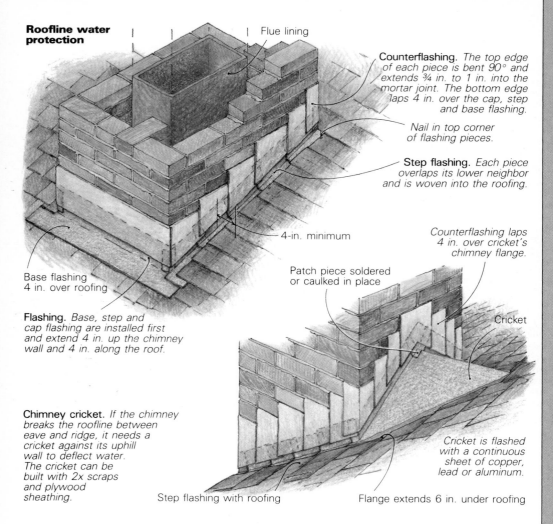

Roofline water protection

Flue lining

Counterflashing. *The top edge of each piece is bent 90° and extends ¾ in. to 1 in. into the mortar joint. The bottom edge laps 4 in. over the cap, step and base flashing.*

Nail in top corner of flashing pieces.

Step flashing. *Each piece overlaps its lower neighbor and is woven into the roofing.*

4-in. minimum

Base flashing 4 in. over roofing

Flashing. *Base, step and cap flashing are installed first and extend 4 in. up the chimney wall and 4 in. along the roof.*

Counterflashing laps 4 in. over cricket's chimney flange.

Patch piece soldered or caulked in place

Cricket

Chimney cricket. *If the chimney breaks the roofline between eave and ridge, it needs a cricket against its uphill wall to deflect water. The cricket can be built with 2x scraps and plywood sheathing.*

Cricket is flashed with a continuous sheet of copper, lead or aluminum.

Step flashing with roofing

Flange extends 6 in. under roofing

chimney work must be steady, strong and roomy. You can rent steel staging or build your own from 2x material. Either way, check the soundness of your staging well before you load it up with bricks and mortar. And before you bring any mud up on the roof, lay down a dropcloth to keep the roof clean.

To get to the staging, I put a ladder on the roof with its upper end supported by a ridge hook. It's a good idea to put some padding between the eave and the ladder rails so that the edge of the roof isn't damaged as you trudge up and down. You'll need a second ladder to get from ground to roof, unless the roof is steeply pitched. In that case, you can simply extend one ladder (if it's long enough) from ridge to ground.

Flashing—You're bound to find generously caulked flashing if you look closely at the chimneys in your neighborhood. I've lost track of the patch jobs I have done trying to stop chimney-flashing leaks. What sometimes happens is that rain gets blown in behind this flashing during a storm. Infrequent leakage doesn't mean the flashing job was poorly done, and caulking is a good stopgap in cases like this. But if your flashing leaks regularly in rainy weather, it probably wasn't installed correctly in the first place.

The Brick Institute of America (1750 Old Meadow Rd., McLean, Va. 22102) recommends flashing and counterflashing at the roofline. This creates two layers of protection,

with the counterflashing covering the base and step flashing on all sides of the chimney.

The Brick Institute recommends the use of copper flashing, but today you'll see more widespread use of aluminum, since it's much less expensive. Through-pan flashing (explained at right) is usually done with lead.

If your chimney straddles the ridge, first install the base flashing against the two chimney walls that run parallel with the ridge. Then step-flash the sides. Each flashing piece should extend at least 4 in. onto the roof, and is held with one nail through its upper corner.

Install counterflashing over the step and base flashing. The bottom edges of the counterflashing overlap the base and step flashing by 4 in. (drawing above left). The top edges of the counterflashing are turned into the masonry about ¾ in. They can be cast into the mortar joints as the chimney is built, or tucked into a slot cut with a masonry blade after the chimney is finished.

If the chimney is located against the side of the house or in the middle of a sloping roof, then you need to build a cricket against the uppermost chimney wall (drawing, above right). Otherwise, water will get trapped here, and you'll eventually have a leak. I use scrap 2xs and plywood to construct the slope, then cover the cricket with building paper and flash it with a large piece of aluminum or copper. The cricket flashing should extend 6 in. under the shingles and be bent up 4 in. onto the masonry, where it's covered by counter-

Through-pan flashing

Driving rainstorms can cause a lot of water to penetrate a chimney through cap and brick. Even when perfectly installed, conventional flashing can do little to stop this kind of penetration. The best way to drain out water that gets between the outer brick wall and the flue lining is to install through-pan flashing. Through-pan flashing is just what it sounds like—a continuous metal pan sloped from the flue lining to the roof. Weep holes between bricks just above the pan provide drainage.

There's some controversy about the effect that through-pan flashing has on the strength and stability of the chimney, since it breaks the mortar connection between bricks in adjacent courses. According to the Brick Institute, "if there is insufficient height of masonry above the pan flashing, wind loads may cause a structural failure of the chimney." This might be a valid warning, but I've never seen this kind of structural problem here on Cape Cod, which has its share of windy weather. And there's no arguing that through-pan flashing creates a more complete water barrier at the roofline than the conventional flashing scheme. Unless you live in earthquake or hurricane country, I can't see any reason not to use this system.

You can use copper or lead for through-pan flashing. John Hilley prefers lead because it's less expensive than copper, more malleable and generally easier to work. A utility knife will cut the stuff. Lead isn't supposed to last quite as long as copper, but I've seen 50 year-old lead pans that are still in good condition.

Building the curb—Before you can install the pan, you've got to build a curb where the chimney walls come through the roof. This is done with a combination of angled bricks or blocks and mortar, set in a form made from 2x lumber (top photo, facing page). The side walls of the curb should match the slope of the roof, and be around 1½ in. above the roofline. The lower and upper walls of the curb should be 4 in. to 6 in. above the roofline, parallel with it, and level. Like chimney walls, curb walls should be at least 4 in. deep.

Top the curb with a smooth layer of mortar, and the next day, rub the surface with a brick and round the corners to soften any sharp edges that could pierce the lead. Then install base, step, and cap flashing against the curb. With through-pan flashing, the pan takes the place of the counterflashing that is usually attached to the chimney walls.

The lead pan—If the joint between flue tiles falls just below or just above the roofline, then one piece of lead works well since you can roll it out on the base, find the outline of the flue, and cut holes for the next flue section to fit through. Alternatively, two or more sheets of lead can be used to make the pan. Just overlap the joints 6 in.

To determine the size of the pan, add 20 in. to the length of the chimney's lower wall and 24 in. to 32 in. to its side-wall measurement. Measure the side wall by following the angled curb with your tape measure. The chimney shown here is 64 in.

wide by 32 in. deep, and it required a sheet of lead 84 in. by 60 in. (7 ft. by 5 ft.). Lead sheets usually come in even foot widths and different lengths. For this job, Hilley trimmed a 6-ft. by 8-ft. sheet to size. These dimensions work for a chimney straddling the ridge. When the front and back walls of a chimney are on the same side of the ridge, the lead on the back or upper side of the chimney should be long enough to extend under two shingle courses plus one inch (more on a steep roof).

With the curb finished and flashed, and the lead cut to size, the next step is to install the pan. Roll the lead out over the curb and position it symmetrically (photo center left), then press down gently to find the outline of the flue tile.

Hilley cuts the hole for the tile about 2 in. smaller than the outside dimensions of the flue. Then he makes relief cuts in each corner and folds up the lead so that the flue can slide through it.

You have to be careful when handling the lead sheet; it's surprisingly easy to tear and puncture. The easiest way to carry a sheet is to roll it up. Never form lead with a hammer. Use your hand, a block of wood or a rubber-handled hammer handle. If you do pierce or tear the pan while installing it, pull the hole up above the surrounding pan so that water will drain away from it. You can also mend a hole by parging it over. As you position the pan, keep in mind that the object is to direct water away from the flue.

Once the flue tile that extends above the pan is in place, pull the lead up the sides of the flue to achieve the necessary outward slope. You can stiffen the top edge of the pan around the flue by folding it over. This helps to prevent sagging. After the lead is formed up tightly around the flues, parge the lead-to-flue joint with mortar.

Laying up the brick—Lay the first brick course over the pan in a thin bed of mortar (photo center right). Be careful to follow the curb under the lead. Make a few weep holes at the lowest points of the pan. You can do this by temporarily inserting a twig or rope in the mortar between bricks, or by leaving a vertical joint between bricks open. You'll have to cut the bricks that fit just above the sloping sides of the pan. Use either a mason's hammer, a chisel or a masonry blade. Once the courses on both sides of the ridge join, finish the chimney just as you would if there were no through-pan flashing.

Finish the through pan by creasing its exposed corners, trimming the pan edges and bending them down against the roofing. Before you start this part of the job, carefully lift the lead and sweep out any loose rubble. Creasing the corners so that they look nice is hard to do. The easiest way to start the crease is by slipping a short length of wood or angle iron under the pan at the corner. Hold the wood or metal edge so that it bisects the 90° corner, and gently start to form the lead over it with your hand. Then, if necessary, use a scrap 2x4 or a similar tool to form the lead into a more defined crease, as shown at right. Sharpening the crease should force the pan down against the roofing. A second crease, close to the first one, will force the pan down even farther.   —B.S.

The curb for the through-pan, above, is a combination of bricks and formed concrete, built directly on the chimney walls where they intersect the roofline. Curb walls should be at least 4 in. thick. The side walls of the curb are sloped to match the roof pitch, and 1½ in. above the roofline. End walls are level, parallel with the ridge and 4 in. to 6 in. above the roofline. At left, a lead sheet is cut to fit over the flue tiles and carefully rolled and bent over the curb. This forms the through-pan that will drain water away from the center of the chimney and onto the roof. Below, the pan is parged to the flue, and the first course of the chimney cap is mortared to it, bearing squarely on the curb.

The last step is to crease the corners of the pan, trim its edges and bend them down against the roof. Wood blocks and gentle hand pressure are used in this final forming.

flashing. Once the cricket is finished and flashed, flash and counterflash the three remaining sides as mentioned earlier.

The cap—The chimney cap covers the airspace between flue tile and chimney wall, stabilizing the flue and directing water away from the rest of the masonry (drawing, below left). It's good to corbel out the chimney's top brick courses or to install a cap that overhangs the chimney walls. This will direct runoff onto the roof rather than onto the brickwork.

You can buy precast caps in a few sizes or cast your own in place. Installing or forming a cap is more complicated than it sounds because of the way the ceramic flue behaves. Cross-sectional expansion of heated flue tile is accommodated within the 1-in. airspace between tile and chimney wall. The concrete chimney cap bridges this gap, and if it's mortared directly to the tile, you're bound to have cracking and breaking problems. Even if the upper tiles stay cool and don't expand widthwise, the tiles near the fire will expand along their length, forcing the entire flue upward.

One way to accommodate vertical movement of the flue is to create an expansion joint between brick courses just below the cap. The topmost flue tile, cast to the cap, forces it upward as the flue stack expands. The cap's weight closes the expansion joint as the stack cools. Hilley creates his expansion joint by sprinkling a thin layer of sand on the brick course before laying down the mortar (photo top left). The sand prevents the mortar from adhering to the brick directly beneath it, but doesn't affect the bond to the bricks above.

The chimney top's first corbeled course creates an inside ledge that will support a form for the concrete cap (photo center left). You can use corrugated sheet metal or scrap wood for the form. If you use wood, as Hilley does, make sure it sits loosely on the brick ledge, and soak the wood in water before pouring the cap. This way, the wood won't swell and force bricks out of their bond.

The last flue tile should project above the level of the last brick course at least 2 in. so the concrete cap can slope down toward the edges of the chimney (photo bottom left).

Forming the concrete cap is the last step. Hilley casts the cap directly to the topmost flue tile, relying on the control joint several courses below to accommodate flue-stack expansion. An alternate method is to cast an expansion joint around the topmost flue tile. To do this, pack some ⅜-in. dia. backing rope around the top flue tile where the cap will fit, and then caulk the space with a flexible, non-oil-base sealant after the cap is cast.   □

The chimney cap

Expansion joint opens and closes with flue movement.

Top three to four brick courses are corbeled out to bring drip edge away from chimney wall.

Surface of cap should be smooth and sloped.

Rough wood contains concrete when cap is cast.

¼ in. to ½ in.

Flue lining

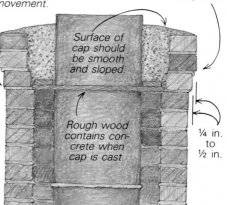

Capping the chimney. **Top, John Hilley creates an expansion joint with a loosely spread layer of sand beneath the mortar. Located several courses below the chimney top, this joint will widen and shrink with the normal expansion and contraction of the flue tile. The wood will be part of the concrete cap's bottom form. At center, the cap is cast between the flue and the top three brick courses. At left, the completed cap slopes away from the flue.**

Energy-Efficient Construction

The Thin-Mass House

Concrete floors and interior stucco walls improve heat storage in passive solar designs

by Max Jacobson

As designers of passive solar homes, our firm has been searching for an elegant yet economical solution to the problem of thermal storage. Instead of concentrating thermal storage in Trombe walls, rockbeds, water pools or eutectic salt arrays, we've now worked out techniques to use thin layers of stucco and concrete both as heat storage and as finished interior surfaces. We're not talking about standard masonry construction. Essentially, our approach takes the typical wood-frame stucco house and turns it inside out: Stucco layers become the finished interior walls, and thin slabs of concrete are used as floors, even in two-story wood-frame houses.

Masonry surfaces warm or cool us by means of radiation and conduction, and a thin masonry envelope can be a very comfortable environment if it's oriented to the sun and properly insulated. Used as finish surfaces, the mass becomes a part of the building's fabric, like the stonework in a whitewashed Mediterranean house or the earthen bricks in our own southwestern adobes. Surrounded in such homes by solid, durable materials, the occupant feels a comfortable sense of permanence.

In addition to these characteristics shared with traditional masonry structures, the thin-mass house also has several advantages.

Lazy mass—Only the first few inches of the interior mass of a masonry structure effectively store heat. The mass in the center of the wall isn't doing much, and in fact, can act as a thermal bridge, conducting the higher indoor temperatures outside. Most of the mass in the center of a concrete wall doesn't have a structural purpose (that's why concrete blocks are hollow), yet it adds to the building's foundation requirements and increases loading during an earthquake. In our thin-mass house, we replace these lazy inches of mass with lighter, well-insulated materials—conventional wood framing and fiberglass insulation.

A little history—Concrete as a finish material has been around for a long time. In 1894, the southern California architect Irving Gill used concrete to duplicate the glossy, hard-packed earthen floors he had found in Mexican adobe houses. He added pigments to the concrete mix to produce neutral, earthy colors.

By 1930, the staining and painting of concrete surfaces was widely practiced, although the concrete was made to look like stone or antiqued wood rather than packed earth. Typically, the surface was painted with a thin wash, followed by darker designs (photo, facing page).

Integral coloring and surface coloring are both still used, but we can also use a new, less expensive method that restricts the color to a 1-in. or 2-in. layer over an existing slab.

Topping over on-grade slabs—When the site permits, an on-grade slab can serve as both the foundation and the ground floor. But clients' first reactions to the suggestion of on-grade slab floors are typically negative. They assume that the floor will look like the inside of a garage and that it will always be cold. But if the slab's perimeter is carefully insulated to prevent thermal bridging, the floor will be comfortable, its

The Lee-Carmichael house (facing page) is a thin-mass building in northern California. It has a heat load of only 205 therms annually, because of its 245 sq. ft. of south-facing glass and the thermal storage capacity of its stuccoed interior walls and concrete floors. Without these features, its yearly heat load would be about 630 therms (1 therm = 100,000 Btus).

temperature fluctuating narrowly around the mean internal temperature.

You can get a gorgeous finish floor by using integrally colored concrete for the entire slab. The pigments are just dumped into the transit mix truck as the load is readied for delivery. The pigment adds about $20 per yard to the cost, so it's important to use it only where it will be visible. In some cases, we prefer to use a two-step system for on-grade slabs, which allows more decorative effects and separates the finish work from foundation frenzy. We first pour a conventional slab, and then finish it with a 1½-in. thick layer of integrally colored concrete.

Because it's nearly impossible to reinforce such a thin layer of concrete (the wire mesh can come to the surface in places) the topping has to be bonded to the reinforced slab below to prevent shrinkage cracks. To do this, you must be sure that the slab has a rough surface. Don't trowel it smooth after screeding. The surface must be absolutely clean. Sweep (or, better yet, vacuum) it free of all dust and debris. Any areas that have had oils spilled on them must be thoroughly washed with detergent and flushed with a hydrochloric-acid wash.

Just before pouring the topping layer, spread a bonding grout of one part portland cement, one part sand and one-half part water over the moist prepared surface. Use a stiff broom to spread it into a $\frac{1}{16}$-in. to $\frac{1}{8}$-in. layer, then pour the topping layer promptly. If the grout dries out, no bond will be formed, and the topping will develop a network of cracks as it dries. The grout spreads easily, so wait until the transit truck is in sight to begin. Any contraction joints in the underlying slab must be repeated in the top layer.

Concrete on wood framing—We often pour thin slabs on wood-framed floors. Joists, of course, must be beefed up to carry the extra load. For a 3-in. slab, we are adding 37 lb. per sq. ft. to the normal value of 50 lb. per sq. ft. of combined dead and live load. This roughly doubles the floor load. You could double the number of floor joists, but we have found that it's more efficient to increase the depth, as shown in the chart below. The chart is for #2 Douglas fir with joists 16 in. o.c.

Span (ft.)	Joists (normal construction)	Joists with 3 in. concrete added
8	2x6	2x8
10	2x8	2x8
12	2x8	2x10
14	2x10	2x12
16	2x12	2x14

Assumptions: live load, 40 psf; dead load, 10 psf; + 37 psf for 3-in. concrete.

When a slab is poured on wood framing, we don't attempt to join the two physically—they remain separate, independent structures. Be-

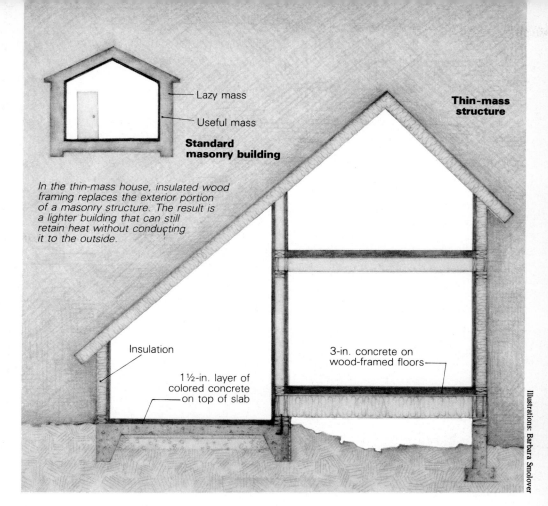

Lazy mass

Useful mass

Standard masonry building

Thin-mass structure

In the thin-mass house, insulated wood framing replaces the exterior portion of a masonry structure. The result is a lighter building that can still retain heat without conducting it to the outside.

Insulation

1½-in. layer of colored concrete on top of slab

3-in. concrete on wood-framed floors

Illustrations: Barbara Smolover

The 1½-in. topping layer of colored concrete is poured onto an underlying slab prepared with a bonding grout. Foam board insulation at foundation perimeters works well in moist locations. Transite panels (cement/asbestos board) placed over the insulation protect it from impact damage.

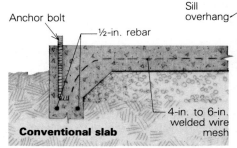

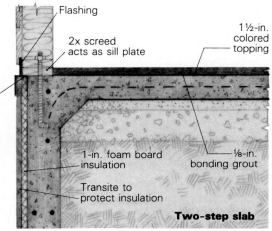

Anchor bolt

½-in. rebar

Conventional slab

4-in. to 6-in. welded wire mesh

Flashing

2x screed acts as sill plate

Sill overhang

1½-in. colored topping

1-in. foam board insulation

⅛-in. bonding grout

Transite to protect insulation

Two-step slab

Mark Johnson

Painted concrete as a form of ornamentation was much in vogue in the 1930s. The geometric designs on this arch in the Los Angeles Public Library resemble mosaics.

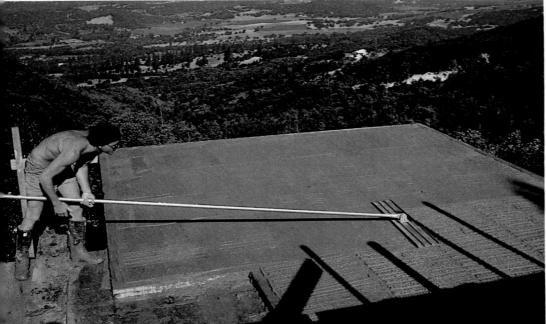

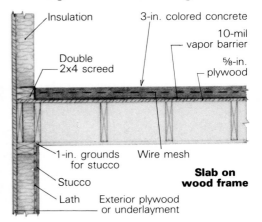

A slab on wood framing requires deep joists to handle the extra load (drawing, below). The plastic membrane keeps concrete from seeping through the plywood substrate. Sheeting wasn't used on the floor above because the underside of this floor is not a finished ceiling. The workers are screeding integrally colored concrete, using the doubled wall plates as guides. Left, the slab has been vibrated into place and is now being floated. Finish troweling will follow.

Insulation   3-in. colored concrete

10-mil vapor barrier

Double 2x4 screed

⅝-in. plywood

1-in. grounds for stucco   Wire mesh

Stucco

Slab on wood frame

Lath   Exterior plywood or underlayment

cause the concrete needs to be reinforced against shrinkage cracking, we use a 3-in. layer with welded wire mesh at its center.

After nailing down the plywood subfloor, we set double 2x screed plates where the walls will go, and lay a 10-mil polyethylene moisture barrier to prevent any moisture in the concrete from seeping through the plywood and discoloring the underlying wood. This step can be omitted if the framing won't be visible from below, but including it will retard the rate at which the concrete slab will dry, and minimize shrinkage cracking.

Spread welded mesh over the floor, and brace the walls with exterior plywood siding or shingle underlayment before pouring the concrete (photos above). To minimize damage to the floor during construction, it's possible to postpone pouring the slab until the structure is nearly complete, but it's difficult to screed the pour level with walls in the way.

Just as the wood floor frames receive a thin coat of concrete, the wall studs are covered on the inside with stucco lath and a 1-in. thick layer of integrally colored cement stucco.

Consistency and pattern—The colored topping layers are thin, so we use small aggregate. As a general rule, the size should not exceed one-third of the slab's thickness. For our 1½-in. slabs, we use ⅜-in. pea gravel.

The recommended proportions of a topping mix are one part portland cement to one part sand to two parts aggregate. Water should be limited to 45 lb. for each 100 lb. of cement. This blend produces a stiff mix with a 1-in. to 2-in. slump. You'll need to increase the water ratio if you have to pump the concrete.

Don't trowel until the sheen has left the surface. This will prevent water and fines (powder and small particles in the aggregate) from rising and causing the floor to "dust" with use.

After troweling, strike expansion joints to control the contraction that occurs during drying. This is especially important in 1½-in. slabs, where we place joints at a maximum of 4 ft. on center, in each direction. Jointing gives you a good opportunity to introduce scale and variety into the floor surface. Using a hand joint edger, you can cut freeform joints or a regular grid using chalklines as guides. Another option is to

use pattern-stamping tools that are pressed into the freshly troweled surface to create brick, stone or tile-like shapes (*FHB* #7, p.6). Ranging from small hand devices to large rolling cages, these tools create a surface of relentless regularity—concrete that looks like it's trying to be something else. We prefer designs that preserve the hand-worked appearance of the jointing. In one example, we began by snapping chalklines in a 3-ft. by 3-ft. grid on the underlying slab. At each intersection, a glazed ceramic tile was placed on a small bed of grout, and tamped down to a 1½-in. height. These tiles became a dispersed screed for the later leveling of the colored concrete. After troweling, joints were struck between adjacent tiles (facing page). We have found, however, that 3-in. reinforced slabs can cover room-sized areas (15 ft. square) without any interior joints and without any appreciable shrinkage cracking.

Curing the finished concrete properly is important because a thin pour can easily dry out too fast. Premature drying will cause hairline cracking throughout the surface. In extreme cases, fissures extend through the pour and

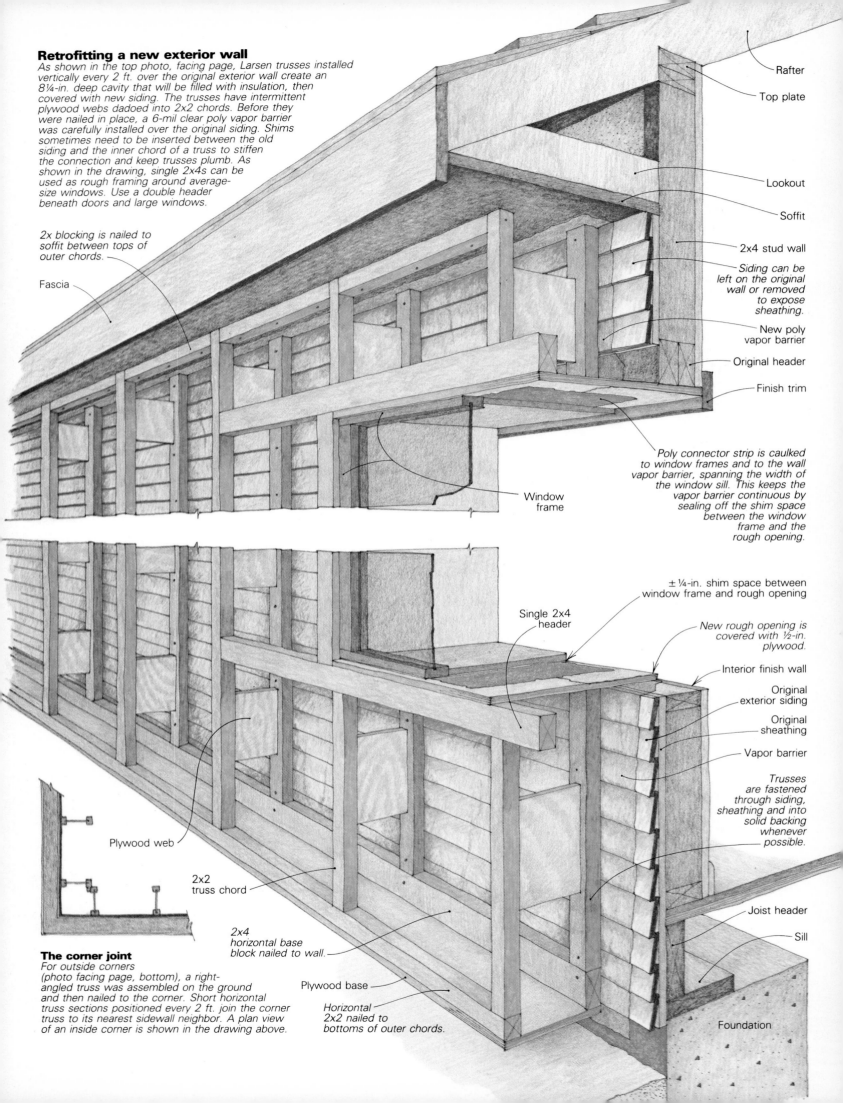

Retrofitting a new exterior wall

As shown in the top photo, facing page, Larsen trusses installed vertically every 2 ft. over the original exterior wall create an 8¼-in. deep cavity that will be filled with insulation, then covered with new siding. The trusses have intermittent plywood webs dadoed into 2x2 chords. Before they were nailed in place, a 6-mil clear poly vapor barrier was carefully installed over the original siding. Shims sometimes need to be inserted between the old siding and the inner chord of a truss to stiffen the connection and keep trusses plumb. As shown in the drawing, single 2x4s can be used as rough framing around average-size windows. Use a double header beneath doors and large windows.

2x blocking is nailed to soffit between tops of outer chords.

Fascia

Rafter

Top plate

Lookout

Soffit

2x4 stud wall

Siding can be left on the original wall or removed to expose sheathing.

New poly vapor barrier

Original header

Finish trim

Poly connector strip is caulked to window frames and to the wall vapor barrier, spanning the width of the window sill. This keeps the vapor barrier continuous by sealing off the shim space between the window frame and the rough opening.

Window frame

± ¼-in. shim space between window frame and rough opening

New rough opening is covered with ½-in. plywood.

Single 2x4 header

Interior finish wall

Original exterior siding

Original sheathing

Vapor barrier

Trusses are fastened through siding, sheathing and into solid backing whenever possible.

Plywood web

2x2 truss chord

Joist header

Sill

2x4 horizontal base block nailed to wall.

Plywood base

Horizontal 2x2 nailed to bottoms of outer chords.

Foundation

The corner joint

For outside corners (photo facing page, bottom), a right-angled truss was assembled on the ground and then nailed to the corner. Short horizontal truss sections positioned every 2 ft. join the corner truss to its nearest sidewall neighbor. A plan view of an inside corner is shown in the drawing above.

convenient thickness because the plywood used to line the rough openings for windows and doors can be ripped from 4x8 panels with little or no waste.

Use your layout lines to align the trusses, and nail them securely to the old wall. Your nails have to be long enough (about 3½ in.) to extend through the 2x2 chord, the old siding (if it's been left on) and sheathing, and preferably into solid backing such as joist headers and top plates. It's easiest to start nailing the trusses at the top and work down. This way, if you have to trim the bottom edge of a truss slightly to make it level with those of neighboring trusses, you can do it standing on solid ground. Nail into the studs in the original wall wherever possible, because solid backing will provide a better hold than the old siding and sheathing alone.

It's important to keep the outer truss chords plumb and vertically aligned so that your siding will go up true. This usually isn't a problem if you're nailing trusses directly to sheathing. But if the trusses are going up over clapboards, as shown in the photo top right, you'll probably have to insert shims here and there between the old siding and the inner chords to keep trusses plumb and in line.

At the base of the wall, the inboard chords of the trusses are cut back 3½ in. and butt against a horizontal 2x4 that's been face-nailed to the wall. On the retrofit shown in the photo, the 2x4 was fastened to the foundation rather than nailed to the bottom edge of the wall. This gives the vapor barrier a chance to stop infiltration around the sill, and also brings the new wall down below the joist headers, a principal heat-loss point in conventional stud-frame houses.

Once the first few trusses are up, you can use them instead of ladders or light scaffolding. Be careful, though. Oversized carpenters or undersized nails might lead to disaster.

At the top of the wall, trusses can simply butt into the soffit or eave if you decide not to nail them directly to rafters or eave lookouts. As long as each truss's connection to the sidewall is solid, there's no need to fuss over eave or gable nailing. But you will need some sort of blocking between the tops of the outer chords (we use 2x2s or 2x4s). This serves as a nailing surface for the siding and also adds lateral rigidity to the top of the wall (drawing, facing page).

Nail a horizontal 2x2 to the bottoms of the truss's outer legs at the base of the wall. This strengthens the lower part of the wall and combines with the inner horizontal 2x4 to serve as a nailing surface for the plywood that boxes in the underside of the new wall.

Detailing at corners—Nailing trusses to the wall is fairly straightforward; making corner joints takes more time. To cope with outside corners, we've found it easiest to prefabricate a corner truss from two straight trusses. One side of this right-angle truss is then nailed directly along one edge of the corner of the house (photo below right). This leaves a gap on the other side of the corner that we bridge

with short truss sections nailed horizontally to the wall every 2 ft.

Inside corners are easier. Basically, we make a right-angled truss and face-nail it vertically to the corner (see the plan-view drawing on the facing page, bottom left). We use vertical 2x2 nailing blocks to connect the outer chords of the truss.

Windows and doors—In most cases I recommend replacing old windows with good-quality insulated units. Using leaky, outdated sash in an extensive energy-conserving remodel just doesn't make much sense, in spite of the initial savings you realize.

Rough openings for windows and doors are framed with headers and rough sills, which are nailed between the outer chords of the trusses on both sides of the opening. Then the entire width of the rough opening from the original interior finish to the outer chord of the truss is covered with ½-in. or thicker plywood. Unless the windows are unusually heavy triple-pane units, you can use single 2x4s as rough framing members. It may be necessary to alter the rough framing in the original wall in order to accommodate the new windows. Allow the usual ¼ in. or so of clearance between all sides of the plywood box and the door or window.

As the window is installed, you need to connect its frame with the vapor barrier that's been lapped around the inner edge of the rough opening. Otherwise, the continuity of the vapor barrier will be broken, and you'll have air leaks in the shim space around the window frame. Caulk a connector strip (about 12 in. wide) to the outside edge of the window frame, locate the window in the rough opening with shims, then seal the free edges of the connector strip to the wall vapor-barrier flaps with caulk and staples. This detail is shown in the drawing, facing page.

Complete the installation by covering the plywood with whatever finish trim you want. It's possible to locate windows and doors anywhere on these broad sills, but most people like to keep windows flush with the new exterior wall. This creates a wide sill inside the house, looks good on the outside and eliminates the necessity of installing a broad, weather-resistant exterior sill.

Insulation and siding—Once the windows are in, it's best to insulate and side the new wall fairly quickly. Unfaced fiberglass batts are the best insulation to use, and you can stuff in as many layers as you need to fill out the depth of the trusses. We often use 24-in. wide by 8½-in. thick fiberglass batts to insulate the new wall cavity. They have to be compressed slightly, but not enough to reduce their insulative value significantly. Install the batts vertically between the trusses. On gable ends, take your wall insulation only as high as the top of the attic insulation.

If you've got easy access to your attic or eaves, it's possible to fill the wall with blown-in insulation after the exterior siding has been installed. If you decide to go this route, be

sure to insulate the spaces beneath windows and doors before sheathing and siding the trusses. Otherwise, these areas may not get enough blown-in insulation.

Some sidings such as horizontal or diagonal boards can be nailed directly to the trusses. Vinyl or aluminum siding, stucco, shingles and vertical boards require backing in the form of horizontal strapping or sheathing. In all cases, use a good wind barrier beneath the siding. Builder's felt is fine; Tyvek (a vapor-permeable wind barrier made by DuPont) is even better, although it's more expensive. □

Illustrations: Frances Ashforth

Retrofit Superinsulation

Old walls serve as a base for a new vapor barrier and a frame of vertical Larsen trusses

by John Hughes

More and more of the new houses being built in Canada are tight and superinsulated. But most of our 6 million or so single-family and duplex dwellings are older, and until recently there's been much less interest in making them energy efficient. The major reason for this is that retrofitting extra insulation and new vapor barriers is far more complicated and frustrating than installing them when a house is being built.

One recently developed approach to this problem relies on straight trusses attached vertically to a house's existing exterior siding. The trusses, which can be fabricated on site or ordered from a lumber dealer, support a new exterior wall and create a cavity for new insulation. In western Canada, they are called Larsen trusses, since John Larsen of Edmonton started to build these truss-based walls some three years ago. Larsen designed his trusses for new construction, but our design firm has found that they are ideal for superinsulation retrofits as well. They'll even work over brick or stucco walls, though fastening the trusses to these materials calls for horizontal wood strapping or masonry nails.

Research in Saskatchewan has shown that cold-air infiltration can be as severe a problem as inadequate insulation when total heat loss is calculated. In most cases, air infiltration can be traced to an absent or inadequate vapor barrier. When retrofitting a wall with trusses, you can install a new vapor barrier over the existing exterior siding before nailing up the trusses. Wrapping the walls of a leaky old house with 6-mil poly usually puts infiltration problems to rest. The 8 in. to 12 in. of insulation in the truss cavities (you can build trusses in almost any width) holds winter heat in and keeps summer heat out.

Constraints—Not every house is an ideal candidate for retrofitting with Larsen trusses. You'll have to check with your building inspector to make sure that making your walls thicker won't violate local setback regulations. And cedar shakes, uneven clapboards or a dual-surface exterior wall that steps out from shingles to stucco at the second-floor level would make proper truss installation difficult. In situations like these, you'd be better off re-

John Hughes is the owner of Passive Solar Designs, Ltd., a superinsulation design firm in Edmonton, Alberta. Photos by the author.

moving the siding and nailing the trusses directly to the sheathing. This strategy also enables you to re-use your siding if it's in reasonably good shape, and if you can keep it that way when you remove it.

Roof overhang is another factor to consider. Ideally, the existing overhang will accommodate the added wall thickness, both at gable and eave. If this is not the case, you have to extend the roof. If you have only a small overhang, then 6-in. wide trusses might work better than 8-in. or 10-in. ones.

If you're superinsulating the walls of an old house, it makes sense to upgrade your attic and foundation insulation too. This can push up the cost and, consequently, the payback time of a retrofit. But doing a comprehensive retrofit is still a lot less expensive than building a new house. You lose the original facade of the building, but its interior can remain virtually unchanged. In some cases, an exterior retrofit is a good way to upgrade the appearance of an older house that is structurally sound but has a dilapidated exterior.

One more thing: The new vapor barrier that is installed around the house will decrease the building's air-change rate substantially. You might need to contact a home-energy consultant who can measure how tight your retrofit is with special pressurizing equipment. If your air-change rate is 0.4 per hour or less, you should install an air-to-air heat exchanger to exhaust stale, indoor air and bring in fresh air from outside.

A new vapor barrier—Installing a continuous, unbroken polyethylene vapor barrier is the best way I know to minimize winter energy waste due to cold-air infiltration, and it's the first step in a Larsen-truss retrofit job. Positioning this vapor barrier against the existing exterior siding or sheathing is bound to raise some eyebrows, since standard construction procedures call for the vapor barrier to be located on the warm side of the insulation. But research in Canada has shown that as long as there's at least twice as much insulation (in R-value) on the cold side of the vapor barrier as on the warm side, no significant condensation will occur in the wall under normal circumstances. This is because the new vapor barrier is still well inside the point in the wall where interior air will cool sufficiently for condensation to occur (the dew point). Even if the house already has vapor-barrier

paint on the interior wall or a layer of poly just beneath it, condensation won't be a problem if the new vapor barrier is located well inside the dew point.

Begin by snapping vertical lines on the exterior walls every 24 in. This gives you the spacing of your trusses. Remove any sharp protrusions from the wall surface that might tear or pierce the vapor barrier. If rafters are also spaced on or close to 24-in. centers, it's a good idea to remove the soffit so you can nail the tops of the trusses directly to rafter ends.

Next, the new vapor barrier is applied. I recommend 6-mil poly that's ultraviolet resistant, since the sheet may be exposed to sunlight for a while after you install it. Use nonhardening sealant to stick the poly sheet to the edges of the house and to seal the joints between the poly sections, which should be lapped at least 6 in. Run a continuous bead of sealant around the top of the house's walls, just below where they meet eaves or gables. Seal the vapor barrier along these edges by stapling through the bead every 6 in. Stapling through a strip of filament tape (preferably with strands running both ways) will keep the poly from being torn loose in windy weather. Repeat this procedure at corners and wherever sheets join. You'll also need to seal the vapor barrier to the foundation, just below where the trusses will stop.

At window and door openings, cut an X in the poly, running each diagonal cut into a corner. This will give you four triangular flaps that should be folded into the house against the old rough opening, after the original windows have been removed. At each corner, there will be gaps that aren't covered by the vapor barrier. Bridge these with small poly strips, caulking and stapling them over the poly flaps and into the rough sill. For more on this technique, see _FHB_ #9, p. 57.

Truss installation—There's no need to use an engineered structural truss to hang the new wall if you can get lighter-weight, less expensive ones, since all you need to support is the weight of the new siding and the insulation. The trusses we use have 2x2 chords and intermittent webs of ⅜-in. plywood. The plywood webs are dadoed into the chords. Most builders in our area use a 210-mm (8¼-in.) deep truss. Adding these to a sheathed, 2x4 stud wall yields a frame that's close to 12 in. thick, not including the new siding. This is a

205

Solar orientation and thermal mass

Type of building	Heat load (therms per year)	Fuel cost (first year)	Value of future heat savings (current dollars)
Non-solar Average orientation, insulation, infiltration	630	$397	—
Solar without mass Proper orientation, insulation, weatherstripping	434	273	$2,416
Solar with wall mass only 1 in. stucco over 2,258 ft.²	334	210	1,232
Solar with floor mass only 3 in. concrete over 1,500 ft.²	248	156	2,296
Solar with floor and wall mass (Lee-Carmichael house)	205	129	2,822

This chart compares the savings achieved through solar orientation and added thermal mass in a house heated by a natural gas-fired furnace working at 70% efficiency, burning fuel that costs 44¢ per therm (100,000 Btus) input. The Lee-Carmichael house, with both floor and wall mass, has a heat load of 205 therms per year. With solar orientation minus the added mass, the heat load would have been 434 therms. Projected over a 30-year period (and adjusted for inflation), the savings in fuel costs amount to $2,822. The source for this analysis is the Passive Solar Design Handbook *(USDOE, Jan. 1980).*

Installation costs for typical floor and wall finishes

	Finish option	Cost of base	Cost of finish	Total cost	Total cost, adjusted for future savings Moderate climate (Calif.)	Severe climate (Wis.)
Slabs on grade 4½-in. slab, 6-in. by 6-in. welded wire mesh over 2 in. sand, over membrane, over 6-in. rock. Edge insulated.	Pre-finished ¾-in. hardwood flooring	$3.97	$4.60	$8.57	$8.57	$8.57
	Mortar-set tile	3.97	4.58	8.55	7.02	6.21
	Mastic-set tile	3.97	4.25	8.22	6.69	5.88
	Carpet	3.97	3.33	7.30	7.30	7.30
	1½-in. integral color topping and tooling	3.97	1.55	5.52	3.99	3.18
	Powder-dusted color and tooling	3.97	.81	4.78	3.25	2.44
	Integral color throughout and tooling	3.97	.39	4.36	2.83	2.02
Framed floors ⅝-in. CDX plywood machine-nailed to 2x8 joists (increased to 2x10 for 3-in. concrete topping), includes blocking.	Pre-finished ¾-in. hardwood flooring	2.15	4.60	6.75	6.75	6.75
	Mortar-set tile	2.15	4.58	6.73	5.93	5.51
	Mastic-set tile	2.15	4.25	6.40	5.98	5.76
	Carpet	2.15	3.33	5.48	5.48	5.48
	3-in. integral color topping	2.41	2.13	4.54	3.01	2.20
	3-in. topping with dusted-on color	2.41	2.61	5.02	3.49	2.68
Framed walls 2x6s 16 in. o.c. includes blocking and plates.	½-in. gypboard taped and textured; two coats of paint	1.37	1.55	2.92	2.92	2.92
	⅜-in. premium redwood paneling over ½-in. gypboard	1.37	2.68	4.05	4.05	4.05
	⅞-in. stucco over lath, integral coloring	1.37	3.70	5.07	4.52	4.48

This chart compares the 1982 installation costs per square foot (San Francisco Bay area) for typical grades of floor and wall finish treatments. Unlike wood flooring, carpet or gypboard paneling, thin-mass finishes help to reduce the heating load, lowering total installation costs. Savings are even greater in a more severe climate, such as Madison, Wis.

tially reduces the need for supplementary heat (top chart), but the incremental worth of each additional pound decreases. This is the law of diminishing returns.

The value of increased comfort levels caused by the all-surrounding nature of the mass is more difficult to measure. A remarkable sense of temperature uniformity results, and heat loads can be so low that it becomes practical to use a woodstove for backup heat.

Costs—Although cost is only one criterion in the choice of a building material, colored concrete is invariably an economical option for a finish floor—especially when future savings on heating bills are subtracted from installation expenses (chart above). Over a slab on grade or a wood platform, a concrete finish can cost less than half the price of a carpeted floor.

Concrete is also less subject to chipping than is ceramic tile. It doesn't require refinishing like hardwood or endless maintenance and eventual replacement like carpet.

The economics of interior stucco are not as compelling. Gypsum wallboard remains more economical, even when the present value of the heat savings is deducted from the cost of installing the stucco. However, if alternative wall finishes (such as plywood paneling) are considered, the stucco is only slightly more expensive, and it will never require painting, varnishing or restaining.

Acoustics and solidity—A thin-mass building is remarkably quiet. This comes as a surprise to some people because they assume the hard surfaces will create an echo chamber. Yet, in terms of reflected sound waves, this building is no different from one with wood floors and gypboard walls. Assuming that each type of room contains soft elements, such as furniture, curtains and occasional area rugs, noises will bounce around in a similar manner.

So where does the sense of unusual quiet come from? The key is the ability of the mass to damp out impact noise almost totally. Gypboard will ring like a drum when struck; the interior stucco will give off a gentle thud. Similarly, footsteps upstairs will transmit readily through a frame floor with a wood or carpet finish. A 3-in. slab upstairs will absorb such impact sounds, creating an effective noise barrier between floors.

Along with the improved acoustics, there is a comfortable sense of solidity. The 3-in. slab on the wood frame creates a much stiffer assembly, something you can sense as you walk over it. This is equally true of the stucco walls; they don't give so much when struck. One of the most frequently voiced complaints about current frame construction is that it's cheap and flimsy. Thin-mass buildings make the opposite impressions. The problem seems to be with the finish materials rather than the wood frame.

Retrofits—Can the benefits of thin-mass construction be built into an older home? Probably not. Most existing homes don't have sufficient south-facing glass for thermal storage to make sense, and few existing homes are insulated and weatherstripped well enough to make the added mass worthwhile. Conservation dollars should first be spent on insulation, weatherstripping, caulking and double glazing. There can also be potential structural difficulties. A 3-in. slab is heavy, and shouldn't be added to wood framing without increasing the number of joists or intermediate beams. Problems can also crop up in earthquake-prone areas if the original structure doesn't have adequate shear bracing. Wall outlets may wind up closer to the floor than the code requires, and changes in elevations at stairs become awkward.

A glance ahead—The concepts we're proposing are best suited for new structures and additions to existing buildings. Economically, thin-mass construction makes sense (especially for floors), and the thin masonry sheath improves acoustics and makes for a feeling of solidity. Finally, the thin-mass technique opens up new aesthetic possibilities and challenges us to introduce vigorous new color and texture into our buildings. □

Max Jacobson is a principal architect in the firm Jacobson, Silverstein and Winslow, Berkeley, Calif.

cause the slab to curl slightly at the cracks. Keep the slab wet and covered with burlap, plastic or waterproof paper for seven days. You can walk on it the second day, but be careful not to tear the protective material. Covering work areas with flattened cardboard boxes is a good hedge against damage.

Coloring options—Masonry suppliers have color charts of a wide variety of pigments. While integral coloring produces the most uniform results, dusting coloring powder over a freshly poured slab and working it into the surface also works well. The dust-on mixture consists of one part portland cement to one part sand and a variable amount of pigment. It is scattered like grass seed at a rate of 60 lb. to 125 lb. per 100 sq. ft. of floor, and it's pressed into the fresh concrete with a wooden float. More pigment gives a more intense color, and spreading the mixture in two separate casts (two-thirds first, then the rest) will result in more uniform coloration.

Painting and staining are options to consider if you don't plan on integral or dust-on color. Many paints and stains have been developed specifically to be used on concrete. Traffic will gradually wear through these finishes, and you'll have to renew them periodically, but protective coatings of clear varnish or shellac will help them last longer and produce a shiny glazed surface. Another trick is to mop the painted floor with a protective coating of starch. When the floor is washed, this coating will come up along with accumulated surface dirt. Mop on a fresh coat of starch to begin the process again.

Thermal performance—Because we use these thin layers of concrete as thermal storage in passively heated buildings, it's crucial that the building be oriented correctly to the sun, insulated well and weatherstripped. Even without any additional mass, such a building will store and release heat better than the average structure, assuming that it can be successfully vented during periods of overheating.

Exactly what are the thermal benefits of adding mass to a building that is already oriented correctly, insulated and weatherstripped? To answer this question, we analyzed a recent design of ours, the Lee-Carmichael residence, in Glen Ellen, Calif. (photo p. 200).

This area has a mild climate, only 2,900 degree days annually. We designed for a low temperature of 29°F. Any benefit we can demonstrate for the thermal mass in this house will be greatly magnified in more severe climates.

The conclusions drawn from our analysis show clearly that it pays to design a house to be passively heated, regardless of whether thermal mass is used. Adding thermal mass substan-

Above right, a 1½-in. thick topping layer of integrally colored concrete poured onto an underlying slab is being worked into a finished surface. Tiles on 3-ft. centers set in mortar before the topping was poured act as a dispersed screed. The finished walls are protected by plastic. After cleaning, the finished floor was sealed (right); a coat of tinted wax will follow.

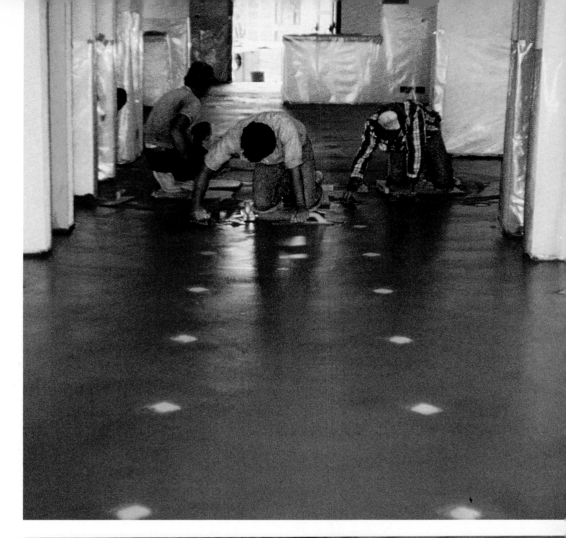

Max Jacobson

The Double Envelope

An architect applies the convection-loop concept with a simple, affordable design

by Joseph Burinsky

The concept of building a house within a house for the sake of solar heating and energy conservation has been called several things since it was introduced in the late 1970s. Terms like envelope, double-envelope, thermosiphoning convection loop and double-shell solar have all been used to describe this idea. All these names refer to the south-facing sunspace and the continuous plenum or airspace that wraps around the house's living space.

The double-envelope concept is based on convection—the natural rising and falling of an air mass as it gains and loses heat. By building two shells, one within the other (thus the double-envelope name), a continuous plenum is created around the house to contain the convective flow of air. During the day, air is heated by the sun in a sunspace on the south side of the house. As this air expands and becomes less dense, it rises into an attic plenum, moving toward the top of the house. As this air cools and is pushed by more rising sunspace air, it sinks down the plenum on the north side of the house, and then is drawn by negative pressure underneath the house and eventually back into the sunspace to be reheated. Along the way, it gives up some of its heat, transferring it through the inner shell of the plenum and raising the temperature of the storage mass beneath the house. At night, the convection loop reverses: Greenhouse air becomes chilled and sinks down into the basement plenum, forcing warmer air to rise up through the north plenum, the attic and back into the greenhouse.

There isn't always detectable air flow in the plenum, but this doesn't mean that the system isn't working. Even when convection slows or stops, the envelope of air that surrounds the inner house continues to buffer the living space from the temperature extremes outside the house.

In spite of the somewhat greater expense of building and insulating a two-shell structure, many people fell in love with double-envelope designs after reading about them in solar and building magazines. The idea of a usable sunspace and a naturally circulating, insulative layer of near-tropical air around the house was very appealing.

But as architects and home owners soon discovered, double-envelope houses don't always perform as they are supposed to. The sunspaces in many houses overheat severely in summer and can't be occupied. In other

The permanent shade hood above the sunspace gives this double-envelope house an unusual appearance, but it protects the sloped glazing and also prevents overheating in the summer. Telltale signs of double-envelope design aren't visible until you get inside the sunspace (facing page). Wood slats in the floor along the wall cover the opening that connects the plenum under the house with the sunspace. Air heated in the sunspace begins a convection current through the attic, down between the double north walls, under the floor slab and back to the sunspace, where the cycle repeats.

houses, daily sunspace temperature extremes during the winter range from below freezing to above 100°F. Insufficient convective air flow in the plenum because of poor loop geometry is another common problem.

Before developing my own double-envelope design, I studied the concept thoroughly, visited several double-envelope houses, talked with designers and owners, collected performance data and attended a number of solar conferences. What I learned was that double-envelope designs must be kept simple so that they can both function well and compete economically with superinsulated and single-wall solar designs.

The Kepner house, shown here, is a slightly modified version of my design study. Built between May and October of 1981, this double envelope is performing very well. The sunspace (photo facing page) is usable year round, and the convective loop has worked well enough to minimize the need for supplemental heat. With three bedrooms (one is still unfinished) and 1½ baths, it was built for around $29 per sq. ft. Even if you consider the good prices the Kepners got on building materials at the time, their house cost them quite a bit less than some of the superinsulated and double-envelope designs I'm now working on (they're closer to $45 per sq. ft.).

Shortly after I met Frank and Tina Kepner, they bought land on the south slope of a ridge overlooking Berwick, Pa. This 6,155 degree-day climate has an annual average of only 49% of possible available sunshine (see *FHB #21*, pp. 60-61 for more on climate and site analysis). Frequently, during the heating season, there's even less sun available—about 20%. These conditions are far from ideal for a largely solar-heated home, so I knew at the outset that this would be a tough test for any double-envelope design.

The crawl-space plenum—Building this house didn't differ greatly from conventional stud-frame construction. As in many double-envelope designs, there are single walls on the east and west sides of the house. The south-facing sunspace windows are double glazed on both inner and outer walls. The north side of the house has double walls, and connects with the sunspace through an open attic space and a plenum beneath the house, which was constructed from 12-in. double-core concrete blocks that are stacked on end with their cores aligned north to south (photo bottom right).

Most of the builders I asked to bid on the house were apprehensive about constructing the plenum, the floor and the foundation that would surround it. Actually, it wasn't very complicated to build. Zane Parker, a local contractor, built the below-grade heat-storage mass with no problems greater than sore hands from lifting the plenum's 1,600 concrete blocks.

First, a concrete-block foundation was laid up atop a reinforced concrete footing. Then Parker installed a 6-mil moisture barrier over the compacted earth between the walls and laid down 2-in. thick extruded polystyrene insulation. A 2-ft. layer of crushed stone followed, and Parker compacted and leveled this heat-storage mass carefully, using a transit, a 28-ft. wood screed and power-driven tampers.

The concrete blocks for the plenum were set in place directly over the crushed-stone base. As shown in the photo center right, the blocks stop 12 in. from both north and south perimeter walls to connect the plenum's sub-floor air passageways with the north-wall plenum and the sunspace.

Directly on top of the plenum blocks we poured a 4-in. thick reinforced concrete slab. This stabilized the block plenum and served as a base for the hardwood finish floor that was later installed over sleepers nailed to the concrete. We insulated the outside walls of the foundation with 2-in. foamboard. The boards are covered with parged-on acrylic cement that's reinforced with fiberglass mesh.

Some designers choose to link the below-grade heat-storage mass with the ground it rests on, tapping on the earth as a source of low-temperature heat (reliably 45°F to 50°F). This is certainly possible, but I feel that this same earth mass can act as an infinite heat sink, actually reducing the temperature of the concrete blocks through which the air circulates. So I chose to isolate the block, slab and stone mass from the earth beneath it with 2 in. of rigid foam insulation.

The double wall—The east, west and north exterior walls are all framed with 2x6s, 24 in. o. c. On the north side of the house, there is an inner 2x4 wall, which forms the 12-in. wide plenum at the back of the house. In theory, even a 2-in. wide plenum should allow for convection to take place. But a 12-in. wide plenum is far easier to build, and can more easily transport a large volume of very slowly moving air.

An important way to encourage convection through the plenum is to minimize barriers or points of friction in the airspace. This means that you don't want many windows or doors in the north side of the house, because the air turbulence caused by their heat loss will disrupt the air flow in the convection loop. I designed only two windows in the north wall of this house, and their jambs and sills are made of slats, rather than of solid wood (top photo, p. 213), so that air can pass through. The only other connections between the walls are four

attic floor joists and four second-floor joists that extend across the plenum to anchor the 2x6 outer wall. All mechanicals and utilities are inside the inner 2x4 wall.

The 2x4 inner wall had to be framed first, since nearly all the 2x10 second-floor joists would bear on it. The inner wall's bottom plate is anchor-bolted to the edge of the 4-in. slab above the block plenum. The 2x6 outer wall bears on the block foundation and was built after the inner wall, like a typical platform frame. As shown in the drawing, facing page, top right, the rafters bear on the outer wall's double 2x6 top plate.

The 2x12 rafters hold R-30 batt insulation between them, with the poly vapor barrier stapled to the rafters and then covered with ⅝-in. drywall. Since the attic floor represents the inner envelope in this design, it too must be insulated, in this case with 6-in. batts. The most important function of the insulation in the inner shell is to reduce night-time heat loss to the envelope.

To finish the walls, Parker and his crew started outside the house and worked inward. First they sheathed the 2x6 wall with 1-in. thick foamboard and nailed up the diagonal T&G cedar siding. Then they friction-fit nominal 6-in. thick fiberglass batts between the outer wall's studs and stapled a 6-mil vapor barrier to the inner face of the 2x6 wall.

Next, the plenum was lined with drywall. In double-envelope construction, the question of fire hazard is always troublesome. Conventional fire blocking in the plenum would destroy the convection loop. One sophisticated and costly approach is to install fire dampers in the plenum. These are fire-resistant flaps hinged and held open by fusable links that will melt at a specific temperature, closing the flap to delay the spread of fire.

Approval of double-envelope designs depends largely on the attitude of the local fire marshall or building inspector. In some instances, officials have interpreted the inner wall as the building boundary and the outer wall as part of the solar-mechanical system, eliminating the need for fire dampers. My approach was to cover the attic ceiling and the plenum walls with ⅝-in. fire-code gypsum wallboard, creating a one-hour fire-resistive wall. Smoke and heat sensors should be installed in the plenum to provide an added margin of safety.

The plenum's finished width of 12 in. allowed the crew a limited degree of maneuverability when positioning and nailing the drywall to the inner face of the 2x6 wall and the outer face of the 2x4 wall (photo top right). On the second floor, Parker left out alternate studs in the 2x4 wall so that material could be passed through more easily. These studs were toenailed in place after all the north wall's drywall was up.

Sunspace and shade hood—Since the sunspace drives the convection loop in a double-envelope house, its design is important. Locating the glazing as low as possible in the loop improves convection, since solar-heated

Double north wall. The 12-in. wide plenum on the north side of the house is created by building two insulated walls. The vapor barrier is stapled to the inside face of the outer 2x6 wall, and the plenum will be lined with drywall.

The hollow-core slab. Concrete blocks, with their cores aligned north to south, form the plenum beneath the house, above. They are dry set between foundation walls on 2 ft. of crushed stone that's been leveled and tamped. Top, a 4-in. reinforced concrete slab is poured directly on top of the blocks. The blocks stop 12 in. from the north and south foundation walls, where the plenum connects with the sunspace and the north-wall space.

Photos this page: Joseph Burinsky

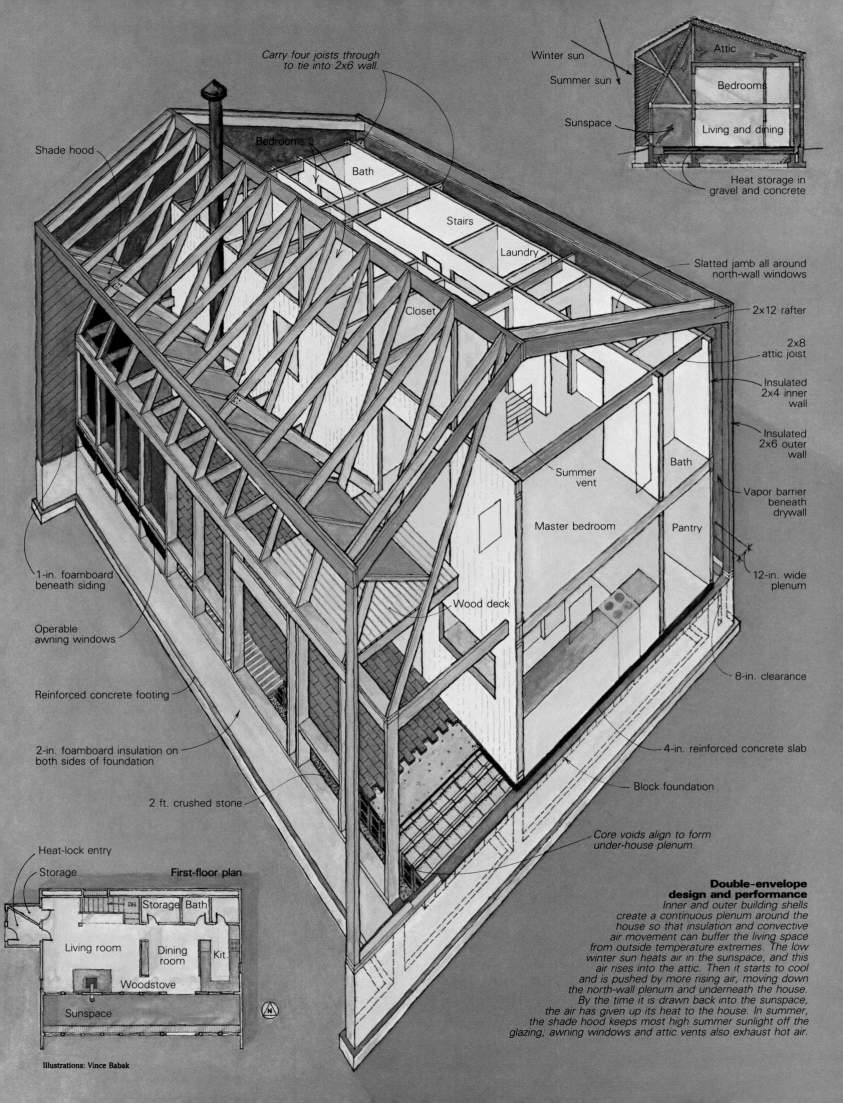

Carry four joists through to tie into 2x6 wall.

Shade hood

Bedrooms

Bath

Stairs

Laundry

Closet

Winter sun

Summer sun

Sunspace

Attic

Bedrooms

Living and dining

Heat storage in gravel and concrete

Slatted jamb all around north-wall windows

2x12 rafter

2x8 attic joist

Insulated 2x4 inner wall

Insulated 2x6 outer wall

Vapor barrier beneath drywall

Summer vent

Bath

Master bedroom

Pantry

12-in. wide plenum

Wood deck

1-in. foamboard beneath siding

Operable awning windows

Reinforced concrete footing

2-in. foamboard insulation on both sides of foundation

2 ft. crushed stone

8-in. clearance

4-in. reinforced concrete slab

Block foundation

Core voids align to form under-house plenum.

Heat-lock entry

Storage

First-floor plan

Storage | Bath

Living room

Dining room

Kit.

Woodstove

Sunspace

Illustrations: Vince Babak

Double-envelope design and performance
Inner and outer building shells create a continuous plenum around the house so that insulation and convective air movement can buffer the living space from outside temperature extremes. The low winter sun heats air in the sunspace, and this air rises into the attic. Then it starts to cool and is pushed by more rising air, moving down the north-wall plenum and underneath the house. By the time it is drawn back into the sunspace, the air has given up its heat to the house. In summer, the shade hood keeps most high summer sunlight off the glazing; awning windows and attic vents also exhaust hot air.

Photo: Joseph Burinsky

Left, two X-shaped braces help support the shade hood above the sunspace. One leg of the X is a 3x8 framing member for the sloped glazing; the other leg is a pair of 2x8s that spans between the hood's framing and the second-floor platform. A small platform, above, overlooks the sunspace.

air will have a good distance to rise, gaining velocity and creating strong convective flow. On a sunny day, you should be able to detect air movement in the plenum.

In this part of the country, there are many winter days when sunlight is more diffuse than direct. The sloped glass (its angle is 65°) performs particularly well on these cloudy or overcast days. Overhead glass also allows plants in the sunspace to grow up and not just out, as they usually do when you've got only vertical glass.

If you decide to build a sunspace with sloped glass, you've got to do something to prevent severe overheating during the summer. In this design, the south-facing roof extends over the sloped glass to shade it during the summer, when the sun is high overhead. My plans for a permanent shade hood caused no small amount of controversy when I was going over the design with the owners. Parker's estimate for extending the south roof, framing the overhang and siding the hood was $2,500. Without the hood, the owners would have to install a hand-operated curtain or awning system, which would cost $1,500 to $3,500. Considering the manual operation and shorter life of the latter option, the shade hood turned out to be a wiser choice. And since the house's completion we've found that its sheathed surface provides additional winter solar gain through second-order reflections off the snow.

The structural support for the shade hood is provided by two X-shaped frames, which can be seen in the photo at left. One leg of each X is actually a 3x8 Douglas fir framing member for the sloped glass. It's fastened to the header above the sunspace's vertical glazing and extends all the way to the ridge, where it's nailed to the ridgeboard. The other leg of the X was made by through-bolting a pair of 2x8s scissors-fashion to the 3x8 with ½-in. dia. bolts. These twin 2x8s are bolted to the second-floor platform and extend to the 2x framing for the south-roof overhang, where they are also bolted in place.

The rest of the sunspace was framed with 3x8 Douglas fir. I designed both the vertical and sloped sections of the frame so that standard insulated patio-door replacement glass could be used for all sunspace glazing. Each tempered-glass panel is 46 in. by 76 in., so the finished opening for each panel should measure 46½ in. by 76½ in. This allows ¾ in. for bearing against stops and ¼ in. for expansion all around the glass panel.

Vertical-grain, all-heart redwood battens compress the ⅝-in. thick glass panels against glazing tape stuck to the redwood stops. The 1x3 battens are fastened with aluminum screws to the glazing stops, which are nailed down the middle of the 3x8 framing members. Sill battens and stops drain through weep slots cut before the wood was installed.

Just above the sunspace floor line and just below the fixed glass, operable awning windows were installed to bring fresh air into the sunspace during warm weather. Stormproof louvers in the attic are opened during the summer to exhaust rising sunspace air that would otherwise cause the house to overheat.

At right, the slatted jamb of this window in the master bathroom spans the north-wall plenum without blocking the envelope's convective air flow. The living room, below right, is framed with roughsawn beams and has a ceiling of tongue-and-groove pine boards. Behind the louvered doors, a wall of insulated glass separates the sunspace from the living room and the dining room.

Finishing up—On the west side of the house, there's a small heat-lock entry with an insulated 2x4 frame. It rests on an insulated slab. The heat lock has two weatherstripped doors—one leads outside and the other opens to the living room. When it's cold outside, very little warm interior air is lost, and not much cold air gets into the living space. Apart from conserving heat, the entry works well as a mudroom.

The owners are fond of roughsawn beams, and we used locally cut roughsawn hemlock beams in the living room (photo below right). Parker designed and built the stairway to the second floor from white oak. Finished flooring on the first floor is also oak. In the sunspace, bricks were dry set with tight joints directly over the concrete slab. A grill of oak slats was built to cover the plenum opening.

Performance—On an average winter day, sunspace temperatures typically range from 50°F to 88°F, while temperatures in other parts of the house are between 66°F and 74°F. This is with moderate sun and no supplemental heat (the house has a woodstove in the living room and electric baseboard heat). So far, the all-time low for the sunspace is 42°F, following four or five subzero days with almost no sun. The owners are both out of the house during the day, but often fire up the woodstove in the morning and evening. They use about 1½ cords of wood per year.

Because of the shade hood, daily highs in the sunspace during the summer average about 5°F lower than outdoor highs. Actually, the highest temperature recorded in the sunspace is 102°F. This occurred in mid-January of 1983, and was due in part to solar gain reflected off the snow. With the awning windows and attic louvers open during the summer, overheating doesn't occur.

It's always possible to improve a project, and if I hadn't been constrained by budget limitations I would have specified night insulation and triple glazing for north, east and west-facing windows. On current versions of this design, I'm improving wall performance slightly by using 1½-in. foamboard insulation against the inside face of the 2x6 wall and plywood underlayment just beneath the exterior siding. And an easy way to increase winter solar gain would be to paint the broad soffit of the shade hood white. In this part of the country, reflective gain off the snow and other surfaces can improve solar performance by as much as 45%. □

Joseph Burinsky practices architecture in Hazleton, Pa., and teaches at the Worthington-Scranton campus of Pennsylvania State University.

The Vapor Barrier

A builder and researcher offers some advice on a controversial and misunderstood subject

by Dan Desmond

In 1807, Charles H. Parry, M. D., of Circus, England, published a scientific paper entitled "On Preventing the Decay of Wood." An obscure subject at first glance, it was of considerable importance to the home owners of the day because it dealt with moisture damage in houses. And Dr. Parry's comments are still relevant because they describe problems that many of today's builders and home owners have to contend with. Parry noted:

"We are frequently mortified by seeing in our houses...that casing, plaistering [sic] and painting fail. The...[interior] moisture penetrates into the [wall cavity], and deposits [water] by condensation on that surface. There is afterwards no current of air to evaporate the water. Such is the process of dry-rot...." Dr. Parry concludes: "The dry-rot may in all cases be infallibly prevented where it is practicable to cover the surface of the...[wall] with a varnish which is impenetrable...by water."

The "impenetrable varnish" that Parry recommended was, in fact, a vapor barrier. And today, the move toward energy-efficient housing has focused a tremendous amount of attention on vapor barriers. Many designers, architects and builders have come to insist on near-perfect vapor-barrier installation, and many questions as to "what kind," "where," "how much" and "by what method" have arisen. The result has been a great deal of confusion, especially among novice builders, who may have read any one of a number of articles predicting an in-the-wall Niagara for those who aren't sufficiently zealous in their adherence to vapor-barrier doctrine.

How did all this come about? Certainly, the lack of a proper vapor barrier can cause damage to a house, as Dr. Parry observed. But who is to be believed, and how do we know that this or that vapor barrier really works? Moreover, how is it that so many fine older homes have vapor protection that is marginal, non-existent or downright backwards, yet stand firm and dry?

I will attempt to answer some of these questions by sharing some of my experiences. For the last 15 years, I have been involved in different kinds of renovation and construction work, and I've seen numerous vapor-barrier/insulation systems. Recently, I've had the opportunity to do specific wintertime testing and data analysis on two public housing units built to HUD minimum property standards. This project involved the computer-assisted measurement of temperatures, dew points, heat flow and condensation in insulated wall sections with and without vapor barriers. More than anything else, this work has taught me to be cautious in assuming that what should happen in theory will in fact happen in practice. Real houses in real-world environments are complicated systems, and concerns over the effects of single systems, like vapor barriers, need to be put in perspective.

Two examples of the need for careful analysis come to mind. Both involve houses I owned which had been retrofitted with insulation some 40 years before I purchased them. The first was a Vermont Victorian farmhouse with mineral-wool insulation blown into stud cavities from the outside, resin paper over the original clapboard siding, and shingles over the resin paper. It had been a busy guest house for many years, so interior humidity from cooking, washing and bathing was continuously high. One would think that after four decades without a vapor barrier, the structure would be on the verge of crumbling away. But when I opened several wall sections during renovations, there was no sign of moisture damage at all; the 2x4 studs were firm and unmarked, nails were only lightly rusted, and the galvanized electrical fittings were as good as new.

The second home, in New Hampshire, had been built in 1840. A section of the kitchen floor had been cantilevered out 6 ft. from the foundation, with the spaces between joists filled with 8 in. of glass wool and covered with tightly stapled tar paper. Again, there was no vapor barrier save for the tar paper, which was on the wrong side of the floor anyway. Opening the cavity to run an outside faucet, I expected the worst, but instead found that the joists and subfloor were in excellent condition, with no sign of moisture damage. While it's risky to generalize from one or two cases, I have found similar conditions while working on many other houses built over the last two centuries, many of which had retrofit insulation or exterior siding.

Am I suggesting that vapor barriers are not really important? No, I'm not. But we do need to look at the problem in a more selective fashion. What these houses and many others have in common is that they leak substantial volumes of air during the winter heating season. In many conventionally built houses, moist interior air can take any one of a thousand paths toward the outdoors. These include leaks in the vapor barrier and cracks in and around windows, doors, floorboards, chimney dampers and attic hatchways. The rate of air change is high in most houses because of these many leaks. Just as warm, moist interior air finds its way outdoors, cold, dry winter air infiltrates, keeping indoor humidity levels low and the potential for damage at a minimum.

Research papers that I've recently read indicate that if you live in a climate with less than 8,000 degree days, any reasonably installed vapor barrier should work just fine. Below 5,000 degree days, there is little evidence that a vapor barrier is necessary, except to control cold-air infiltration.

Leaky construction due to poor vapor-barrier installation isn't limited to pre-1970s architecture. Both job-site experience and pressurized fan tests done by home-energy analysts and consultants tell us that few houses are built as well as they are designed on paper. Vapor barriers aren't easy to install, and in their conventional location just beneath the interior finish wall or ceiling, they need to have openings cut in them for electrical outlets, plumbing stacks and chimneys. Once in place, vapor barriers are more often than not punctured, torn or otherwise damaged unintentionally by electricians, drywallers, carpenters and other workers in the course of construction. This is why the theoretical performance of a vapor barrier usually differs so much from actual performance.

Of course, there are exceptions, and some houses have vapor-transfer problems from day one. It's ironic that the houses built or renovated with the greatest care—supertight, superinsulated new houses or old houses that are retrofitted with new insulation and vapor barriers—are the ones most liable to suffer moisture damage due to an inadequate or improperly installed vapor barrier. Unless these houses have a reliable air-to-air heat exchanger, indoor moisture levels can rise quickly, creating abnormally high vapor pressure. This is just the situation you want to avoid, since it means that minor water-vapor escape routes through vapor-barrier punctures or leaks may become major ones.

Unless you are building a house to exceptional standards of energy efficiency (better than R-24 insulation in the walls and less than

½ air change per hour), you needn't lie awake nights wondering if there is a pinhole behind the walls.

How vapor barriers work—Since water vapor is a gas, it can pass through (permeate) even seemingly dense materials such as plaster, gypsum, wood and masonry. Vapor barriers work by slowing (not eliminating) this vapor transmission, thereby limiting the volume of water that might condense in a structural cavity. Vapor transmission is usually a concern of builders in temperate climates because vapor transfer is caused by the difference between the interior and exterior dew-point temperature, and the difference is not sufficiently large in warmer regions to warrant the use of vapor barriers. In fact, in hot climates, moist, hot outdoor air will diffuse toward the air-conditioned indoors, so north-ern-style vapor barriers would be on the wrong side of the wall.

Up north, the difference between indoor and outdoor winter dew-point temperatures can drive considerable quantities of water vapor from inside the house into the stud cavities, where it will condense into droplets when it reaches any point at or below the dew-point temperature. Where serious condensation occurs, it can saturate and weaken, or corrode some materials. It will promote the growth of mildew and other micro-organisms, and cause paint to blister and peel. Ice crystals will expand in saturated materials, further damaging the structure. In addition to material damage, condensed water can cause a considerable heat loss by conduction through saturated insulation.

The extent to which damaging concentrations of vapor will accumulate in a building cavity depends on the quantity of vapor in transmission, the dew-point temperatures, the amount of cavity ventilation and the degree to which materials can absorb moisture before becoming saturated. For example, dry wood can absorb a considerable amount of water (about 20% by weight) before becoming saturated and prone to damage. Certain insulating fibers, such as fiberglass and mineral wool, are virtually non-absorptive, so moisture won't damage them at all but will only reduce their insulative value.

Cellulose, on the other hand, can absorb a great deal of water. Once it gets wet, it stays wet for a long time. When it does dry, it has a papier-mâché consistency with fissures and voids that easily pass convected air. Needless to say, it's not much good as insulation in this state. I offer this caveat from first-hand experience opening wall cavities filled with cellu-

Glossary

Water vapor
Air is a mixture of gases, and water vapor is one of them. Keep in mind the distinction between water vapor and visible (liquid) moisture. Many materials that appear to repel moisture are nonetheless permeable to water vapor.

Vapor barrier
This term is actually a misnomer, since all vapor barriers are permeable to some degree. In practice, a vapor barrier controls moisture by retarding the rate of vapor transmission, and not by physically blocking every last water molecule it encounters.

Vapor pressure
Every gas in the mixture we call air exerts a force of its own, called vapor pressure. If a volume of moist air is on one side of a permeable wall and a volume of dry air is on the other, the water molecules in the moist air will attempt to reach an equal concentration on both sides of the wall because there is a differential in vapor pressure between the two bodies of air. Diffusion is another term for this process. The concepts of "moist" and "dry" can be confusing, however, and if you are trying to decide where to place a vapor barrier, it is best to apply it nearest the "warm" side of the room. That is because vapor pressure is a function of

temperature as well as vapor content. For example, if the air near a woodstove is 80°F and only 20% relative humidity, it still has better than double the water-vapor pressure of outdoor air at 20°F and 100% relative humidity.

Relative humidity (Rh)
A measurement of the water-vapor content of air at a given temperature. It is relative in that it measures the ratio between the amount of water vapor in the air and the maximum amount of water vapor the air could hold without condensation occurring. The maximum amount of vapor is designated the saturation point of air and equals 100% relative humidity.
A given volume of air can hold more water vapor as its temperature rises. For example, at a given relative humidity, 70°F air can hold about eight times as much water as 20°F air.

Dew point/frost point
Dew point is the temperature to which the air must be cooled to cause some of its water vapor to condense to a liquid. It is important to realize that air doesn't drop all its vapor at once and become bone dry when it reaches a dew-point temperature; it merely drops that quantity of moisture it can no longer hold at that temperature. If the sun

then warms the outer reaches of the wall, the water droplets may again become vapor.
Water vapor always travels from the zone of higher vapor pressure to that of lower vapor pressure. Dew-point temperature is a direct component of vapor pressure, and the difference in dew-point temperatures on both sides of a wall is one way of measuring vapor-pressure differential. The frost point is 32°F, where water vapor condenses to ice crystals.

Permeance
Permeance is the vapor-transmission rate of a material multiplied by its thickness. Permeance is measured in Perms. A vapor barrier with a rating of 1 Perm would allow the passage of one grain of water per sq. ft. of barrier per unit of vapor-pressure differential (expressed in inches of mercury). Water contains 7,000 grains per pound (or pint).
To be considered an adequate vapor barrier, a material must have a Perm rating of 1.0 or less. Four-mil polyethylene has a Perm of .08; insulation-foil backing a Perm of .02, and insulation-coated paper backing a Perm of .2 to .3. One vapor-barrier paint has a Perm rating of .6. Perm figures for other building materials are listed in the chart at right.

Perm ratings for common building materials

Perm	Material
	Vapor barriers
0.00	1-mil aluminum foil
0.06	6-mil polyethylene
0.03	10-mil polyethylene
	Paint and wallpaper
0.5	Aluminum paint (2 coats)
0.6	Vapor-barrier paint (1 coat)
1	Vinyl wallpaper
1	Oil paint on wood (3 coats)
2	Oil paint on plaster (2 coats)
10	Latex paint on wood (3 coats)
20	Ordinary wallpaper
	Foam insulations
1	1-in. urethane
1	1-in. Styrofoam
4	1-in. beadboard
9	4 in. urea formaldehyde
	Fibrous insulations
30	4 in. mineral wool
30	4 in. cellulose
30	4 in. fiberglass
	Masonry
1	4-in. brick
2	8-in. concrete block
3.2	Poured concrete, per inch (1:2:4 mix)
0.12	Glazed tile
	Papers
0.2	Builder's foil
18	15-lb. tar paper (builder's felt)
40	Builder's sheathing paper
0.24	Coated roll roofing
	Other
0.5	½-in. CDX plywood
3	¾-in. thick fir
20	Plaster
50	Gypsum drywall
11	Hardboard
0.06	5-mil acrylic sheet

lose, and from observations made in the aftermath of several fires. If cellulose doesn't get wet, it will work fine, so I'd use it only behind or above a first-rate vapor barrier.

Vapor barriers are generally applied on the warm side of house cavities because these are generally the highest vapor-pressure areas. There are exceptions, however, such as a floor over a damp cellar with ambient temperatures within, say, 15°F of the upstairs. In this instance, the vapor-pressure differential between the cellar and the living space would be small, so you could either eliminate the vapor barrier or put it on the cold side of the joists, below the floor insulation. (If this is your choice, local fire codes may require a barrier with a minimum flame-spread rating.)

The sheet barriers used as ground covers beneath slabs and in basements and crawl spaces aren't really vapor barriers. Their purpose is to isolate wet ground from the building. One thing you have to remember about this type of barrier is that it can work against you if water gets on top of it for any reason.

Attics pose another problem. I use a vapor-barrier paint on the ceiling just below the attic rather than a membrane vapor barrier in the floor cavities of the attic. All roofs and chimney flashings will eventually leak in time, and I don't relish the idea of heavy, wet insulation or a small lake lying unnoticed on the membrane barrier overhead.

Unless you are building a house to exceptional standards of energy efficiency (better than R-24 insulation in the walls and less than ½ air change per hour), you needn't lie awake nights wondering if there is a pinhole behind the walls.

Different vapor barriers—Any vapor barrier with a Perm of 1.0 or less (for an explanation of this and other vapor-barrier terms, see the glossary on the previous page) will be adequate in most situations. Perm ratings for different vapor barriers and building materials are listed in the chart alongside the glossary. Remember that the critical performance factor has been shown to be the quality of installation, and not merely a low Perm rating. Membrane barriers such as foil and polyethylene are usually stapled in place. Some builders overlap the seams where one sheet meets another, and caulk this joint carefully to keep the vapor barrier as continuous as possible. I think this is difficult to justify unless the designer is using the vapor barrier as a primary means of controlling air-change rates.

There are some misconceptions about the effectiveness of foil-backed insulating rolls. The paper "ears" are anything but a gas-tight seal even when carefully stapled in place, yet

there is no evidence whatsoever that houses insulated with this material are prone to moisture damage, and these products have been in use for several decades. I have also had success using a vapor-barrier paint in place of conventional sheet barriers. It is applied directly on interior finished surfaces such as plaster or drywall. The only such interior paint I've used is Glidden Insul-Aid (900 Union Commerce Bank Building, Cleveland, Ohio 44115). It's a water-base primer that contains a vapor-resistant resin formulation and has a Perm rating of .2 to .3. The Enterprise Companies (1191 S. Wheeling Rd., Wheeling, Ill. 60090) manufacture a vapor-barrier paint that can be used as an interior finish paint and has a Perm rating of .6.

Many builders do a good job of sealing up the inside of the house with the polyethylene sheet, but deliberately build the exterior sheathing loose in order to avoid trapping any transmitted vapor inside the cavity. This is all well and good, but heavy, dense winter air can then infiltrate and settle in stud cavities and remove heat by means of convection loops in the wall where warm and cold air rise and fall as though on a vertical conveyor belt. I have measured the heat flow in heavily insulated walls built in this manner, and despite textbook assurances of dead-air space, I have found surprising turbulence and chilling from infiltrated air.

One possible solution to this problem may lie in the use of a specialized housewrap that is permeable to moisture but still able to restrict convection behind the exterior finish wall. One such product is Tyvek, recently developed by DuPont (Wilmington, Del. 19898). It's basically a spun sheet of olefin fiber that is highly permeable to moisture but not to convected air. According to DuPont, Tyvek has a Perm rating of 94 and can transmit more than 70 times as much moisture as 15-lb. builder's felt, but has about half the air porosity of the felt. If products like this come into widespread use, it may be possible to reduce the risk of damage from condensation simply by keeping exterior wall cavities warmer.

Another helpful practice would be the elimination of conventional electrical outlet boxes in vapor-barrier walls, since a polyethylene vapor barrier has to be cut to accommodate these boxes. A surface-mount electrical system can be integrated into woodwork and molding, thereby preserving the integrity of the vapor barrier.

Whatever kind of vapor barrier you use, don't try to make a terrarium out of your house. Supertight houses with no provision for air change can be a serious threat to the health of their occupants. The outgassing from synthetic materials used in construction and furnishings can cause the accumulation of formaldehyde gas, radon gas and other hazardous substances. Humidity, which is largely controlled by air exchange in conventional designs, can rise to uncomfortable levels, promoting the growth of mold and bacteria, and causing moisture damage to the interior surfaces of the house.

Controlling vapor pressure—Regardless of the state of your vapor barrier, there are a few steps you can take to minimize the amount of vapor that might condense in the structural cavities of your house. First, try to keep average house temperatures in the 55°F to 65°F range during cold weather. Cooler air contains less moisture at a given relative humidity, and cannot transmit as much vapor. For example, at 55°F, the water content at a given relative humidity will be about half that of the air in a 70°F room. If you have a special need for a warmer room, limit it to one with a supplemental heater.

Houses without vapor barriers can also suffer damage from excessive humidity brought on by mechanical humidification. Unless absolutely necessary, the relative humidity in such houses should be kept below 40%.

Ventilate attics and crawl spaces. Minimum standards for vents in an attic with a gable roof are 1/300 of the floor area beneath the roof. Crawl spaces require 2 sq. ft. of vents for each 100 linear feet of exterior wall plus ⅓ sq. ft. for each 100 sq. ft. of crawl space. In calculating vents, you must use Net Free Vent area, which is the size of all the openings in the vent, and not the size of the vent itself. Most manufacturers of vents supply a chart listing the Net Free Vent area.

In new construction, do not seal joints of exterior insulating boards like foil-faced urethane. The foil can act as a second vapor barrier, trapping moisture in the wall. Most of the other products used as or with exterior sheathings are permeable enough to allow accumulated wall-cavity moisture an escape route. Resin building paper, felt building paper, tar-impregnated fiberboard and CDX plywood are all acceptable materials, providing the exterior siding is vapor permeable. Permeable siding includes brick, clapboard, wood shingles, stucco and ventilated aluminum or vinyl siding. When in doubt, ventilate the top of the wall cavities. If the exterior of the house shows signs of moisture accumulation—blistering paint, staining or discoloration—consider installing 1-in. circular (button) vents at the tops of the stud cavities.

If you're retrofitting with vinyl or aluminum siding, do not use foil-faced backer board with alleged insulating qualities unless it is vapor permeable. I wouldn't install any siding product that lacks a means of ventilation.

You'll need an air-to-air heat exchanger if your calculated air-change rate is less than ½ per hour. If your heating contractor cannot provide an air-change figure, there are energy analysts who use a doorway-mounted fan to pressurize a house and thereby measure the air-change rate.

Finally, if you're not sure that your house has a vapor barrier, or if you are installing retrofit insulation in an old house, try using a vapor-barrier paint on the warm side of all exterior walls and ceilings. This will be a lot easier than ripping off interior finish walls to install a poly sheet. □

Dan Desmond lives in Lancaster, Pa.

The Balloon-Truss System

An economical superinsulating technique

by John Amos

A few years ago, I became convinced that standard platform framing on top of a concrete foundation was not the way to go if I wanted to build an energy-efficient house. Too many cracks to plug. I also felt that double-wall and double-shell techniques were wasteful and fussy, and, as a result, too expensive. I started looking for a building system that could economically provide a deep, uninterrupted cavity for the easy application of continuous insulation. Just as important, I also wanted to join the above and below-grade walls into one to provide a smooth, unbroken interior surface for the simple installation of a continuous vapor barrier.

What I came up with is a wall-framing system based on modified parallel-chord trusses (available as floor trusses). I use trusses made of two 2x5s, with steel "space-joist" webs that leave plenty of room for insulation between the chords. These trusses are 5 in. wide, 11 in. deep and 18 ft. high. The lower eight feet are of pressure-treated foundation lumber.

A number of builders and designers have begun looking at trusses as a way to get thick wall cavities that can be heavily insulated (see *FHB #5*, pp. 54-57, *FHB #19*, pp. 69-71, and pp. 205-207 of this volume). My technique goes a bit further, though, because I'm after more than good insulation. My system borrows spacing (32 in. o. c.) from post and beam, allowing windows and doors to be hung without the need for headers and jacks.

As its name implies, the balloon-truss system takes from balloon framing the concept of running studs continuously from sill to roof with no interruption at the second floor. These trusses have to be 18 ft. high because they extend all the way down to the footing—a simple pressure-treated sill resting atop a level gravel pad. From platform framing, I took the idea of assembling the walls on the ground. For speed and efficiency, I even install windows, doors, siding and basement sheathing before calling in a crane to lift the walls into place (photo right).

The first house I built using this system went up in 25 days and cost $23 per sq. ft. through drywall. It has 1,536 sq. ft., and a

Raising the walls of a truss-framed house is a one-day project. The doubled cable from the crane is hooked onto a 3-in. angle iron 20 ft. long, which is secured by blocking nailed to the inside of the wall trusses.

Photo this page: George Dashner

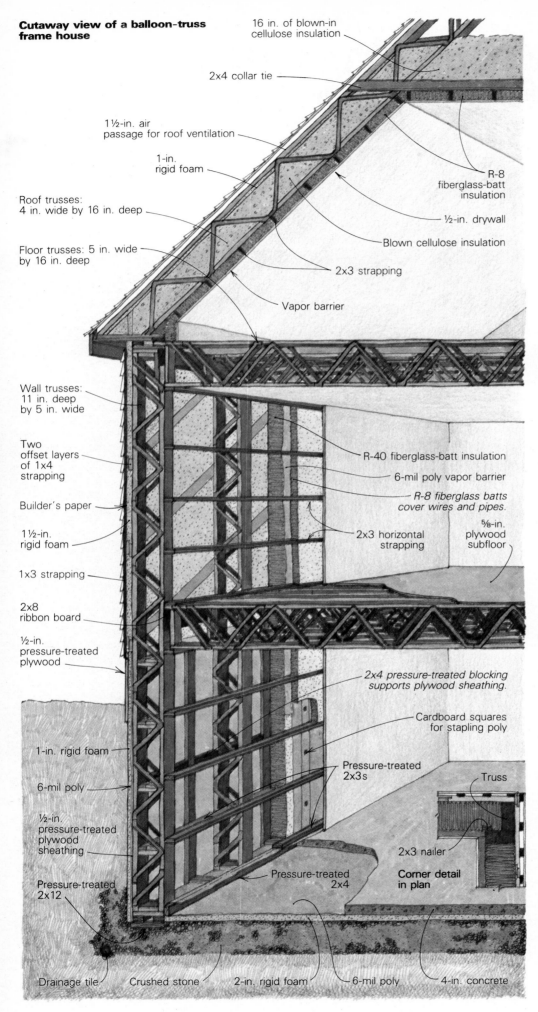

Cutaway view of a balloon-truss frame house

16 in. of blown-in cellulose insulation

2x4 collar tie

1½-in. air passage for roof ventilation

1-in. rigid foam

R-8 fiberglass-batt insulation

Roof trusses: 4 in. wide by 16 in. deep

½-in. drywall

Floor trusses: 5 in. wide by 16 in. deep

Blown cellulose insulation

2x3 strapping

Vapor barrier

Wall trusses: 11 in. deep by 5 in. wide

R-40 fiberglass-batt insulation

6-mil poly vapor barrier

Two offset layers of 1x4 strapping

R-8 fiberglass batts cover wires and pipes.

Builder's paper

2x3 horizontal strapping

1½-in. rigid foam

⅝-in. plywood subfloor

1x3 strapping

2x8 ribbon board

2x4 pressure-treated blocking supports plywood sheathing.

½-in. pressure-treated plywood

Cardboard squares for stapling poly

Pressure-treated 2x3s

Truss

1-in. rigid foam

6-mil poly

½-in. pressure-treated plywood sheathing

2x3 nailer

Corner detail in plan

Pressure-treated 2x4

Pressure-treated 2x12

Pressure-treated 2x4

Drainage tile   Crushed stone   2-in. rigid foam   6-mil poly   4-in. concrete

maximum heating load of 7,000 Btus. It cost $104 to heat with electric resistance at $.04 a kwh in an 8,900 degree-day heating season.

Since then, I've built five balloon truss-frame houses. I contract my services as a house designer and project manager to the home owner, who becomes the general contractor. There's nothing fancy about the work, so we usually hire low-cost, unskilled local labor to assemble the house—sometimes the owner's friends and relatives. We normally complete the house in 30 days to the drywall stage, using the suppliers to finance the construction. The mortgage draw due at this stage pays the suppliers. Only the relatively low labor cost has to be financed or paid for out of pocket. Each house is custom designed, and the basic frame components are fabricated by a truss manufacturer, who delivers them to the site.

The houses don't have to look unusual. Last July, we built a 2,736-sq. ft. house for Hal and Millie Stone in Oxford Junction, Nova Scotia. Although some of the work-in-process photos in this article are of other houses going up, this is the house I'll be talking about.

The Stones chose a rectangular design because of its inherent energy efficiency—that is, its low ratio of window and wall surface area to interior floor space. We calculated that heat from passive-solar gain, mechanical systems, lights, cooking and the residents themselves would total 4 kilowatts in an airtight version of the Stones' house. Even at a design temperature of −10°F, we could reduce heat loss to the same figure by installing R-55 insulation in the walls, R-70 in the ceiling and R-10 below the floor, and by allowing for the 70% recovery rate of the air-to-air heat exchanger. A hundred and fifty dollars worth of portable electric heaters spotted where they were needed around the house would maintain the house at 70°F.

Our excavation was 6½ ft. deep, and its perimeter was several feet larger than the 28-ft. by 40-ft. dimensions of the house. We leveled all the excavated dirt and rocks around the hole so we'd have a relatively flat platform to set up on to build our wall sections. A 12-in. deep bed of ¾-in. to 1½-in. crushed stone was compacted over the undisturbed soil at the base of the excavation. Then we installed our footing—a simple pressure-treated 2x12 leveled and squared on top of the stone to act as a locator for the walls, as shown in the drawing at left. Finally, we buried 4-in. drainage tile in the crushed stone 1 ft. outside and below the footing.

Building the walls—We staked floor trusses to the ground to act as a big worktable where we could build the wall and gable-end assemblies (photo facing page, top). We set wall trusses on 32-in. centers across them. The end-wall trusses were laid out so that there would be one at the end of each wall. Front and rear-wall trusses were held back so that sheathing and siding would overlap them by 11 in., and the front and back walls would overlap the side walls (detail, drawing at left,

Assembling the truss walls. **Wall sections are put together on the ground, top, with floor trusses spiked in place to act as an assembly table. Wall trusses are spaced 32 in. apart. Windows, doors and siding are all installed before the walls are lifted into place. Front and rear-wall trusses, left, are held back 11 in. so the corners can fit together as shown. The footings, above, are pressure-treated 2x12s squared and leveled on top of a foot-deep bed of crushed stone. The walls on top of them are tugged together with come-alongs. The photo above also shows the pressure-treated shoe at the bottom of the wall, and the blocking that supports the basement sheathing.**

and photo above left). We also adjusted the spacing for doors and windows where necessary, and then secured a plate to the top of the truss studs and a shoe to the bottom. Both were of ½-in. plywood 11 in. wide, spanning two 2x4s, but the shoe was pressure treated to match the lower 8 ft. of the wall trusses. We didn't use trusses for the triangular gable ends. They were framed with double walls of horizontal 2x4s on 32-in. centers.

To prepare for the basement sheathing, we installed pressure-treated 2x4 and 2x3 blocking and cribwork as support and nailers (photo above right). We checked the walls for square by measuring diagonally from corner to corner, and we straightened the plate and shoe using a string as a guide. Then we

sheathed the below-grade portion of the wall with 4x8 sheets of ½-in. pressure-treated plywood, and sealed joints between the sheets with acoustic sealant.

Above grade, most of the windows and doors were designed to fit between the wall trusses, so they didn't need any headers or jacks. However, the double doors and bay window had to be framed up in the traditional manner. After this was done, we sheathed the walls diagonally with 1½-in. thick 2x8 rigid foam sheets. In Canada, these come with ½-in. shiplapped edges. To provide lateral bracing for the frame, we staggered pairs of 1x4s, and ran them along the foam's shiplaps. Builder's paper and vertical 1x3 strapping on 10½-in. centers (nailers for the siding) went over the

foam. Then we attached the 6-mil poly vapor-barrier sleeves to the window and door frames (see *FHB* #9, p. 58). We next installed the airtight triple-glazed windows and the insulated steel doors with magnetic seals. Finally, the vinyl double 4 siding was installed.

House raising—After the main walls and the gable ends were built on the ground, a crane was brought in to erect them. In one afternoon the four walls, second-floor joists, and gable ends were raised. To snug the corners together we used come-alongs at the top and bottom of the walls to slide the front and rear walls the last few inches along the sills (photo above right). We nailed the walls together at their tops and bottoms, and secured the bases

Illustration: Frances Ashforth; Top photo: Don Hemmings

Insulation. The whole interior of the house from footings to second floor is insulated at once, and sealed with an uninterrupted vapor barrier, left. The first-floor trusses (facing page, left) are installed on a ribbon board nailed to a plywood strip inside the vapor barrier. Under the roof on the second floor (facing page, right) a single 32-ft. by 56-ft. sheet of poly is stapled in place under cardboard squares. Later, 2x3 strapping will be installed and insulation blown into the spaces between the vapor barrier and the roof sheathing.

of the gables to the end walls beneath. (Sometimes, a small gap—never much larger than ¼ in.—remains because of a bow in a wall. When this happens, we reposition the come-along and tighten things up.) Once everything was in place, we simply nailed the corners together with 16d nails every 6 in. or so.

Because you can't tie up a crane at $60 an hour while you get things just right, we left fine-tuning until the next day. We covered the outside of the foundation sheathing with 6-mil poly. One-inch rigid foam over the plastic protected it from damage during backfilling.

Floors and roof—With the walls up and plumb, on raising day we installed the second-floor trusses (these are similar to the wall trusses, but 16 in. deep), also with the crane. To ensure the integrity of the vapor barrier, we installed a 2-ft. wide piece of 6-mil poly all around the top of the wall before these floor trusses went in. This strip of plastic would later be lapped over poly in the roof, the gable ends and in the rest of the wall. The trusses, which we installed 16 in. o. c., were sized to span the entire 26-ft. width of the house, obviating the need for interior bearing walls, and minimizing connections to the exterior walls.

The roof went up next, a piece at a time. Rafters are also space-joist trusses, 16 in. deep on 32-in. centers. We installed collar ties 8 ft. 4 in. above the second floor. Then the soffit and fascia board were cut and placed. Because the walls were already sheathed and sided, we didn't need to erect staging for this. We sheathed the roof with ⅝-in. T&G plywood, and installed foam insulation stops to ensure ventilation at the eaves.

Insulation—With the house weathered in, we stuffed the walls below the second floor with insulation before installing the first-floor trusses. Like the second-floor trusses, those on the first floor wouldn't penetrate the insulation cavities between the wall trusses. The crew wove 8-in. by 24-in. pieces of R-20 friction-fit batts between the webs. The 27-in. by 11-in. cavity between the trusses was filled with two layers of 24-in. by 48-in. R-20 batts and a 12-in. by 48-in. batt placed on edge. Over this, the crew ran 20-ft. wide poly all around the 18-ft. high, 136-ft. inside perimeter of the structure, and used Tremco acoustic sealant along the seam with the 2-ft. wide poly band around the second-floor joists. Insulation and vapor-barrier installation took a day.

The next morning, we installed a 1-ft. wide band of ¼-in. plywood around the walls over

the vapor barrier where the main floor was to be attached. The plywood protected the plastic from punctures. We nailed a 2x8 ribbon board over the plywood with eight 3½-in. nails into each wall truss. The floor trusses were set on this board, every other one aligned with a wall truss to brace it against the thrust of the backfilled earth (photo above left).

Next, we installed the vapor barrier in the gable ends, slope, and ceiling of the second story. To install and support a single 32-ft. by 56-ft. sheet of plastic overhead, we chalked a line, corresponding to the centerfold of the poly, on the gable ends and across the ceiling. The fold of the plastic was stapled, through 2-in. square pieces of cardboard, along this centerline. The plastic sheet was then spread out and stapled over the slopes and ceiling (photo above right). After nailing 2x3 strapping over the poly on 16-in. centers perpendicular to the rafter trusses, we insulated the roof and the gable end above the second floor with blown cellulose. The strapping supports the insulation's weight in the roof.

The crushed-stone floor on the lower level was dealt with last. We covered it with a layer of 2-in. rigid foam, then unrolled a 28-ft. by 40-ft. piece of 6-mil poly over it, running its edges up the walls. Next, we set 2x4s on edge around the perimeter of the floor and poured our slab level with their top edges. When the concrete set, we folded the edge of the floor plastic back on itself over the exposed 2x4 edge, ran a bead of Tremco acoustic sealant along it, pressed the long edge of the wall vapor barrier into it, and sandwiched the resulting seam between the 2x4 below and a 2x3 above. The sleeves of 2-ft. wide poly that had been sealed to the window and door frames before their placement in the walls were all sealed to the main-wall vapor barrier, as were the air-intake and exhaust pipes for the heat exchanger and the sewer, water and electrical pipes when the time came.

After we installed horizontal 2x3 strapping

on 2-ft. centers over the 6-mil poly, a blower door was used to depressurize the house. Used in this way, the blower motor has a vacuum effect, so that easily noticed jets of air are drawn in through holes in the membrane. After these holes were patched, the house registered .38 air changes at 50 pascals.

A house this tight needs a heat exchanger. I had one built for the Stones' house that uses two two-speed squirrel-cage fans. The high speed of the fans is used to vent the clothes dryer and the Jenn-Air range top, and to power the central vacuuming system. The Jenn-Air has its own fan, and requires the fresh air brought in by the heat exchanger. The low speed is used to vent the house as a whole. The heat exchanger is activated by relays from the dryer, range and vacuum system, and by single-pole humidistats placed in the bathrooms and kitchens.

Wiring and plumbing on exterior walls— The 2x3 strapping over the vapor barrier created a 2½-in. cavity where we could install electrical boxes and 1½-in. dia. plumbing vent pipes without penetrating the poly. We installed wiring and pipes simply by pressing in the vapor barrier and sliding them behind the strapping. We then pressed 23-in. by 48-in. R-8 friction batts in the 2½-in. cavity. Besides added insulation, the fiberglass protects the poly from punctures. Finally, we screwed the drywall to the strapping.

Labor and economics—Throughout early construction, the eight workmen worked in groups of two. Each pair was given a task, and would complete it on each of the six major sections of the house. Since the truss frame members arrived ready-made, most construction work was limited to measuring and nailing. Because much of the work was done at tabletop height, it proceeded with speed, precision and safety.

Hal Stone and his son Philip set aside six

weeks of summer vacation to work on their home. Their contribution reduced the average wage to $5.50 per hour. In all, $10,000 went for labor up to and including the installation of the drywall. Floor troweling, wiring, blown insulation and interior finish were extra.

Costs different from those of a standard platform-frame house were as follows. The complete truss frame—wall studs, floor joists, roof rafters, gable ends, wall plates and shoes, and all the pressure-treated wood for below grade—was delivered for $11,500. (By comparison, a poured-concrete foundation alone would have cost $7,000.) The total insulation cost was $6,200. This included the exterior sheathing, below-grade protection against dampness, and insulation stops in the rafters. The rolls of 6-mil poly and the acoustic sealant cost $325. The use of the crane added another $420. The total cost for the house, including land, services, garage, appliances and the very high-quality trim the clients asked for, was $104,000, or about $38 per sq. ft.

For the first year, the space-heating load for the house was 5,900 kwh, which cost the Stones $370 (Canadian). The cost should be about ⅓ less in subsequent years, after the materials in the house have dried out.

The "space-joist" balloon-truss frame permitted me to build a high-performance house faster and more economically than other methods of energy-efficient construction. Moreover, the system eliminates having to wait for a foundation contractor, having to carry a skilled crew, and having to pay for, rig and work on staging. The frame, consisting of four truss components—rafter, gable, wall and floor—simplifies the materials list and reduces the number of subs and suppliers. The system is like a giant erector set that requires little more than a pencil, a saw, a hammer, and some help from the neighbors. □

John Amos is a builder in Sackville, N. B. Photos by the author, except where noted.

Ridge-Vent Options
Two site-built alternatives
to the standard aluminum fixture

Editor's note: Until a few years ago, roof venting was pretty much confined to open attics. The target then was heat buildup in the summer. Fixed-louver gable-end vents of wood or sheet metal, or mushroom (eyebrow) vents and turbine ventilators were used on the roof to get air flowing through the attic on days when there was a breeze. Moisture accumulation in the winter wasn't a serious problem in most climates, so venting a cathedral ceiling (where the finished ceiling is nailed directly to the rafters) was seldom even considered.

But with the demand for tighter construction, a growing awareness of the problems trapped moisture can cause, and the threat of ice-damming in heavy snow areas, roof venting has become an essential part of building. Model-code requirements call for a total Net Free Ventilation Area that's ⅟₁₅₀ of the total square footage of the ceiling below the roof. The Net Free Ventilation Area, or NFVA, is the total opening of the vent, usually measured in square inches, after the area taken up by

screening or louvers has been subtracted from the total coverage of the vent. The ⅟₁₅₀ ratio can be reduced to ⅟₃₀₀ if a vapor barrier is used on the underside of the ceiling joists or rafters, or if the vented areas are split equally between low (soffit) vents and high (ridge, gable or roof) vents.

Venting-hardware manufacturers use a ventilation rate expressed in cubic feet per minute flow as a measure, and call for an even greater NFVA: .4 to 1.0 cfm per sq. ft. of floor area for winter moisture removal, and 1.5 to 2.0 cfm per sq. ft. of floor area for summer heat removal (see FHB # 7, p. 52). But many builders feel this much venting isn't necessary, given the widening use of vapor barriers and the leaky nature of even the tightest wood-frame house.

Our mail shows that our readers, both novice and professional builders, are trying to get a handle on when venting is necessary, how much should be supplied, and how best to build it in. The most effective way to vent a gable roof is to combine strip soffit vents and a continuous ridge vent. This

system makes use of the fact that warm air rises, establishing a convection current from the eaveline to the ridge. It is also the only way to vent a shed or gable roof when the finished ceiling is nailed to the bottom of the rafters.

The standard fixture for the ridge is an extruded-aluminum vent that runs continuously along the peak. But a lot of our readers think manufactured ridge vents are irritatingly conspicuous as the crowning touch of a roof, especially when the vents are dented—a not infrequent occurrence considering the light-gauge metal (about .019 in.) from which they are made.

Two such critics are Doug Amsbary, a building contractor in Franconia, N. H., and Eric Rekdahl, an architect and builder in Berkeley, Calif. Amsbary remarked that he had seen too many vents bent and battered by snow shovels and chimney sweeps' ladders. He was looking for an alternative that was more attractive and durable for the custom houses he builds.

In the mild climate of California,

where continuous ridge vents are less common, Rekdahl found himself needing to vent cathedral ceilings without using add-on hardware that would detract from his designs. Both of these builders have come up with alternative ridge vents made from standard lumber and roofing material, and have used them on houses they built last year. Their designs, quite different in style and structure, are detailed here. Like most site-built solutions to new problems, these vents are a first-generation response. Neither contains a full wind baffle, which can be important in high wind areas to prevent the convection currents from the soffit to the ridge from reversing, and to keep blowing snow and rain out. However, both Amsbary and Rekdahl are satisfied so far that insulation and sheathing are staying dry. There are lots of ways to solve venting problems; and these two designs will undoubtedly be refined as more are built. One reason for presenting them is to offer others a starting point for their own solutions. —Paul Spring

---

by Doug Amsbary

Manufactured aluminum ridge vents are easy enough to install, and they do their jobs well. No argument. But I take a lot of care with the custom homes I build, and I set about last summer to design a ridge vent that I could make on site that would be sturdier and more attractive. I started with blocks that are toe-nailed to the rafters at the peak, and incorporate a commercial aluminum strip vent 2¼ in. high. The rest is just sheathing, drip edge and asphalt shingles. The materials cost just a little more than an aluminum ridge vent (see the comparison chart on the facing page), but there's quite a bit more labor in my design. After fabricating four of these site-built vents this summer, though, I decided it was worth the extra time.

These vents can be used on either new work or renovation. Make sure that there will be room (1½ in. to 2 in.) between the insulation and the sheathing for a continuous flow

of air from the soffit vent at the bottom of each rafter bay to the ridge on top.

On a retrofit, the first task is to strip off the ridge shingles and the first few courses below on either side of the peak. Then, carefully lay out where the bottom edge of the blocks will be toenailed to the rafters (the blocks I have been using are 7¾ in. from the long point of the pitch cut at the ridge to the square cut on the downhill end). Add ⅟₁₆ in. to ⅛ in. to your calculations when cutting the sheathing back so that there is enough room to slide the vent strip into place. Take several measurements along the ridge in case the peak isn't straight. Now snap a line and make the cut on the sheathing. If the sheathing isn't plywood, small nailers might have to be added to support the sheathing boards where the vent terminates next to a rafter. On new construction, wait to sheathe the upper portion of the roof until you've installed the vent strip. Otherwise the sheathing will get in the way of nailing the bottom flange of the aluminum venting.

Ridge blocks—I use 2x softwood for the ridge blocks. They are nailed directly on top of the rafters, and must extend above the sheathing exactly 2¼ in. Before cutting all of the blocks, give the first pair a trial fit at several spots on the ridge so that you can adjust the plumb cut if necessary. Use fairly clear stock for the ridge blocks so that split-out is minimized. Make a few spares while you're at it. In fairly dense framing wood, predrilling may be necessary.

The blocks should be toenailed into the existing rafters. I use four nails for each block—one on each side of the block, one on the downhill end, and the last one driven down from the top of one block into the other. It is a good idea to install the end sets of blocks first and then string a line between them. I found that if the rafters didn't join the peak either at the same height or along a straight line laterally, the precut blocks tended to pick up the sweeps and dips. This makes it frustrating to install both the vent sheathing and the vent

Material cost breakdown	
Standard ridge vent	**$22.49**
10-ft. length of aluminum vent	18.00
1 connector	1.90
1 end cap	.80
10 oz. asphalt roof cement	1.79
Site-made ridge vent	**$26.09**
2 pcs. 10-ft.1x10 #3 pine	6.64
Ridge blocks—12-ft. 2x3 spruce	1.92
2½ 8-ft. pieces #SV202 @$1.45	4.10
2 pcs. 5-in. galvanized drip edge	2.90
8d and 16d nails @ $.40 lb.	.20
1-in. galv. roofing nails @ $.65 lb.	.38
1 bundle ridge shingles	9.95

Based on suppliers' prices in 1983 for materials bought in quantity. Comparison based on 10 lin. ft.

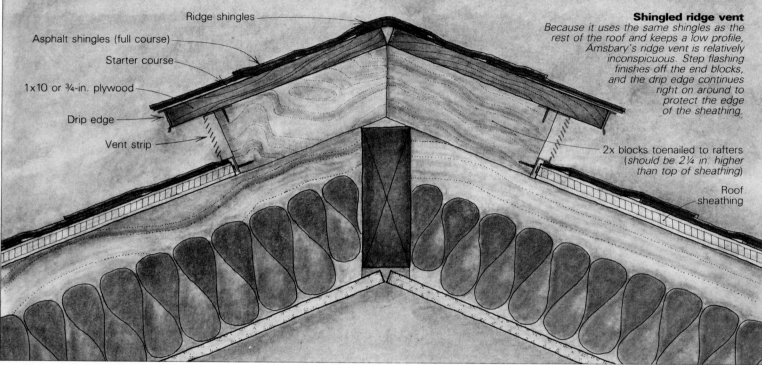

Ridge shingles

Asphalt shingles (full course)

Starter course

1x10 or ¾-in. plywood

Drip edge

Vent strip

Shingled ridge vent
Because it uses the same shingles as the rest of the roof and keeps a low profile, Amsbary's ridge vent is relatively inconspicuous. Step flashing finishes off the end blocks, and the drip edge continues right on around to protect the edge of the sheathing.

2x blocks toenailed to rafters (*should be 2¼ in. higher than top of sheathing*)

Roof sheathing

strip itself. A bit of planing or ripping on the blocks may be necessary at this point if the ridge is a real roller coaster. In any remodeling, compromise is the key, and a string line will help you define that compromise here.

Vent strip—The aluminum vent strip I use— Model SV 202 by Air Vent Inc. (6907 N. Knoxville Ave., Peoria, Ill. 61614)—is intended for soffits, but with its low profile and small (⅛-in.) louver openings, it works well for the ridge. This vent strip comes only in a mill finish, but the pieces can be spray-painted to match the roof color.

You can use either staples or 4d box nails to secure the top and bottom flanges of the vent strip. On new construction you will be able to nail the bottom flange directly to the block ends, if you hold the sheathing back until you're finished. On a retrofit job, you will only be able to catch the edge of the roof sheathing by nailing from inside the vent between the blocks.

Vent strips come in 8-ft. lengths. Lapping the pieces isn't necessary, but do butt them over the nearest set of ridge blocks. Make

sure your vent strip extends slightly (1/16 in.) beyond the end set of ridge blocks so the step flashing you'll be using as a finished end cap will fit tightly beside the vent strip.

Sheathing and roofing—The next step is sheathing the little gable you've created. You can use either ¾-in. plywood or 1x softwood solid stock. So far I've cut these pieces 10 in. wide to the long point of the peak, but 12 in. might help cut down on any possibility of driving rain working its way through the vent strip. It would also cut down on the number of times I accidentally kicked the vent strip after it was installed.

Whatever width sheathing you use, make sure that it overhangs the blocks by at least 1 in. around the entire perimeter of the vent. To protect the edge of the sheathing, install either galvanized or aluminum drip edge using ¾-in. roofing nails. This kind of attention to detail during construction will mean that the vent won't require any more maintenance than the roof it sits on. Next apply a starter course of asphalt shingles on both sides of the vent—use the top portion of the shingle

ripped down to include the self-sticking asphalt adhesive. Then nail on the first (and only) course of shingles on each side of the vent, being careful to line up the cutouts with the pattern already established on the roof. Cap this peak with standard ridge shingles laid toward the prevailing wind, and you're finished with the top of the vent.

As you approach the bottom edge of the vent from each side of the main roof, be sure that you hold back the next-to-last course of shingles about an inch so that the last course can fit tightly beneath the ⅜-in. offset at the vent strip's bottom edge. These last shingles will have to be face-nailed. Use a spot of roof cement on the top of each exposed nail. Also run a small bead of plastic roof cement along the juncture between the vent strip and the short top course of shingles to keep everything dry underneath.

To seal off the exposed outside face of the last set of blocks, use step flashing, or make up the necessary pieces using colored coil stock, aluminum or copper. Cut this flashing so that it fits tightly against the underside of the ridge-vent sheathing. □

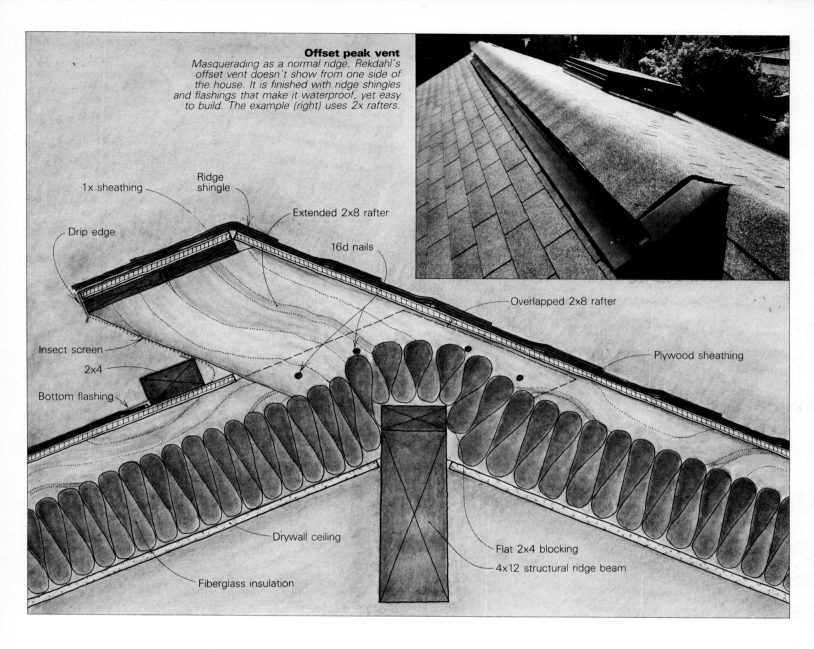

Offset peak vent
Masquerading as a normal ridge, Rekdahl's offset vent doesn't show from one side of the house. It is finished with ridge shingles and flashings that make it waterproof, yet easy to build. The example (right) uses 2x rafters.

1x sheathing
Ridge shingle
Drip edge
Extended 2x8 rafter
16d nails
Insect screen
2x4
Overlapped 2x8 rafter
Bottom flashing
Plywood sheathing
Drywall ceiling
Flat 2x4 blocking
4x12 structural ridge beam
Fiberglass insulation

by Eric Rekdahl

Last winter I was faced with designing a two-story addition whose roof was limited in height from the outset by the height of the existing one-story roof nearby. Even using a slab on grade for the addition's ground floor to stay as low as possible, I had to nail the gypboard ceiling in the addition's partial loft directly to the underside of the gable roof rafters to get the required head clearance. This plan left me no choice but to vent the roof from the soffit up through the ridge. I began experimenting at my drawing board with a way to incorporate a ridge vent without breaking the plane of the gable roof on at least the one side that faces the yard, and without adding a lot of conspicuous hardware. The solution I came up with creates a continuous ridge vent by lapping the rafters coming from each side, and letting one side run long beyond the normal roof peak. These long rafters are cut off at the same angle as the return side of the roof. Since the top edges of these rafters are higher than anything else on the roof, they create an offset ridge. With stan-

dard ridge shingles capping this peak, it looks like a normal gable from one side of the ridge. But on the other side, the extra rafter length creates a miniature clerestory, whose intervening spaces provide the venting.

A nice side benefit of the overlapping roof rafters is the increased resistance to structural separation. Each rafter has full bearing on top of the structural ridge beam, and by spiking them together at the top, they interlock nicely around the beam. To allow an unimpeded flow of air from the soffit vent at the bottom of the rafters to the ridge vent at the top, we nail 2x4 blocking flat on top of the ridge beam, and use metal crossbracing to hold the rafters in position at midspan.

The vent detailing is simple and inexpensive. First we nail 1x sheathing to the end of the extended rafter (this is the short return pitch of the offset peak). Once the screening is stapled down, this area is built up with ½-in. plywood. A 1½-in. by 2-in. drip edge is then nailed to the top of the plywood to protect its edges and the sheathing below it.

The screening is simply ¹⁄₁₆-in. insect screen (we used a dark-colored fiberglass screening)

available off the roll from most hardware stores. Code in my area now requires ¼-in. insect screen to meet Net Free Ventilation Area needs, but this size mesh will keep only large insects out. We start the screening high on the 1x sheathing of the little return, and stretch it down across the opening and onto the slope of the main roof. We use a staple gun or hammer tacker, and tack it every few inches.

The bottom of the screened openings requires a flat 2x4 nailed to the slope of the lower, main roof and pushed tightly against the bottom edge of the extended rafters, as shown in the drawing. The undercourse of 15-lb. felt I use with asphalt shingles on top of the roof sheathing is wrapped up over this 2x4. It's capped with a simple flashing that begins at the top with a ¾-in. lip that sits in front of the screening. This prevents blowback—water that has fallen on the flashed surface being driven back toward the vent opening by the wind. The flashing then slopes down the face of the 2x4, folds over its bottom edge and lies flat on the asphalt shingles below. We had ours made up by a local sheet-metal shop in 10-ft. lengths. □

Designing for a Temperate Climate

How a Maryland architect dealt with seasonal extremes using an arsenal of passive-solar and mechanical systems

by William Bechhoefer

Although cherry blossoms may come to your mind when you think of the climate in our nation's capital, for anyone who lives here, as I do, it's the extremes between summer and winter that are a strain on comfort and pocketbook. As a native of Washington, D. C., I can remember a few 0°F days each damp, gloomy winter, although temperatures in the 20s and 30s are the norm. Blizzards are unusual, but Washington does average 20 in. of snow each season. Houses need heating from October through early April, for an average of 4,300 heating degree-days.

I remember the summers as being even worse. People who could afford it sent their families to more pleasant parts of the country, away from the oppressive heat and sticky humidity. Cooling degree-days average 1,500, but with air pollution resulting from the high concentration of automobiles and very high humidity, comfort requires air conditioning from June through September. In a poorly designed house, cooling can be even more expensive than heating.

As I was growing up, dealing with the climate meant applying liberal doses of cheap energy. But energy isn't cheap anymore, and finding economical ways to make houses more comfortable is one of my main concerns these days, both as a practicing architect and as a teacher helping students understand what design tools are available to them. Designing an energy-efficient building for a temperate climate is difficult because of the conflicting conditions each season presents. To cope with summer heat and humidity, you need shading and good air movement through the house. A lightweight structure is ideal for summer conditions because it doesn't hold much heat.

Winter requirements are just the reverse. The heating season demands that solar gain be stored in the mass of the house, and that air infiltration be kept to a minimum.

There is a whole palette of energy strategies that can be used in this kind of climate. But their application must be carefully integrated with the peculiarities of the site and with a broad range of other concerns if the house is to become a home. I recently designed a house in Olney, Md., that is a good example of how workable solutions come from compromise, and from using a variety of cooling and heating schemes.

The site—The house was to be built on a 200-acre farm, on which cattle would be raised and a stable of horses maintained. My clients were a doctor, an artist and their four children. On the one hand, they wanted a working farmhouse that would fit in with the surrounding woods and pastures. On the other hand, they needed an efficient house they

Various heating schemes are evident in this view from the dining room into the kitchen. The return-air ducts behind the Kalwall tubes collect the stored heat and distribute it to all parts of the house. The woodstove on the right contributes supplemental warmth; the grate above it allows excess rising heat to be absorbed by the Kalwall tubes above. The floor plans, below, show how the house is divided into zones with the living areas on the south.

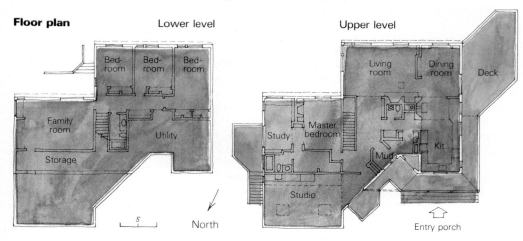

Floor plan   Lower level   Upper level

Bedroom   Bedroom   Bedroom

Living room   Dining room   Deck

Family room   Utility   Study   Master bedroom   Kit

Storage   Mud

Studio

5

North   Entry porch

could entertain in easily. They wanted good views to the outside, and good light within.

We surveyed possible sites for the house on horseback, a first for me. This gave us easy access to woods, pastures, creeks and ravines. Most important, though, the quiet and ease of horseback riding let us absorb the beauty of the land without distraction.

We rejected a dramatic site overlooking a creek because it was too close to the barns and exposed to severe northwest winter winds. There were several other nice sites, but they were too far from roads and utilities.

We eventually decided to build on a wooded knoll overlooking pastures and a creek. The views of the farm were great, and from an energy standpoint, the deciduous trees would provide vital shade in the summer. I sited the house just below the crest of the knoll to get some protection from winter winds. The views of pasture land were to the southeast, which is also the solar orientation I like best because it minimizes overheating in the afternoon, a problem in all seasons. It also squeezes out a little extra solar gain on winter mornings.

The owners had decided to dam the creek for a cattle pond (photo previous page). The pond would enhance the view, and would play a significant role in the energy strategy for the house. The approach to the house would be from the northwest under a natural arbor of trees. The road, which was almost half a mile long, would be easy to build since most of it was on a dedicated right-of-way.

What the clients wanted—Energy concerns are an important aspect of all good design, but are not necessarily the first needs to be addressed. I think that the design of a good home begins with the wishes of the owners and the requirements of the site. Since one of the owners was an artist, a large studio with good natural lighting was near the top of the list. They also asked for a separate living area for the children. The kitchen had to be large enough for several sets of hands to make dinner, yet compact enough to be used easily by one person. It needed to be open and informal, and still feel detached from the living and dining rooms when they were used for formal entertaining. Several rooms were to be trimmed out with black walnut that had been cut on the farm and air dried in the barn for over a year, and large expanses of wall were needed for hanging paintings.

How energy efficient?—My clients wanted an energy-efficient home. But they didn't want a house that looked solar. A greenhouse was out because they didn't want to grow plants indoors; furthermore, they didn't want anything obstructing their views. Nor were they interested in monitoring the temperature of the house and making daily adjustments. I certainly couldn't blame them for thinking that moving insulation into place in a large house would become an unpleasant chore in the winter, as would adjusting windows for precise ventilation patterns in the summer.

In short, my clients were not enthused by

Good natural light was important to the owners. Kalwall tubes in the open solar attic reflect and transmit light to all major rooms on the upper floor. The black walnut trim in the living room was cut from trees on the farm.

the dictum, "passive houses are for active people." I was not disturbed by this. There should be room for different degrees of concern about energy usage, and people have a right to expect technology to serve lifestyle rather than the other way around. Experience, too, has taught me that most kinds of movable insulation are not cost effective in a well-designed house in this climate with current energy costs. Where conditions are more severe, movable insulation makes better sense.

With all this in mind, I established the broad outlines of the house. First, I decided to take advantage of the slope of the site to make a two-floor house, with the parents on the upper level and the children on the lower, largely earth-sheltered level. On the upper floor, I placed the master bedroom, its study, and the living and dining rooms on the south side (drawing, facing page). On the north, where the light would be best for painting, is the studio. The kitchen, which generates heat, and the entry airlock and mudroom are also on the north. On the lower level, the bedrooms and a recreation room are to the south, and mechanical and storage rooms to the north. I think that this kind of zoning—dividing the house into southern living areas and northern buffer areas—accomplishes more than any other fuel-saving device.

Energy conservation—Along with the north-south zoning, the orientation of the house is the most important factor in energy conservation. Most of the glass is on the southeast, with overhangs calculated to permit both summer shading and the penetration of winter sun. I based my calculations on the sun-angle charts in Edward Mazria's *The Passive Solar Energy Book, Expanded Professional Edition* (Rodale Press, Emmaus, Pa., 1979), and shortened the overhangs somewhat because the site is so heavily shaded. While some sun does enter the house from late March through early September (when there is the greatest danger of overheating), it doesn't strike any of the heat-storage devices, and it just brightens the interior.

Earth-sheltering the lower level and siting the house below the crest of the hill provides protection from northwest winds. The roof is pitched to deflect these winds over the house.

Insulation is an important part of any conservation strategy. I used 2x6 studs 24 in. o. c. for the exterior walls and fitted them with foil-faced fiberglass batts. Including the exterior sheathing and cedar shingles, this gave us R-22 walls.

The roof was framed with 2x12 rafters, also on 24-in. centers, which allowed room for 9 in. of kraft-paper faced fiberglass and approximately 2 in. of airspace between the insulation and the sheathing. The roof is ventilated with soffit vents and ridge vents. Includ-

ing the extra-heavy asphalt shingles, the roof's insulation rating is R-33.

I did not use a supplementary vapor barrier. In the Washington area, condensation in the walls doesn't amount to much. While infiltration could be reduced with a polyethylene membrane, I don't like to seal a house too tightly because of interior air pollution. Most of my clients share this concern. Although construction is purposefully tight, I count on a little infiltration. If that sounds a bit casual, it's not. Last winter, the whole-house ventilator fan was turned on by mistake, and smoke began to fill the house immediately. It became evident that with the windows closed, the only available source of new air was a reverse draft down through the woodstove chimney. So the house is tight. It's just not too tight.

Heating—Heating is best approached in three stages: collection, storage and distribution. Collection was not a big problem in this house, because of its excellent orientation. The windows on the south admit all the solar gain needed. Heat distribution wasn't a problem either. Since most of my clients don't want to live with the temperature variations of a totally passive house, I usually provide backup heating and air-conditioning. The ducts and fan for these easily do double duty for the primary systems.

Heat storage was the greatest challenge. The choices are limited. Masonry floors and walls make an excellent heat sink. But because the house is wood-frame construction and all the floors were to be wood, the masonry was mostly below grade, in the walls of the lower level. The hollow block was filled with sand to increase its mass, and the walls were insulated on the outside with Dryvit. This

way, the heat-absorbing mass of the wall remains exposed inside the house.

I had drywall glued directly to the concrete block in the downstairs bedrooms, making it part of the mass and eliminating the usual air space that comes with mounting drywall on furring strips (which would have defeated the radiation of heat from the wall). This required that I use surface-mounted electrical outlets, so I designed a large base molding to cover them. This molding became a design element in all the downstairs rooms.

However, I needed a great deal more heat storage than the lower-level walls could give. I was intrigued by the aesthetic possibilities of water-storage tubes. Kalwall Sun-Lite tubes (Solar Components Corporation, P.O. Box 237, Manchester, N. H. 03105) have been used successfully for a number of years for solar mass. They are made of translucent fiberglass, and I was especially drawn to the possibilities of filtering and reflecting light into the rooms of the house with tubes located in the attic.

My calculations (see p. 229) showed me that 32 tubes 12 in. in diameter and 8 ft. high would give the house the mass it needed if they were incorporated in an open solar attic that ran through the center of the house. I designed the attic with large clerestory windows for heat collection. Light passes through the tubes into the studio and kitchen on the north side of the house. On the south side, light is reflected off the tubes into the master study and bedroom, and into the living and dining rooms (photo facing page). A return-air duct (drawing, top of next page) draws air across the warm tubes and distributes it through the conventional backup air system. This duct is automatically controlled, and doesn't open until the temperature in the solar attic

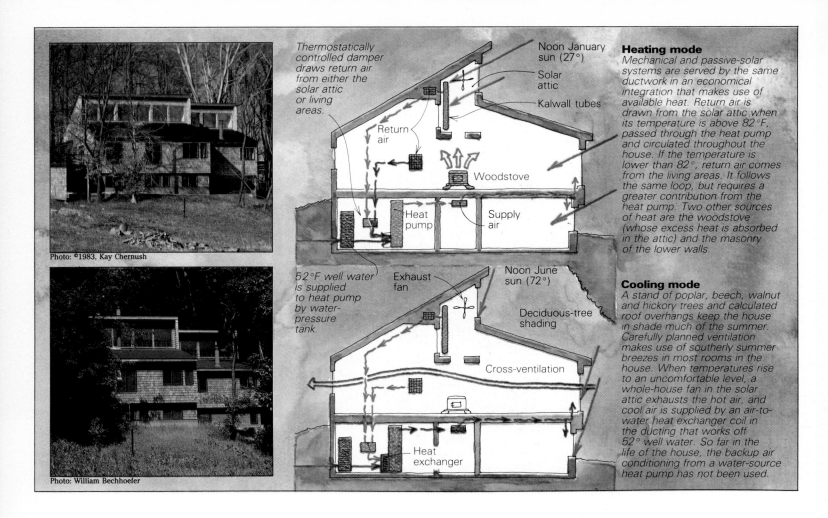

Photo: ©1983, Kay Chernush

Photo: William Bechhoefer

Thermostatically controlled damper draws return air from either the solar attic or living areas.

Return air

Noon January sun (27°)

Solar attic

Kalwall tubes

Woodstove

Heat pump

Supply air

52°F well water is supplied to heat pump by water-pressure tank.

Exhaust fan

Noon June sun (72°)

Deciduous-tree shading

Cross-ventilation

Heat exchanger

Heating mode
Mechanical and passive-solar systems are served by the same ductwork in an economical integration that makes use of available heat. Return air is drawn from the solar attic when its temperature is above 82°F, passed through the heat pump and circulated throughout the house. If the temperature is lower than 82°, return air comes from the living areas. It follows the same loop, but requires a greater contribution from the heat pump. Two other sources of heat are the woodstove (whose excess heat is absorbed in the attic) and the masonry of the lower walls.

Cooling mode
A stand of poplar, beech, walnut and hickory trees and calculated roof overhangs keep the house in shade much of the summer. Carefully planned ventilation makes use of southerly summer breezes in most rooms in the house. When temperatures rise to an uncomfortable level, a whole-house fan in the solar attic exhausts the hot air, and cool air is supplied by an air-to-water heat exchanger coil in the ducting that works off 52° well water. So far in the life of the house, the backup air conditioning from a water-source heat pump has not been used.

reaches 82°F. Since the attic acts as a plenum for warm air that rises into it by natural convection, little heat is wasted.

I began to think of the solar attic as a long core in the middle of the house, which could be used not only to collect heat and light for the entire structure but also to integrate the separate parts of the house. I used its walls as main structural supports for the roof and as dividers between major spaces. Bathrooms and closets fit in below it. So does the woodstove, which is adjacent to both the kitchen and dining room. Excess heat from the stove passes through a grill into the solar attic and further heats the tubes.

Finally, the solar attic provides heat for a domestic hot-water preheating system. Cold water is passed through three 29-gal. black-painted steel tanks on its way to the conventional water heater. The tanks replace three Kalwall tubes in the center of the attic, which is hidden from view. The system is virtually maintenance free and very economical.

There is significant heat loss from the clerestory windows in the solar attic. However, the entire heating bill for the winter of 1982-83 was about $400, very low for a 3,600-sq. ft. house in this area. Movable insulation can be added to these windows if rising energy costs warrant it.

Cooling—My cooling strategy for the house was quite simple. The shading from trees is very important, as is the earth-sheltering of the lower level. Except for two lower bed-

rooms, every room has cross-ventilation. The southeast orientation of the house opens it to prevailing summer breezes.

Excess heat rises naturally into the solar attic. The water tubes absorb some of it, preventing the air from overheating. Ideally, the windows in the solar attic would be operable to allow hot air to escape naturally. But that was rejected for its inconvenience and relative cost compared to fixed glazing. Instead, the excess heat is exhausted by a whole-house ventilator fan.

Well water provides another source for cooling. Air is passed over a heat-exchange coil in the ducts located in the basement utility room, which lowers its temperature and humidity. This air is then delivered to the house through the ducting system (drawing, above). The whole-house fan, natural ventilation, and the cool air from this coil were all that was needed to make the house comfortable during the summer of 1982. The backup air conditioning wasn't used at all.

Backup systems—My clients suggested a water-source heat pump for their mechanical heating and air-conditioning needs. They knew of several successful installations in the area, and asked me to gather some information on them. Water-source heat pumps (see *FHB* #14, p. 16) use water rather than air as a heat source. The advantage is that the water, which comes from a well, is a constant 52°F, whereas air temperatures can drop well below the minimum 26°F to 32°F required for a con-

ventional heat pump. The potential gains in efficiency are enormous.

The only data we needed was the rate of flow from the well. Twelve gallons per minute was the minimum requirement, and the well came in at more than 20 gallons per minute. Our mechanical contractor chose Carrier equipment because its quality is good and parts are always available. He downsized the system, recognizing that the house would get much of its heat from the sun, and that the heat pump would be activated only when the temperature in the solar attic dropped below 82°F. The idea of scaling down the size of heating and cooling equipment when it's to be used as a backup is a difficult concept to get across to most mechanical engineers.

Water taken from the ground must be returned there to keep from lowering the water table significantly. Since water passing through the heat pump is either warmer or colder than the original well water, it must be returned to the ground at some distance from the well to replenish the water table without affecting the temperature of the source well. The cattle pond proved an ideal spot, and we avoided the usual necessity of drilling a second well by using it.

The only other backup system is the woodstove. The wood box nearby can be filled from the outside. All wood is cut on the farm, which makes the stove very cost-effective. □

William Bechhoefer teaches architecture at the University of Maryland.

Illustrations: Frances Ashforth

Working with Kalwall tubes

Kalwall Sun-lite water-storage tubes were an early entry in the explosion of solar products in the last decade. They remain popular because of the their effectiveness and simplicity. Made of translucent fiberglass .040 in. thick to transmit natural daylight, they are simply containers for one of the best mediums for heat storage—water.

Kalwall tubes come in a variety of sizes, with one closed end, and a friction-fit cap to seal the top against evaporation once they have been filled in place. They are easily installed and take 60% less space with 80% less weight than an equivalent rock or masonry heat-storage system. Even the largest tubes weigh less than 20 lb. empty.

The water can be dyed to increase solar absorption by up to 35% over undyed water, but light transmission is reduced considerably. The dyes available are black, yellow, bronze and blue. Even tubes with clear water will have some color due to the refraction of light. The tubes can also be used in applications where light transmission is not a factor, as in a Trombe wall or an enclosed solar attic. The chart below right gives the necessary design data.

Calculations—As with any form of heat storage, finding the compromise between too much mass and too little is important. Too much mass means that the material won't get warm enough to make much of a difference in heating the house. Too little mass, and some of the incoming solar energy will be wasted by overheating the air. As a result, the house will get too much heat during daylight hours, and not enough at night.

To determine the size and number of water tubes I needed in the house in Olney, I turned to Edward Mazria's passive-solar text. Using the tables, I calculated the amount of solar gain that would be available in the attic on Dec. 21, assuming clear conditions. I began with the square footage of clerestory glass and subtracted an allowance for shading by nearby trees and the house itself. Deductions were also made for transmittance of the wall of the water tubes. The final figure was 203,300 Btu/day.

I wanted to use 8-ft. long, 12-in. dia. tubes, since they best fit the space. Each of these would hold 6.25 cu. ft. of water, and have a heat capacity of 7,800 Btu, assuming that exposure to sunlight would cause a 20° rise in the water temperature. Simple arithmetic told me I'd need 26 tubes to heat the house. As a check, I used Mazria's rule of thumb for this climate, which calls for about 1 cu. ft. of water for each square foot of solar window. Figuring things this way, I would need 48 tubes to avoid overheating the air, but this would result in a rise in water temperature in each tube of only 10° or 11°, not enough Btus to heat the house.

Ultimately I compromised with 35 tubes. These would fit comfortably in the solar attic, and I would be able to get about a 15° temperature rise in the water without serious risk of overheating. By replacing three of these tubes with steel tanks for pre-heating domestic hot water, I reduced my order to a total of 32 tubes.

There are more precise methods for solar

Design specs for Kalwall tubes								
Diameter (in.)	Length (ft.)	Volume (gal.)	Weight (lb.)		Heat capacity	Heat-storage capacity	Floor loading (lb.)	
			Empty	Full	(Btu) 20°F rise	(Btu°F-sq.ft. glazing)	Per lin. ft.	Per sq. ft.
12	4	23.5	8	204	3,900	49 *	204*	260*
12	8	47	12	404	7,800	49 *	404*	514*
18	5	66	16	567	11,000	73.5**	378**	321**
18	10	132	19	1,122	22,000	73.5**	748**	635**

* Spaced 12 in. o.c.   ** Spaced 18 in. o.c.

design than Mazria's, but I have found that his tables and rules of thumb are simple and reliable when combined with intuition and experience. Computer energy-analysis programs that are now becoming available make greater accuracy even easier to achieve for the designer.

Installation—The filled tubes are heavy. They can sit directly on a concrete slab on grade with no problem. However, if they are mounted above grade, you've got to beef up your floor construction.

The tubes are stable and stand by themselves. It's still a good idea to support them at the top in earthquake areas, or if they form a wall that people would be tempted to lean against. A wood or steel kickplate near the bottom to protect against impact also

makes sense where foot traffic is heavy. It is advisable to set the tubes in a pan or on a drainable waterproof surface, just in case they leak. In the house I designed in Olney, Md., I used an elastomeric membrane under the tubes with drains to the DWV system in the bathroom below.

The tubes should be unpacked and examined carefully upon receipt. According to the manufacturer, leaks are rare, and shipping damage is almost always the cause. Fill the tubes before you install them to check for leaks and to clean out any dust and dirt.

Once the tubes are in place, fill them with a garden hose. If you're not planning to use any dye, add a little swimming-pool chlorine to the water to inhibit the growth of algae. The friction-fit cap simply fits on top and completes the installation.   —W. B.

Index